Globalisation and Ecological Integrity in Science and International Law

Globalisation and Ecological Integrity in Science and International Law

Edited by

Laura Westra, Klaus Bosselmann
and Colin Soskolne

CAMBRIDGE
SCHOLARS
PUBLISHING

Globalisation and Ecological Integrity in Science and International Law,
Edited by Laura Westra, Klaus Bosselmann and Colin Soskolne

This book first published 2011

Cambridge Scholars Publishing

12 Back Chapman Street, Newcastle upon Tyne, NE6 2XX, UK

British Library Cataloguing in Publication Data
A catalogue record for this book is available from the British Library

ISBN (10): 1-4438-2734-7, ISBN (13): 978-1-4438-2734-8

This book is dedicated to William Rees:
Our inspiration for a long time, and our host for 2010

TABLE OF CONTENTS

Part IV – Issues in Global Governance

Part V – Ecological Integrity and Indigenous Peoples: Cases and Issues

LIST OF FIGURES AND TABLES

Figures

Tables

INTRODUCTION

LAURA WESTRA, KLAUS BOSSELMANN AND COLIN L. SOSKOLNE

The Global Ecological Integrity Group (GEIG) has been meeting annually since 1992. In 2010, the meeting took place at the University of British Columbia, hosted by William Rees. For this meeting, the GEIG returned to its roots. The chosen focus for 2010 was precisely the initial impetus of the Group's research from 1992 to 1999, that is, the quest for a fuller understanding of the role of ecological integrity in the various fields of human endeavour.

The GEIG quest started with an attempt to find a more thorough and explicit understanding of the scientific meaning of ecological integrity. The most recent thinking of the GEIG is presented in this book and, for ease of access, is divided into seven parts.

We were fortunate to have with us, James Karr, who joined us to address, once again, several of the issues that were central to our research at the start, focussing on the complete dependence of the human enterprise on the integrity of all living systems. Part I of this book is devoted to the "Science of Ecological Integrity", starting with a chapter by James Karr. The work of Pavel Cudlín then traces the interface between ecosystem services and citizens' responsibility for just governance. In the next chapter, Robert Goodland explains the importance of ecological integrity in relation to some of the gravest problems of our times: climate change and hunger. Both this chapter and the next one, by Heather McLeod-Kilmurray discuss the necessity for a radical change of diet on a planetary scale, to ensure sustainability and ecological justice.

In Part II, we return to yet another aspect of ecological integrity, that is, its relation to ethical norms. Donald Brown analyses the ethical implications of the Copenhagen Accord regarding climate change. Peter Brown goes even further, seeking a moral foundation for all governance based on ecological principles. Helmut Burkhardt outlines the main imperatives to be followed to achieve sustainability, and Sheila Collins traces possible scenarios to preserve the "global commons".

After 1999, the GEIG's main focus shifted somewhat, to consider the physical integrity and health of individuals. This required both exploration and explication of the legal dimensions of integrity in relation to human rights. Thus, in Part III, Joseph Dellapenna discusses existing forms of international law, and Sara Seck proposes a turn to "Third World Approaches to International Law" (TWAIL) for the better protection of ecological integrity. Klaus Bosselmann discusses the possible presence of "global constitutionalism", followed by Owen McIntyre who traces the relation between human rights and water in international administrative law. Kathryn Kintzele then presents her research on the ecological aspects of the constitutions of various countries around the world.

In Part IV, we return to the theme pursued in Florence in 2009, the problem of appropriate, ecologically sound governance. Timmer, Kissinger and Rees consider the importance of the regional aspects of governance, necessary in order to diminish the ecological impact of current policies. Alex and Sabina Lautensach then discuss human security in the context of rights, while Michael Schröter exposes the conflicts between liberalism and sustainability. The next two chapters examine the question of the interface between globalization and governance through two regional case studies: in the first one, Philippe Crabbé examines the effects of globalization in sub-Saharan Africa, while in the final chapter of Part 4, Vicky Karageorgu considers inter-basin water transfers in Greece.

In Part V, we further examine one of the major themes our group has studied, especially since 2007 at our Halifax, NS, meeting: the relation between ecological integrity and indigenous peoples. Linda Te Aho and Mimi Lam discuss, respectively, Maori and Saami approaches to the protection of ecological integrity in local landscapes, while Jack Manno explains the forms of governance in the Ononodaga Nation, in the State of New York.

In Part VI, we return to another major theme our group has studied since 1998: the importance of biological and ecological integrity to public health. These three chapters, range from the general problem of Peak Oil, discussed by Donald Spady, to Colin Soskolne and Shira Kramer's analysis of an ongoing legal case regarding decades of industrial pollution around the harbour of Sydney, NS. A general discussion of the interface between cancer and environmental epidemiology follows by Vladimír Bencko, pointing to technological promise for preventative public health.

In Part VII, the final section of this book, we consider issues which have been a more recent focus of GEIG, that is, the role of the media in reporting and presenting unsustainable practices (Rose Dyson), the role of the Internet in general (Robert Rattle), and in reporting environmental disasters (Doug Daigle). This Part speaks to the foundational role of media and communication technologies in influencing cultural values for a sustainable future.

As usual, the papers presented at the meeting included a broad interdisciplinary range of areas and concerns, connected only by the foundational role that ecological integrity plays across the breadth of social and legal concerns.

In his Concluding Chapter, J. Ronald Engel captures the overall sense of the 2010 GEIG meeting, underscoring its prophetic vocation. One of the defining characteristics of GEIG, he notes, is its "prophetic stance" toward the world. While this stance is chiefly rooted in the prophetic heritage of the Western religious traditions, it also has roots in other cultures such as those of indigenous peoples. This deep prophetic theme underlies the motivations for participation in the Group. As the papers in this volume demonstrate, it is with prophetic eyes that GEIG sees the world and assesses what changes are needed and possible.

GEIG is an island of prophetic truth-telling for those who come together each year for a week of rigorous interdisciplinary exchange and mutually supportive interpersonal relationships; and, also for those who turn to its published proceedings for greater understanding of our global problematique. GEIG is a place where we can pursue the prophetic vocation of making the arguments which alone can give us genuine hope for our future. Each step in the prophetic argument that James Karr sets in motion in Chapter One is repeated and developed in a fresh way by each of the subsequent authors of this volume. Each author has an allegiance to one or another expression of the higher law — the unwritten, universal principles of fairness, morality, and justice that hold in judgment all instances of political and economic decision-making.

The 2010 Vancouver conference left participants with the great unanswered question of how, by prophetic argument, we can bring about the changes in the world that our faith in the higher law requires. No prophet has yet given an adequate answer to this question, and nor has GEIG been able to make the argument that will bring about the changes between humans and nature, and between nations and peoples that the world so desperately needs. This question must be an essential part of the future agenda of the Global Ecological Integrity Group if it is to remain true to its prophetic vocation.

PART I

THE SCIENCE OF ECOLOGICAL INTEGRITY

INTRODUCTION

LAURA WESTRA

In this part of the book, we return to the original intent leading to the formation of the Global Ecological Integrity Group (GEIG): the quest for a thorough, scientific understanding of ecological integrity and its role in human society. James Karr's introductory Chapter refers to it, as "an essential ingredient for human's long-term success". The present development of most of the world, fostered by globalization and the unsustainable thrust toward "growth", goes apace with "growing biotic impoverishment" minimal to the presence of ecological integrity. "However", Karr points out, "it is not only non-human species that fall prey, increasingly, to extinction, thus decimating natural systems and the support they lend to all life."

An additional irreparable harm comes from the loss of indigenous/traditional ecological knowledge (TEK), and the diverse cultures and languages from which that knowledge originates. Karr also lists the many faces of biotic impoverishment, leading to both the direct and indirect depletion of living systems, human and non-human, and for both of which ecological integrity is absolutely necessary. Therefore, Karr maintains, it is necessary to ensure that this understanding is present in all political decisions, and that it is used to shape public policy globally.

Pavel Cudlin et al., argue, in Chapter 2, that "ecological citizenship" requires the appropriate valuation of the ecosystems that have been sustaining human life for millions of years. Over the last hundred years, "expanding human populations", and their "insatiable interests" have combined to cause massive destruction in those systems. The main cause of this disastrous result, Pavel argues, is that only "direct ecosystem provisioning services" are valued by markets, while the "depletion and degradation of natural and environmental resources" (i.e., ecological services), are not perceived as valuable. This chapter concludes with several proposed methods for arriving at an appropriate valuation of ecosystems.

Perhaps the most obvious result of the disintegrity that prevails globally, and of the degradation of natural systemic functions, is the presence of climate change. In Chapter 3, Robert Goodland and J. Ahnang decry the inconclusiveness and lack of political will that characterized the last two meetings regarding climate change (i.e., in Copenhagen (2009), and in Mexico (2010)). They propose a simple and inexpensive way to eliminate the greenhouse gas emissions attributable to livestock's supply chain, from forests cleared to supermarkets. The facts regarding land-based livestock speak for themselves. As more than 50 billion animals will be raised in 2010, one quarter of land worldwide is not used for grazing livestock, and one-third of all arable land is being destroyed, mostly from livestock and feed production.

It is particularly significant that even the World Bank Group's extensive peer review of Goodland and Ahnang's work, by their climate change specialists, approved the publication of the article on which this chapter is based, and that other UN Agencies have also adopted their findings. Yet, while the energy industry is closely monitored (with no significant reduction in the use of fossil fuels), the food industry is neither monitored nor curbed. The authors propose that living "Green" must start with "Eating Greenfully".

In Chapter 4, Heather McLeod-Kilmurray reprises the argument about food governance. She asks why laws, policies and individual actions are not moving in this direction?. Although most environmental protection Acts, Acts related to Agriculture, and other instruments relating to various aspects of food production, speak of "sustainable development", the laws governing food production in the Western world do not oppose "the massive concentrated power of the agricultural industry" despite the fact that "the global adoption of a low meat diet for the next 20 years" would halve the costs "of mitigating climate change up to 2050". This chapter concludes with "principles" and "recommendations" to move forward in that direction, clearly necessary for the survival of humanity and for the attainment of ecological and social justice.

CHAPTER ONE

ECOLOGICAL INTEGRITY: AN ESSENTIAL INGREDIENT FOR HUMANS' LONG-TERM SUCCESS

JAMES R. KARR

Introduction

Human history, like evolution itself, has been marked by relative stasis punctuated by periods of rapid change. Harnessing fire, making and using tools and weapons, and inventing the wheel were early mileposts signaling, we are told, an unbounded human ingenuity. These and other innovations allowed humans to tap natural capital and spread virtually throughout the world, living year-round from sea level to mountain tops, from equatorial heat to polar cold. The success of humans in these diverse natural settings resulted directly from the ability to tune culture and religion to diverse regional conditions. This progress—especially the specialization of nineteenth-century science and the hubris of twentieth-century technology—has led us to believe that we can repeal the laws of nature and forget the connections between society and its life-support systems.

Throughout evolutionary time, the success of living things has depended on the accumulation of information passed from generation to generation in the genetic blueprints of DNA. Humans, though, perfected another connection: the legacy of knowledge and culture passed from parents to their children and their children's children across hundreds, even thousands, of generations. During early human evolution, important knowledge was primarily biological—how to find food and shelter, escape from predators, avoid disease. Humans, like all other organisms, had to know their regional environments and how to support their families within these environments.

But with the agricultural revolution, these connections began to fray. By the nineteenth century, scientific and societal specialization combined with rapid, massive industrialization and free-market economics and seemed to promise escape from dependence on, or even connections with, other living systems. Now the "information age" gives us "virtual reality," completing our isolation from the rest of the living world and, some claim, clinching an end to human need for the biological knowledge so important to our ancestors.

Touting the uncanny ability of humans to be innovative and thus improve their lot, optimists—including economists, technologists, and futurists—see improvement in the human condition as an inevitable outcome of human ingenuity. But the world we have created may not be the ideal world we intended to create. Nature continues to challenge us through the very by-products of our own ingenuity: human ingenuity has had serious unintended consequences, which can no longer be ignored.

The most serious unintended consequence of this attitude is growing biotic impoverishment, the systematic reduction in Earth's ability to support living systems (Woodwell 1990), which extends from the degradation of the global physical and chemical environment to the impoverishment of human culture itself (Chu and Karr 2001). Concern over the implications of this trend for the quality of human and nonhuman life is now widespread, and the concepts of ecological health or ecological integrity are being invoked as guiding principles for policymaking. The multifaceted concept of integrity requires the integration of disciplines from science to philosophy and adds "a totally new note in the discourse of environmental concern" (Westra 1994).

Unintended Consequences

A major consequence of human "progress" is the homogenization of global society; human language, technology, and culture are becoming more homogeneous as we seem to become more independent of the idiosyncrasies of local natural systems. The rich diversity of human cultures is disappearing even more rapidly than the natural systems that nurtured that diversity. English is becoming a global language, and linguists are predicting that at least half of the world's languages will go extinct in the next century. These disappearing languages "are beyond endangerment," says Michael Krauss; "[t]hey are the living dead" (Haney 1995). Knowledge of indigenous medicines and other indigenous cultural adaptations, too, are fast being lost while cellular telephones, gasoline-powered engines, and computers spread to every corner of the globe.

Twenty-first century society is reluctant to acknowledge that ancient wisdom matters in the modern world (Davis 2009).

Another consequence of human "progress" is the impoverishment of Earth's life-support systems (Woodwell 1990; Millennium Ecosystem Assessment 2005, Chu and Karr 2011). Biological diversity declines as natural systems are degraded and destroyed, and ubiquitous pests and weeds homogenize the biological surroundings of humans and their industrial society. The loss of diversity reveals humans' flawed planetary stewardship, but, more important, it represents a loss of the unique life-support systems, including human culture itself, that the human species needs for survival. The integrity of the entire biosphere is threatened.

Global biotic impoverishment, including the homogenization of human culture, means that we are losing the adaptive complexes that once tied each human culture to the geographic region where it evolved. Our ingenuity and hubris let us forget the importance of these connections. It chained us instead to clever ways of extracting resources from environments that are too often depleted by our actions and all too often impoverish local human communities. We inherit and pass on to future generations a legacy of toxic effluents, destroyed and fragmented landscapes, depleted forests and fisheries, and collapsing cultures. The failure to maintain human bonds with place, biology, and culture—our connections to living systems—is likely the single most important challenge that future human generations will face.

We Must Learn from History

We first saw planet Earth from space more than forty years ago and suddenly realized just how isolated we were and how dependent on a small, yet unique, piece of space debris. Until then, we had only seen ourselves up close. From the distance of space, we see ourselves and our planet exposed in an unexpected fragility and vulnerability.

We would do well to remember that, like Earth alone in space, Easter Island is an isolated place, separated from other land by more than 1800 kilometers of ocean. When first settled 1000 to 1500 years ago by some two dozen Polynesian explorers, Easter Island was densely forested, with ample natural resources (Ponting 1991). By the seventeenth century, the population of Easter Island had burgeoned to about 10,000 people, but less than a century later, the island's human society had collapsed. When Western Europeans arrived in 1722, they found a treeless island and a small population living in primitive conditions. These Easter Islanders had no cultural memory of the society—their own ancestors—that only a few generations before had carved and placed the island's famous massive stone monoliths.

Easter Island is not a unique situation. Environmental changes in soil, vegetation, and water caused by humans are primary factors in the decline of many local and regional civilizations: Angkor Wat, Mesopotamia, Indus, Greece, Rome, Egypt, Maya, Inca, Aztec, and the Moche. Like Easter Island, all are a metaphor for human society on the globe today. In contrast to the optimists' view of an inevitable and continuous advance to human society, many others believe that humans have overshot Earth's carrying capacity—as a thriving Easter Island society overshot that of the island—and that nothing short of substantial change in human behavior will reverse this trend. Such concerns are not merely extremist hand wringing; current public policy and legislative initiatives will not protect either natural or human environments.

The complex reasons for this inadequacy lie in the unrelenting hubris of a society that behaves as if it could repeal the laws of nature. Plans generated by economists, technologists, engineers, and ecologists have too often assumed that lost or damaged components of ecological systems are inconsequential or can be repaired or replaced. Yet we see the consequences of this attitude everywhere: In the Pacific Northwest, hatcheries are expected to sustain salmon stocks while little is done to restore degraded river and coastal environments, curtail harvests, or protect seasonal river flow. Throughout the world, expensive fertilizers are expected to replace depleted soil nutrients. Groundwater is depleted to supply unsustainable amounts of water to crops, livestock, and people. These consequences and many others point to the folly of maintaining the status quo.

Multidisciplinary initiatives seeking to improve environmental policy *are* cropping up in many contexts, driven by goals such as environmental justice (Bullard 1994), protection of biodiversity (Millennium Ecosystem Assessment 2005), and control pollutants (Colborn and Clement 1992). They are grounded in concepts such as ecological economics (Jansson et al. 1994), sustaining capitalism (Hart 2005), conservation biology (Groom et al. 2006), and industrial ecology (Ayers and Ayers 2002).

Fruitful use of these concepts requires the human species to recognize its fundamental dependence on living systems and to develop a core societal vision capable of integration—a vision that should be similar to the Socratic vision of medicine. This larger goal has been variously expressed as the protection of biological integrity (Karr 1991), ecological integrity (Karr and Dudley 1981; Nash 1991; Westra 1994), or ecological health (Costanza et al. 1992). Although the terminology varies, all of these visions focus on the reality that healthy biological systems are critical to the success, and survival, of the human species.

Growing Concern

Concern about the integrity of life-support systems has evolved over nearly two centuries. For most of the twentieth century, the most visible demonstration of "environmental awareness" was the conservation movement in the developed world. But voices now coming from all corners of society draw attention to the severity of present ecological crises (Karr 2002: Appendix). A Health of the Planet Survey by the Gallup Organization (Dunlap et al. 1993) showed "strong public concern for environmental protection throughout the world, including regions where it was assumed to be absent."

Scholars too are calling for shifts in human behavior. A worldwide collection of 1575 scientists, including 99 Nobel Prize winners, noted that "human beings and the natural world are on a collision course. . . . A great change in our stewardship of the Earth, and life on it, is required if vast human misery is to be avoided and our global home on this planet is not to be irretrievably mutilated" (Union of Concerned Scientists 1992). In the same year, the National Academy of Sciences and the Royal Society of London (1992) issued a joint statement recognizing the need for industrial countries to modify their behavior radically to avoid irreversible damage to the Earth's capacity to sustain life. A 1993 Population Summit held in New Delhi (Science Summit 1993) called for action to turn 1994 into "the year when the people of the world decided to act together for the benefit of future generations."

Universities and governments have also joined the chorus. In the 1990 Talloires Declaration, the leaders of hundreds of universities from throughout the world expressed their deep concern "about the unprecedented scale and speed of environmental pollution and degradation, and the depletion of natural resources." Business and labor also recognize the need for change. Forty-eight international industrialists and business leaders from more than 25 countries called for renewed efforts by business and government to make ecological imperatives part of the market forces governing production, investment, and trade (Schmidheiny 1992). Stuart Hart (2005) maintains that to be profitable in the future businesses must "simultaneously raise the quality of life for the world's poor, respect cultural diversity, and conserve the ecological integrity of the planet for future generations."

The United Steelworkers of America (1990) overwhelmingly endorsed a report that says, "We cannot protect steelworker jobs by ignoring environmental problems." Further, the "greatest threat to our children's future may lie in the destruction of their environment," and "the environment outside the workplace is only an extension of the environment inside." At the August 1993 Parliament of World's Religions (Briggs 1993), the leaders of Christianity, Buddhism, Islam, Judaism, Hinduism, and other faiths developed a "global ethic." Among other things, that ethic condemns environmental abuses. In an age of unparalleled technological progress, poverty, hunger, the death of children, "and the destruction of nature have not diminished but rather have increased."

The Commission on Life Sciences of the National Research Council (1993) concludes that society possesses many of the "tools to address environmental problems of enormous consequence to our social and economic well-being. But we are not using those tools most effectively." This demonstrates an important modern paradox: As scientific understanding of Earth and human effects on it expand, the threats to Earth's living systems—human and nonhuman—worsen (Karr 2008).

Biotic Impoverishment Goes Beyond Extinction

These organizations and the constituencies they represent recognize that all is not well on planet Earth, but they do not explicitly define the main problem: biotic impoverishment (Woodwell 1990). Biotic impoverishment is visible today in three major forms: indirect depletion of living systems through degradation of the chemical and physical environment; direct depletion of nonhuman living systems; and direct depletion of human systems (Table 1; Karr 1995b, Chu and Karr 2011).

Table 1. The Many Faces of Biotic Impoverishment, with Examples.

A. Indirect Depletion of Living Systems

1. **Soil depletion and degradation** (erosion, degradation of soil structure, salinization, desertification, nutrient leaching, loss of soil biota)

2. **Degradation of water** (pollutants, flow alteration, wetland drainage, depletion of surface and groundwater, homogenization of aquatic biota)

3. **Alteration of global biogeochemical cycles** (nutrient enrichment, acid rain, alteration of water cycle, outbreaks of pathogens and red tides)

4. **Chemical contamination** (land, air, & water pollution by pesticides, heavy metals, and others; bioaccumulation, ocean acidification, fish kills)

5. **Global climate change and ozone depletion** (global warming, alteration of rainfall distribution and amount, effects on health)

B. Direct Depletion of Nonhuman Living Systems
1. **Renewable-resource depletion** (depleted populations of fish and trees; altered food webs; extinctions)
2. **Biotic homogenization** (extinction and invasions)
3. **Habitat destruction and fragmentation** (biotic homogenization, loss of landscape mosaics and connectivity)
4. **Genetic engineering** (homogenization of crops, antibiotic resistance)

C. Direct Depletion of Human Systems
1. **Emerging and reemerging diseases** (occupational hazards, pandemics, AIDS, Ebola, Hantavirus, Lyme disease, nutritional and stress diseases)
2. **Loss of cultural diversity** (genocide, ethnic cleansing, loss of knowledge and linguistic and cultural diversity, loss of knowledge)
3. **Reduced quality of life** (environmental refuges, malnutrition and starvation, failure to thrive, poverty)
4. **Environmental injustice** (environmental discrimination, economic and generational inequity, racism, gender inequity)
5. **Political instability** (resource wars, civil violence, international terrorism, environmental refuges)
6. **Cumulative effects** (surprises, collapse of civilizations, "boom and bust" cycles, "natural" catastrophes, disease and biodiversity interactions)

Indirect Depletion of Living Systems

The primary physical systems that humans depend on are air, soil, and water. The productive potential of soils is degraded by erosion, salinization, desertification, and compaction. But soil is much more than its physical constituents: depletion of the organic activity in soil is also serious. Degradation of water resources--including chemical pollution, surface and groundwater depletion, and flooding--is pervasive. Norman Myers (1993) rightly notes, "Our future will be deeply compromised unless we learn to manage water as a critical ingredient of our lives."

Chemical contamination of air, soil, and water has for many years been the primary focus of government and the public; the primary concern has been the threat to human health from a diversity of chemical pollutants. Special attention has been directed to the narrow problem of contaminants that induce cancers in animals and humans. Other contaminant red flags include bioaccumulation; immunological and developmental deficiencies; and a growing number of reproductive and intergenerational effects.

Historically, the consequences of human activity were limited in space and time, but the increase in toxic chemicals and radioactive materials during the twentieth century has created global problems with legacies that will be present for thousands of years. Some of these affect people directly; many will have long-term indirect impacts on biological systems by altering biogeochemical cycles, global climate, and ozone concentrations.

Direct Depletion of Nonhuman Living Systems

Humans directly deplete renewable natural resources by harvesting fish, timber, and other products. Habitat destruction and fragmentation associated with harvest, urbanization, and other activities have perhaps the farthest-reaching effects on biological systems. Yet relatively little attention has been paid to habitat loss except when it threatens species with extinction. Human activities may even be responsible for increased frequency of red tides in coastal environments and insect and disease outbreaks in forests. Especially devastating to regional living systems is the homogenization of plant and animal communities through extinction and the spread of nonnative species, particularly commensals of human society.

Direct Depletion of Human Systems

The advances of modern medicine over the past three decades have lulled us into a false sense of security about human health and the environment. To be sure, antibiotics to control pathogens and pesticides to control pests have helped check many diseases. Yet, with a few exceptions like smallpox, diseases have not been conquered. Virulent forms of *E. coli*, tuberculosis, influenza, yellow fever, and malaria are becoming more difficult to control. In addition, emerging "new" diseases caused by bacteria (legionnaires' and Lyme diseases), viruses (Ebola, Hantavirus, HIV/AIDS), and parasites (*Cryptosporidium*) are cropping up. Human population growth and behavior, global travel patterns, resistance to antibiotics, reductions in natural immunity in human populations stressed by other environmental degradation (e.g., global warming), and destruction of natural habitats all contribute to this trend.

The impoverishment of human systems is also manifest as reduced cultural diversity (genocide and loss of knowledge), reduced quality of life and economic deprivation (failure to thrive in infants, malnourishment in 20% of people), and environmental injustice (racism, economic exploitation, lack of intra- and intergenerational equity).

Collectively, this broad sweep of issues illustrates the magnitude of the environmental challenge facing all members of the human community. It also reminds us of the close association and common underpinning of environmental and social concerns, and it provides an opportunity to exercise the ingenuity that has brought us this far. The loss of species; the destruction of agricultural lands; and the differential exposure to environmental hazards of economically disadvantaged people, often people of color, degrade the quality of human life. As human influence expands, the limits of technology, especially unintended consequences of technology, become more obvious. Depletion of water supplies cannot be "fixed" by engineers making water to refill aquifers; lost salmon spawning grounds cannot be "fixed" by adding gravel or "large woody debris."

Citizens and political leaders, humanists and scientists must work together to develop creative solutions. Failure to do so will relegate the world to continued biotic impoverishment and threaten the sustainability of human society. Ecologists' participation in these partnerships is critical to their success. In the same sense that medical doctors must be trained to recognize and understand the attributes of a healthy human, ecologists and environmental scientists must understand the attributes of healthy biological systems--systems that must be sustained over the long term, in the service of humans and for their own sakes.

Ecological Integrity and Ecological Health

If biotic impoverishment is the problem, then protecting the integrity of living systems must be the goal. But how do we define *biological integrity* in a world that is increasingly altered by the actions of humans? How do we reconcile the inevitable changes required to accommodate a growing human population and the proliferation of modern technology while guarding the planet from irrevocable biotic impoverishment? Answering these questions in clear and explicit terms is especially important as we seek to bring scholars from diverse disciplines together to focus on common problems.

What do *health* and *integrity* mean? How do we integrate concepts of integrity in their philosophical, legal, biological, cultural, and ethical senses (Westra 1994)? What kind of health or integrity do we seek? Are we seeking "environmental health," or is that phrase too narrowly associated with the effects of toxic substances on human health? As a societal goal, biological integrity suggests a meaning beyond human health. And the sum of physical, chemical, and biological integrity is ecological integrity (Karr and Dudley 1981).

Aldo Leopold (1949) was the first to invoke the concept of integrity in an ecological sense: "A thing is right when it tends to preserve the integrity, stability, and beauty of the biotic community. It is wrong when it tends otherwise." Nearly forty years ago, as the United States Congress drafted the Water Pollution Control Act Amendments of 1972, it sought a broad statement reflecting a vision absent from earlier water resource legislation. "Can we afford clean water? Can we afford rivers and lakes and streams and oceans which continue to make possible life on this planet?" asked the late Senator Edmund Muskie (Congressional Research Service 1972). "These questions answer themselves." Congress explicitly included "integrity"—"to restore and maintain the physical, chemical, and biological integrity of the nation's waters"—as the underlying goal of its legislation. Two major aims are clearly incorporated in this congressional language: active protection of remaining high-quality aquatic systems and a return of the nation's waters to a state of health. Since 1972, the integrity concept has been invoked integrity as a societal goal in diverse ecological and geographic contexts: Great Lakes Water Quality Agreement (1978); amendment to Canada's National Park Act (1988); Kissimmee River (Florida) Restoration Project (1989); National Wildlife System Improvement Act (U.S.; 1997); National Parks (U.S.) Omnibus Management Act (1998); Freshwater Strategy for British Columbia (1999); and European Union Water Framework Directive (2001).

Integrity implies an unimpaired condition or the quality or state of being complete or undivided; it implies correspondence with some original condition. The term most appropriately refers to the condition at sites with little or no influence from human actions; the organisms living there are products of the evolutionary and biogeographic processes influencing that site. Biological integrity (Karr and Dudley 1981; Angermeier and Karr 1994) refers to the capacity to support and maintain a balanced, integrated, adaptive biological system having the full range of elements (genes, species, assemblages) and processes (mutation, demography, biotic interactions, nutrient and energy dynamics, and metapopulation processes) expected in the natural habitat of a region. Although somewhat long-winded, this definition carries the message that: (1) Living systems act over a variety of scales from individuals to landscapes, (2) A fully functioning living system includes items one can count (the elements of biodiversity) plus the processes that generate and maintain them, (3) Living systems are embedded in dynamic evolutionary and biogeographic contexts that influence and are influenced by their physical and chemical environments.

An evolutionary foundation ties the concept of integrity to a benchmark against which society can evaluate sites altered by human actions. The complex biological systems that evolved at a site have already proved their ability to persist in, and even modify, the region's physical and chemical environment. Their very presence means that they are resilient to normal variation in that environment. Species abundance, for example, changes as a function of changing physical environment and changing interactions among species in a local assemblage. But the bounds over which systems change as a result of most natural events are limited when compared with the changes imposed by human activities like row-crop agriculture, urbanization, or dam building.

Human society sets aside extensive areas in parks and reserves to protect their natural state, to protect their integrity. These areas deserve protection because of the diverse values they provide to society. Water bodies, both on the surface and underground, deserve special protection as well, because they provide water to drink and support recreational and other values. Most important, rivers are the lifelines of a continent, reflecting the condition of surrounding landscapes, linking landscapes across great distances.

Because of the demands of feeding, clothing, and housing more than 6.9 billion people (November 2010), few places on Earth maintain a biota with evolutionary and biogeographic integrity. The growth of human populations in the last few centuries has made our species the principal driver of global change. Providing for human needs has required massive alteration of the planet in ways that preclude a return to the pristine environments of the preindustrial era. Thus, biological integrity is lost on a large share of the planet and is unlikely to be regained. Yet loss of ecological integrity for all lands and waters in all regions of the world is unacceptable on scientific, economic, aesthetic, and ethical grounds.

Health implies a flourishing condition, well-being, vitality, or prosperity. An organism is healthy when it performs all its vital functions normally and properly; a healthy organism is resilient, able to recover from many stresses; a healthy organism requires minimal outside care. The concept of health applies to individual organisms as well as to national or regional economies, industries, and natural resources such as fisheries.

Ecological health describes the preferred state of sites modified by human activity—areas cultivated for crops, managed for tree harvest, stocked for fish, urbanized, or otherwise intensively used. At these sites, integrity in an evolutionary sense cannot be the goal. Healthy land use, with or without active management, should not degrade a site for future human use or degrade areas beyond that site (Karr 1995b). Soils, for example, should not be eroded or otherwise transformed in ways that reduce future productivity. Groundwater should not be depleted.

Land use should not have deleterious effects beyond a site; atmospheric contamination should not result in downwind effects, such as tree death or ozone depletion. Healthy sites should not release contaminants or eroded soils that degrade sites elsewhere.

According to these two criteria—no degradation of a site for future use and no degradation of areas beyond that site—most modern agricultural and urban land use, for example, are not healthy. Recent initiatives for sustainable, or healthy, communities and agriculture recognize this reality.

Like losing ecological integrity everywhere, failing to protect the ecological health of intensively used lands is also unacceptable on scientific, economic, aesthetic, and ethical grounds. We must choose to develop a conceptual framework to define "acceptable" and "sustainable" uses.

These concepts are easily applied to small parcels of land; scaling up to large landscapes is another matter. What proportion of a landscape should be protected under a biological integrity goal? The World Commission on Environment and Development (1987) recommended 12%, but that percentage seems inadequate. Why 12%? Should that proportion vary with regional ecosystem type (desert, forest, grassland)? Which sites deserve the highest priority for protection? Water bodies, fragile sites such as steep slopes, or sites with the highest biodiversity? How shall we decide?

Measuring Health and Integrity

Ecologists have not been especially adept at defining or measuring either ecological health or biological integrity. The track record of freshwater management provides an instructive example. Human society depends on fresh water and on the resources associated with freshwater and marine systems. One third or more of modern humans lack access to safe drinking water and sanitation services. Improved water conservation, and treatment and recycling programs, can delay crisis, but human ingenuity will remain on a treadmill trying to keep up with expanding demand created by an expanding population. Even where supplies of fresh water remain adequate, resource degradation continues because society chronically undervalues the products and services, besides water, provided by aquatic ecosystems.

Although rivers are in many ways the lifeblood of society, prevalent attitudes toward rivers reflect disdain for their value and arrogance about our ability to replace or repair them. Despite the mandate in the Clean Water Act for protecting the integrity of the nation's waters, for example, it took nearly two decades to begin to incorporate that concept into water resource protection, largely because appropriate benchmarks were not defined for evaluating success in attaining these goals. That failure led to six realities about the condition of water resources (Karr 1995a): (1) water resources, especially their biological components, are in steep decline; (2) degradation stems from more than chemical contamination, the primary focus of conventional water quality programs; (3) long-term success in protecting water resources requires careful thought about goals and benchmarks, including development and uses of criteria for protecting biological integrity; (4) the legal and regulatory framework in place today does not respond soon enough to continued degradation; (5) the quantitative expectations that constitute biological integrity vary geographically; (6) because ecological and biological systems are complex, multiple components of these systems should be protected.

These realities are a consequence of using the wrong indicators. In water resources, for example, the major indicators have been pollution permits issued, fines levied, and counts of chemical contaminants released into waterways. Managers made little or no effort to evaluate the biological condition of water resources (Karr 1991).

The use and development of biocriteria are altering approaches to water resource planning and decision making throughout the world. Moreover, pioneering efforts are changing the way we express a broad range of environmental trends at international, national, regional, and local scales.

Clearly, we need to stop paying heed to the wrong indicators or to stop ignoring the indicators we have. Complex systems are regulated by negative feedbacks, the way a thermostat regulates a furnace. Society pays attention to economic feedback signals, as it does when it uses monetary policy to reduce interest rates or control inflation; to a lesser extent, it also pays attention to social signals. But under the assumption that we can repair ecological damage or replace ecological systems, we fail to respond to signals that should change our behavior toward those systems. Failure to take biotic impoverishment as a warning to change may thus destroy what we depend on. We have pushed ecological as well as social support systems beyond their self-organizing capacity, and biological and social dysfunction is the result.

Reclaiming Lost Connections

Traditionally, human actions have not been judged in terms of their influence on ecological health or integrity. When human populations were small, resources were abundant, technological skills were less advanced, and environmental degradation was less extreme. As a result, society did not notice or understand ecological integrity or health. Degradation was local and usually short-lived; an undeveloped frontier was always available. But the frontier is gone now—as it was for Easter Islanders. Supplies of many renewable resources have been depleted; chemical pollution is pervasive; and the global atmosphere is changing under human onslaught.

If we are to stem biotic impoverishment and reverse environmental degradation, we must: (1) set societal goals based on broad concepts of ecological integrity and ecological health; (2) forge partnerships among scientists, engineers, policymakers, resource managers, and citizens to develop approaches for attaining those goals; (3) revise the legal framework guiding environmental policy to ensure that both ecological risks and threats to human health are minimized; (4) protect existing resources; (5) restore resources that are degraded; and (6) reduce resource consumption.

In essence, we must rediscover and reclaim the connections to living systems that nineteenth-century science and twentieth-century technology let us pretend we could forget. We must go back in our thinking to the legacy of Baron Alexander von Humboldt, late eighteenth-century explorer, geographer, writer, naturalist, "and the man most responsible for bringing the practice of science into mainstream Western culture" (Sachs 1995). In Humboldt's day, "science" meant "natural science," and it took interdependent relationships in nature, including the place for humans, for granted. Aware, from his explorations, of the diversity in nature and among humans, Humboldt nevertheless believed that "a knowledge of that interdependence was the 'noblest and most important result' of all scientific inquiry" (Sachs 1995). He clearly understood the links between natural resource use, human welfare, and social justice.

For most of the past century, human energies have been directed toward economic growth, to the production and distribution of goods and services, to the production of wealth. Our consumption-centered political economy reflects those goals. Just as advancing technology was viewed as progress, economic growth and the generation of wealth have become not only products but a sign of our species' ability to exercise its self-bestowed birthright to monopolize an expanding fraction of Earth's resources. Political economy reduces everything to social, especially economic, constructions, "blatantly disregarding all that is not human" (Greenberg and Park 1994) and ignoring the dependence of humans and their economies on the larger natural economy. It treats economic systems as self-contained, disconnected from their surroundings, rather than as subsystems dependent upon a finite Earth system (Daly 1991; Karr 2008).

To protect that ecological system--the ultimate source of all human wealth--a "political ecology" needs to take shape as a guiding principle for public policy. Political ecology would incorporate ecological concepts into human culture and politics, and consider culture and politics within ecological realities. Political ecology would recognize the folly of behaving as if price mattered and cost did not. We pay a price at the checkout counter, but we often pay a higher cost when we ignore things of value that cannot be measured in numbers, when we ignore the value of things that can be measured, or when we lose something that we did not know was important until it was gone (Orr 1994). Recognizing the dependence of society on natural resources--explicitly adding political ecology to our thinking--will require us to ask not only if we *can*, but if we *should*. Remembering the connections will at least make us consider more carefully whether price reflects both economic and ecological costs.

We cannot avoid the use of technology, but we can no longer adopt technology without careful evaluation of its ecological effects. We can no longer adopt a technology because of its short-term advantages; we must first evaluate its long-term consequences. We must ensure that protecting ecological health and integrity plays a central role in decisions about consumer goods and development of technologies, including when, where, and how to apply them.

For most of the past century, politics, economics, and engineering have been the driving forces in societal decision making. If the price seemed right to build a dam, for example, and the technology was there to do it, society built the dam in the interest of "progress" even though some questioned the wisdom of that action. In effect, society based its decisions on narrow aims of those who forgot the fundamental lessons Humboldt understood 200 years ago. They did not understand that, like frosting on a layer cake, human economies take their shape from society and the Earth's natural resource base. Society--its formal and informal rules, laws and regulations, and ethics and morality--forms the top layer of the cake; the natural resource base forms the bottom layer. Without that natural resource base--without other living organisms--neither human society nor human economy will flourish.

In effect, we depend on ecology replacing physics as the master science of the 21st century: "A master science is, in part, the dominant scientific discipline of a historical epoch. . . . a master science generates and orders the concepts through which society understands itself and its relation to its surroundings" (Homer-Dixon 2009).

Acknowledgments

Portions of this paper were first presented at a workshop titled Engineering within Ecological Constraints (Karr 1995b) and in an earlier GEIG conference publication (Karr and Chu 1995).

References

Angermeier, P. L., and J. R. Karr. 1994. Biological Integrity versus Biological Diversity as Policy Directives: Protecting Biotic Resources. *BioScience*. 44: 690-697.

Ayers, R.U., and L. W. Ayers, eds. 2002. *A Handbook of Industrial Ecology*. Edward Elgar Publishing, Northampton, MA.

Briggs, D. 1993. World's Clerics Draft Global Ethic: Violence, Sexism, Environmental Abuse Are All Targeted. *Seattle Times*, September 1.

Bullard, R. D., ed. 1994. *Unequal Protection: Environmental Justice and Communities of Color*. Sierra Club Books, San Francisco, CA.

Chu, E. W. and J. R. Karr. 2011, in press. Environmental impact, concept and measurement of. In *Encyclopedia of Biodiversity, 2nd edition*, S. A. Levin, ed. Academic Press, Orlando, Florida.

Colborn, T., and C. Clement, eds. 1992. Chemically Induced Alterations in Sexual and Functional Development: The Wildlife/Human Connection. *Advances in Modern Environmental Toxicology* 21: 1-403.

Congressional Research Service. 1972. *History of the Water Pollution Control Act Amendments of 1972*, ser. 1, 93rd Cong., 1st sess.

Costanza, R., B. G. Norton, and B. D. Haskell, eds. 1992. *Ecosystem Health: New Goals for Environmental Management*. Island Press, Washington, DC.

Daly, H. E. 1991. Elements of Environmental Macroeconomics. In *Ecological Economics: The Science and Management of Sustainability*, R. Costanza, ed. Columbia University Press, New York, NY, pp. 32-46.

Davis, W. 2009. *The Wayfinders: Why Ancient Wisdom Matters in the Modern World.* House of Anansi Press,Toronto, Ontario.

Dunlap, R. E., G. H. Gallup, Jr., and A. M. Gallup. 1993. Of Global Concern: Results of the Health of the Planet Survey. *Environment* 35(9): 7-15, 33-39.

Greenburg, J. B., and T. R. Park. 1994. Political Ecology. *Journal of Political Ecology* 1: 1-12.

Groom, M. J., G. K.Meffe, C. R. Carroll, and contributors. 2006. *Principles of Conservation Biology, 3rd ed.* Sinauer Sunderland, MA.

Haney, D. Q. 1995. Experts Say World May Lose Half Its Languages. *Seattle Times*, February 19.

Hart, S. L. 2005. *Capitalism at the Crossroads: The Unlimited Business Opportunities in Solving the World's Most Difficult Problems.* Wharton School Publishing, Upper Saddle River, NJ.

Homer-Dixon, T. 2009. The Newest Science. *Alternatives Journal* 35(4): 8–11, 38.

Jansson, A., M. Hammer, C. Folke, and R. Costanza (eds). 1994. *Investing in Natural Capital: The Ecological Economics Approach to Sustainability.* Island Press, Washington, DC.

Karr, J. R. 1991. Biological Integrity: A Long Neglected Aspect of Water Resource Management. *Ecological Applications* 1: 66-84.

—. 1995a. Clean Water Is Not Enough. *Illahee* 11: 51–59.

—. 1995b. Ecological Integrity and Ecological Health Are Not the Same. In *Engineering within Ecological Constraints,* P. Schulze, ed. National Academy Press, Washington, DC, pp. 97–109.

—. 2002. What from Ecology Is Relevant to Design and Planning. In *Ecology and Design: Frameworks for Learning. Island Press,* Washington, DC, pp. 133–172.

—. 2008. Protecting Society from Itself: Reconnecting Ecology and Economy. In *Sustaining Life on Earth: Environmental and Human Health through Global Governance.* Rowman & Littlefield, Lanham, MD, pp. 95–108.

Karr, J. R. and E. W. Chu. 1995. Ecological integrity: reclaiming lost connections. In L. Westra and J. Lemons, eds. *Perspectives on Ecological Integrity.* Kluwer, Dordrecht, Netherlands. Pp. 34–48.

Karr, J. R., and D. R. Dudley. 1981. Ecological Perspective on Water Quality Goals. *Environmental Management* 5: 55-68.

Leopold, A. 1949. *A Sand County Almanac.* Oxford University Press, New York.

Millennium Ecosystem Assessment. 2005. *Ecosystems and Human Well-Being: Biodiversity Synthesis.* World Resources Institute, Washington, DC.

Myers, N. 1993. *Ultimate Security: The Environmental Basis of Political Stability.* W. W. Norton, New York, NY.

Nash, J. A. 1991. *Loving Nature: Ecological Integrity and Christian Responsibility.* Abingdon Press, Nashville, TN.

National Academy of Sciences (U.S.), and Royal Society of London. 1992. Population Growth, Resource Consumption, and a Sustainable World. Joint Statement by the Officers of the Royal Society of London and the U.S. National Academy of Sciences, Washington, DC.

National Research Council. 1993. *Research to Protect, Restore, and Manage the Environment.* National Academy Press, Washington, DC.

Orr, D. W. 1994. *Earth in Mind: On Education, Environment, and the Human Prospect.* Island Press, Washington, DC.

Ponting, C. 1991. *A Green History of the World: The Environment and the Collapse of Great Civilizations.* St. Martin's Press, New York, NY.

Sachs, A. 1995. Humboldt's Legacy and the Restoration of Science. *World Watch* 8(2): 28-38.

Schmidheiny, S. 1992. *Changing Course: A Global Business Perspective on Development and the Environment.* MIT Press, Cambridge, MA.

Science Summit. 1993. Population Summit of the World's Scientific Academies. A Joint Statement by Fifty-eight of the World's Scientific Academies. Office of International Affairs, National Research Council, Washington, DC.

Union of Concerned Scientists. 1992. World Scientists' Warning to Humanity. Union of Concerned Scientists, Cambridge, MA.

United Steelworkers of America. 1990. Our Children's World: Steelworkers and the Environment. Report of USWA Task Force on Environment. United Steelworkers of America, Washington, DC.

Westra, L. 1994. *An Environmental Proposal for Ethics: The Principle of Integrity.* Rowman and Littlefield Publishers, Lanham, MD.

Woodwell, G. M., ed. 1990. *The Earth in Transition: Patterns and Process of Biotic Impoverishment.* Cambridge University Press, Cambridge, UK.

World Commission on Environment and Development. 1987. *Our Common Future.* Oxford University Press, Oxford, UK.

CHAPTER TWO

VALUATION OF ECOSYSTEM SERVICES TOWARDS ECOLOGICAL CITIZENSHIP

PAVEL CUDLÍN, JOSEF SEJÁK
AND JAN POKORNÝ

Introduction

Land systems in almost all developed countries call for a new integrated resource management toward greener sustainable landscape that will cover the material needs of the current generation, while optimizing restoration activities toward potential natural vegetation. It is necessity to introduce a new unique integrative methodology for the co-existence of traditional economic land-use with optimal landscape restoration management through integrated ecological and economic valuations.

Earth System Governance initiative consists of such distribution of resources that reconciles the needs of human generations, while surviving and restoring the ecosystem services. Human societies and their economies decisively depend on the life supporting functions and services of earth's ecosystems (MEA 2005). Natural ecosystems protect against harmful cosmic radiation; continually control the composition of atmosphere; produce fertile soil and biomass; clear air and water; mitigate climate extremes; maintain biodiversity; decompose organic waste etc. (deGroot et al. 2002).

The principal activities could be to develop the capability for understanding, assessing, and enhancing human use of terrestrial ecosystems and their multiple services, considering key sectors of rural land use in case study areas, with a view to their interpretation across wider regions. The research activities have to focus on describing and understanding both previous and current ecosystem functions, values and services to society, in rural, urban or peri-urban environments, and linkages to and influences from the natural, economic and socio-cultural environment (Costanza et al. 1997).

For hundreds of millions of years, the landscape ecosystems (vegetation on continents) have been creating the life-supporting conditions for heterotrophic organisms. Self-organizing processes in autotrophic ecosystems tend to climax in vegetation that is characterized by maximal use of solar energy and by maximal ability to produce life-supporting conditions (composition of atmosphere, mitigation of temperature extremes, cleaning air and water, retaining nutrients, etc.) and to keep the nutrients and water inside the ecosystem (Ripl 2003).

However, over the last hundred years, expanding human populations and their insatiable self-interested striving for material wealth have caused a situation in which half of the most important world ecosystems have been destroyed. One substantial reason for the massive destruction of landscape ecosystems is the fact that their life-supporting services have not yet been valued, and physical and legal persons could freely change them into anthropogenic surfaces (traditionally into agricultural lands, recently mainly into construction sites, industrial zones, technical infrastructures, etc).

"Land development" of natural surfaces became one of the most profitable businesses in current market economies and, hand in hand with the belief in a never-ending economic growth, with growing loans, sub-prime debts and overgrowth of financial assets, has led to current financial and economic crises. Developers are changing natural ecosystems free of charge and transfer the negative externalities onto the taxpayers who are not only losing ecosystem services, but also have to pay for ex-post nature restoration activities. Therefore, it is high time to internalize the negative externalities to those who economically benefit from the destruction of landscape ecosystems (Seják and Cudlín 2010).

Simultaneously, at a macroeconomic level, the system of national accounts does not account for the degradation in ecological services and does not record the depletion and degradation of natural and environmental resources. Only direct ecosystem provisioning services (natural resource use: agricultural, forest and urban land and mineral and fossil resource sites) are valued by humans on markets and have been creating severe pressures on all other life-supporting ecosystem services.

Ecological Citizenship as a Driving Force for "Alternative" Sustainable Consumption

There have been several suggestions in scientific literature regarding different ways that citizenship and the environment might be related, for instance "ecological citizenship" (Christoff 1996, Smith 1998, Curtin

2002, Dobson and Sáiz 2005), "environmental citizenship" (Luque 2005, Dobson and Bell 2006), "green citizenship" (Dean 2001), "sustainability citizenship" (Barry 2006), "environmentally reasonable citizenship" (Hailwood 2005) or "ecological stewardship" (Barry 2002). Let us note first two terms in greater detail. Ecological citizenship can be considered as an example of post-cosmopolitan citizenship, in contrast with environmental citizenship, expressed in the language of the two major citizenship traditions; liberal and civic republican (Dobson and Sáiz 2005). It possesses four principal characteristics: i) it deals in the currency of non-reciprocal responsibility; ii) it works with a non-contiguous and non-state understanding of political space, best understood in terms of the "ecological footprint"; iii) it argues that the private arena is as much a sphere for citizenship as the public arena; iv) it stresses the notion "citizenship virtue" requiring ecological citizenship's specific obligations.

Ecological citizenship could be instantiated in the economic sphere, through practices such as ecological modernization (Christoff 1996), social economy, sustainable consumption (Seyfang 2005), ethical investment (Carter and Huby 2005), participatory rights and personal duty (Melo-Escrihuela 2008). Environmental education (Hailwood 2005, Gough and Scott 2006) and activism (Latta 2007) might also help green understandings of citizenship.

Ecological citizenship can be considered as a driving force for "alternative" sustainable consumption, via expression through consumer behavior, such as purchasing local organic food (Seyfang 2005).

The ecological footprint can be considered as the ecological citizenship's version of political space, and global warming as an example of the asymmetrical relations of globalizing cause-and-effect that calls forth post-cosmopolitan obligations (Dobson and Sáiz 2005). The ecological footprint of a specified population can be characterized as the area of land and water ecosystems required on a continuous basis to produce the resources that the population consumes, and to assimilate the waste that the population produces (Rees 2003, Wackernagel and Rees 1996).

The aim of our paper is to collect arguments confirming that valuation of ecosystem services can promote ecological citizenship consciousness.

From Ecosystem Processes to Ecosystem Functions and Ecosystem Services

According to de Groot et al. (2002) ecosystem processes could be defined as a complex of interactions among abiotic and biotic parts of ecosystems (backed with energy and matter) and ecosystem functions as a capacity of natural processes and parts to provide ecosystem services (Figure 2-1). In a broad sense, ecosystems processes involve the transfer of energy and substances (Lyons et al. 2005).

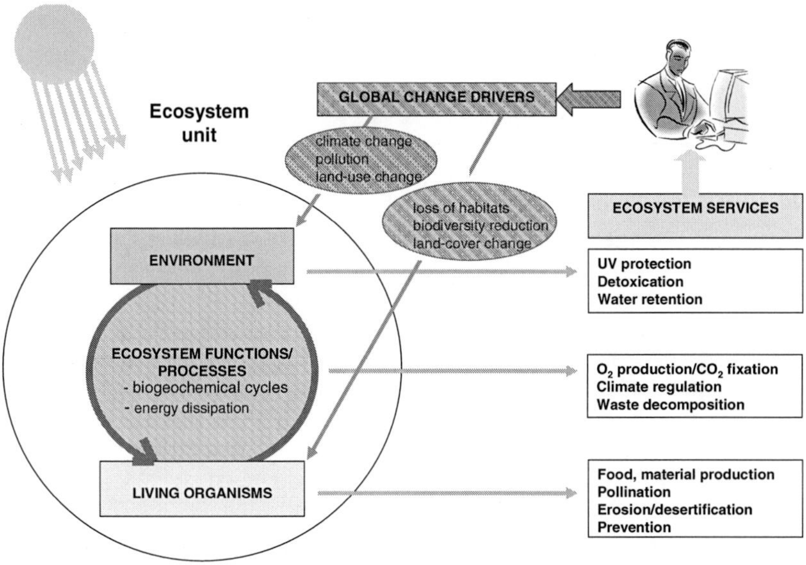

Figure 2-1. Global change effects on ecosystem processes, functions and services.

Ecosystem services are the benefits that humans obtain from ecosystems (MEA 2005) and the concept of them has become an important model for linking the functioning of ecosystems to human welfare (Fisher et al. 2009). Ecosystem services also have present and future dimensions. The patterns in which they are used today will affect their ability to continue to "produce" the outputs that people care about and, thus, people's "option" to use such goods and services in the future (Bingham et al. 1995).

Global driving forces (not only for ecosystem processes) are categorized as demographic, economic, socio-political, cultural and religious, scientific and technological, and physical and biological. Drivers in all categories other than physical and biological are considered to be indirect. Important direct drivers include, among others, changes in climate, plant nutrient use, land conversion, and diseases and invasive species (Nelson at al. 2006). For example, nutrient cycling is a process where one outcome is clean water. Nutrient cycling is a service that humans utilize, but only indirectly.

Clean water provision is also a service that humans utilize, this time directly. Clean water, when consumed for drinking, is a benefit of ecosystem services. The benefit being the point at which human welfare is directly affected and the point where other forms of capital (built, human, social) are likely needed to realize the gain in welfare (Fisher et al. 2009).

Often discussed is a role of biodiversity for intensity of various ecosystem processes and ecosystem services. The functional groups of organisms could be divided into two categories: keystone process species and species fundamental for ecosystem resilience (Folke et al. 1996). It means that at least some minimum number of species is essential for an ecosystem functioning under constant conditions, and a larger number of species is essential for maintaining the stability of ecosystem processes in a changing environment (Loreau et al. 2001). The ecosystems are dynamic and their populations of species rise and fall. One species may substitute for another species, physical processes change. Natural or human effects cause changes in ecosystem structure and function. However, some ecosystems are more desirable to people than others. Some ecosystems provide more habitats for threatened or endangered species than others; some ecosystems provide more water purification service or recreation opportunities than others (Bingham et al. 1995).

The conservation costs and benefits are distributed across markedly different types of stakeholder recipients, from local users deriving short-term benefits in terms of consumption and/or income from locally marketed products, to citizens of other countries deriving welfare from the long-term indirect-benefits from ecosystem services, such as carbon sequestration (Turner et. al. 2003).

Valuation of Ecosystem Services

Non-market valuation approaches

Demand curve (preference) methods

People traditionally value only those parts of nature that provide them with some benefits (utility). Within this utilitarian approach, values of natural assets are derived in mainstream economics as sums of discounted future net benefits from their utilization (Seják and Cudlín 2010).

The utilitarian approach has been used since the mid 19th century for valuing natural resource services (e.g. Faustmann 1849). Natural resources have been valued, as production factors and assets, through their market services in economic activities (production, exchange and consumption) as a sum of the future net benefits (rents) discounted to present value in finite or infinite time horizons (for non-renewable and renewable natural resources).

Since the 1870s the utilitarian paradigm has been supplemented by the neoclassical concept of marginal valuation. The value of a good has been derived from its marginal utility for the user. The total value of a commodity is given by the number of commodity units produced (supplied) multiplied by the marginal utility from the last unit consumed (demanded).

The general expression for the value of an asset, V, in the base year, 0, is simply the sum of the net economic benefits it yields in each year t, over the lifetime, T, of the asset, discounted to present value by the discount rate, i.

$$V = \sum_{t=0}^{T} \frac{r_t Q_t}{(1+i)^t} \qquad (1)$$

where r is the unit rent calculated as revenue minus the marginal cost of harvesting, and Q is the total harvest in a given period (Lange 2004: 74).

The most frequent assumptions in natural resource valuation are of constant rent benefit r over time, as well as constant discount rate i over time, and in the case of renewable natural resources, the assumption of an infinite time horizon for their use (formally it leads to an infinite time series that is convergent; see time series in higher mathematics). With these assumptions, formula (1) can be written in the following simple form

$$V_s = \sum_{t=1}^{\infty} \frac{r}{(1+i)^t} = \frac{r}{i} \qquad (2)$$

where V_S = natural resource value under assumptions of constant *r* and *i* and infinite time horizon.

This is the simplest formula for estimating natural resource value by means of the revenue method. The summation of discounted rents as infinite time series converges to the ratio r/i. It is the well-known formula of capitalized rent that shows the natural resource value V_S as an amount that, invested into a bank at an interest rate of *i*, annually brings an interest of *r*.

During the 20th century, the rapid growth in the use of natural resources resulted in half of the most valuable world ecosystems (mainly natural forests and wetlands) being destroyed and changed into anthropogenic surfaces. Only at the end of the 20th century did scientists realize that humankind and all other living species are crucially dependent on the life-supporting functions of ecosystems. This consciousness has been common for natural nations and has crossed through philosophy history since Platon time. Similarly, as in the case of natural resource valuation, non-market environmental assets started to be valued within the utilitarian tradition from the late 1960s and early 1970s by identifying individuals' willingness to pay for their non-market environmental functions and services. Scientists have started to develop hypothetical markets for the valuation of supporting, regulating and cultural services of ecosystems, asking people (consumers) what they are willing to pay for such services.

A number of imputation methods have been developed that estimate total economic value by revealing the preferences held by human individuals (Moran and Bann 2000, Farber et al. 2002, Turner et al. 2003). Within the utilitarian paradigm, the concept of total economic value does not encompass any value that intrinsically resides in environmental assets (Turner et al. 1994: 109). All these utilitarian, demand-side approaches (stated preference methods, like *Contingent Valuation Method*, *Travel Cost Method*, *Hedonic Method* etc.) suffer from the fact that valuations are done by consumers who are far from integrating ecosystem services into their value systems.

Supply-side methods (cost based and energy embodied)

Alternative theoretical concepts for valuing non-market ecosystem services come from the supply-side. The replacement cost of providing a substitute service that would perform a similar function to an ecosystem (e.g. flood protection service of wetland may be valued on the basis of the costs of building man-made flood defences). As Costanza et al. write:

> "In fact, one additional way to think about the value of ecosystem services is to determine what it would cost to replicate them in a technologically produced, artificial biosphere." (Costanza et al. 1997: 255).

Emergy valuations create the third group of non-market methods, based on emergy analysis, i.e. an analysis of energy embodied in ecosystems, have been elaborated especially in the United States. Emergy (embodied energy) is defined as available energy, required directly and indirectly to make a service or product, and measures both the work of nature and that of humans in generating products and services [units: emjoules]. Emergy theory comes from the presumption that all the real wealth of the environment comes from the work of the geobiosphere in the form of sun, tides, and heat energy deep in the earth (Odum H.T. 1996). Emergy analysis identifies full energy inputs and reflects the supply curve approach (broader then the labour theory). However, up-to-now emergy methods abstain from reflecting the hugest solar energy processes and their embodiment in synergies of vegetation and water phases.

While demand-side non-market valuations can be criticized as unilateral approaches to revealing an economic value of ecosystem goods and services (as they answer only how much some human individuals are – very often hypothetically - willing to pay or willing to accept), similar criticism is often targeted toward a cost-based valuation as insufficient reflection of real ecosystem benefits. Another criticism can be, and is, led against fragmentary characters of preferential valuations as they value some selected cases and do not pretend to a systemic ecological valuation of national territories.

Evaluation of Ecological Injury Caused by Decreased Ecosystem Service Provision

Our effort comes from the intention that biodiversity protection, conservation, and sustainable use should be enlarged from protected areas into the whole national territories. This can be put into practice by the economic instruments counterbalancing the standard economic land uses and the market prices of "land development". Destruction of biodiversity continues due to the fact that, up to now, ecosystems (and biotopes as their carriers), have not been expressed in humans' value systems (as showed in MEA 2005).

Two methods of systemic nature valuations

In the Czech Republic, during the last decade environmental researchers developed two main methods of systemic ecological valuation of national territory.

Biotope valuation method (BVM) identifies the carrying capacity of individual biotopes as specific environments for plant and animal species and reflects mainly biodiversity aspects of biotopes (point valuation of species and biotope type diversity, their rareness and vulnerability). All biotopes of national territory are ranked by eight ecological characteristics (similar as in Hessian state of Germany) that are aggregated into the resulting number of points. Biotope classification moves out of Natura 2000 mapping and more in detail of developed man-influenced biotopes (Cudlín et al. 2005). Biotope monetary values reflect the average restoration costs necessary for maintaining and improving the landscape quality. The monetary value of one point (€ 0.4944) is derived from a set of representative restoration projects as average national cost needed for one point improving of $1m^2$ of the biotope (Seják et al. 2003, summary see: http://fzp.ujep.cz/projekty/bvm/bvm.pdf).

Energy-water-vegetation method (EWVM) has been elaborated recently (Seják et al. 2010) and is based on studies of differences in the sun energy dissipation and water flows, described by the Energy-Transport-Reaction model (Ripl 1995), which produce the main life supporting services for humans (e.g. climate regulation, water and nutrient retention, CO_2 sequestration). Water has the second highest specific heat capacity of all known substances, after ammonia, as well as a high heat of vaporization (40.65 $kJ \cdot mol^{-1}$), both of which are a result of the extensive hydrogen bonding between its molecules. These two unusual properties allow water to moderate Earth's climate by buffering large fluctuations in temperature.

Due to these water properties, and due to the decisive role of vegetation in natural ecosystems that we pointed out, the dominating roles that energetically open natural ecosystems play in the biosphere are climate regulation and water retention services (as shown in the picture below by differences in evapotranspiration). According to real monitored data on differences in the sun energy-dissipation, it is verified that through the self-organizing processes natural ecosystems tend to climax in vegetation, in which the energy use efficiency is maximized, as well as the loss processes minimized.

Coming from the MEA argumentation that "The capacity of ecosystems to provide services derives directly from the operation of natural biogeochemical cycles that in some cases have been significantly modified" (MEA 2005: 33), attention was focussed on the regulating ecosystem services, mainly on climate regulation and water retention services, which have direct impacts on mitigating the global warming processes. The procedure is based on quantifying the efficiency of solar flux dissipation by means of vegetation and water, through evapotranspiration and photosynthesis processes (using monitored data from different types of ecosystems). Air-borne and satellite remote service data are also involved to quantify selected ecosystem functions at landscape level (Seják et al. 2010).

Monetary values of those ecosystem services were estimated by the ***replacement cost method***, i.e. through the cost of providing a substitute service that would perform a similar function to an ecosystem. One example of such valuation is summarized in the Table 2-1.

Table 2-1. Annual ecosystem service values of 1 ha of river floodplain.

Service	Calculation	Value
Flood control service	Investment costs for the retention of 1 m^3 of water by a man-made pond, in the Czech Republic, is € 4. For 1 ha of floodplain with flood control capacity of 5,000 m^3 it is € 20,800 of capital costs, which brings presented annual service (using 5% discount rate).	€ 1,040
Biomass production	5 tons annually x 4,000 kWh x € 0.08/kWh (electricity cost price) x 0.5 (efficiency)	€ 800
Nutrient retention	1 t of base cations and nutrients compared to eroding drained arable lands = 1,000 kg x € 1.4 (price of 1 kg fertilizers)	€ 1,400
Wildlife habitat provision	Alluvial *Alopecurus* meadows are valued according to BVM by 46 points per 1 m^2, per 1 ha it means 460,000 points x € 0.4944 per point = € 227,424 of capital value, which brings presented annual service (using 5% discount rate).	€ 11,370
Oxygen production	3.5 mil. litres O_2 x € 0.01-0.03 per litre	€ 70,000
Climate regulation	Modest estimation is 500 litres of evapotranspired water from 1m^2 during vegetation season. Annual climate regulation service of 1 ha can thus be estimated 500 x 1.4 kWh (0.7 kWh cooling, 0.7 kWh warming) x 10,000 x € 0.08 (electricity cost price).	€ 560,000
Total annual services from 1 ha of river floodplain		€ 644,610

Having the results of the first global demand-side valuation of the world ecosystem services (Costanza et al. 1997) and respecting the definition of economic value of goods and services as a result of demand-side and supply-side aspects, we revealed that the replacement cost value of four selected ecosystem services (i.e. climate regulation, small water cycle support, oxygen production, and wildlife habitat provision) is much higher compared to human willingness to pay for such services (Seják et al. 2010).

Based on demand side methods (preference approach), Costanza et al. (1997) estimated the world annual ecosystem services as 1.8-fold of annual world GDP (USD 18 trillion). Our pilot estimation of annual ecosystem service values in the CR, based on replacement cost methods, assessed that annual ecosystem services at least fifty times surpass an annual GDP.

Conclusion

Nature or near-nature biotopes provide a large array of services (e.g. production of biomass, climate regulation, prevention of floods and soil erosion, maintenance of genetic resources), which are highly sensitive to changes of climate and land use. Science, the community, private business, and policy makers must engage with each other in a robust manner to come up with an evidence-based research agenda which reflects (i) the sensitivity of the supply of ecosystem services to global change, (ii) the socio-economic demand for ecosystem services and (iii) the required design of policy instruments, market mechanisms, and land-use practices to support the adaptation of individual regional ecological economic systems to global change.

Command-and-control and market-based instruments could be reviewed to better understand their role for cost-effective decision-making related to biodiversity conservation and sustainable use at different government levels and support to ecosystem services provision, particularly under conditions of uncertainty created by global change. Guidelines and/ or scenarios could be drawn to apply benefit transfer in a systematic way. Changes in ecosystems services could be evaluated for policy purposes, as well as the distributional effects of changes in biodiversity policies.

Taking into account that a major part of decisive life-supporting ecosystem services has not yet entered into the value system of human individuals, the replacement cost approach and the values derived seem to be the efficient way for quantifying the decisive existential importance of ecosystem services for the human species.

Dissemination of information about the values of ecosystem services could promote ecological citizenship consciousness in a similar way as an ecological foot-print has done.

References

Barry, J. "Vulnerability and Virtue: Democracy, Dependency, and Ecological Stewardship." In *Democracy and the Claims of Nature: Critical Perspectives for a New Century*, edited by B.P. Taylor and B. A. Minteer, 133-152. Lanham MD: Rowman & Littlefield, 2002.

—. "Resistance is Fertile: From Environmental to Sustainability Citizenship." In *Environmental Citizenship*, edited by A. Dobson and D. Bell, 21-48. London: MIT Press, 2006.

Bingham, G., Bishop., R, Brody, M., Bromley, D., Clark, E., Cooper, W., Costanza, R, Hale, T., Hayden, G., Kellert, S., Norgaard, R., Norton, B., Payne, J., Russel, C. and Suter G. "Issues in ecosystem valuation: improving information for decision making." *Ecological Economics* 14 (1995): 73-90.

Carter, N. and Huby, M., "Ecological citizenship and ethical investment." *Environmental Politics* 14 (2005): 255-272.

Christoff, P. "Ecological Citizens and Ecologically Guided Democracy." In *Democracy and Green Political Thought: Sustainability, Rights and Citizenship*, edited by B. Doherty and M. De Geus, 151-169. New York: Routledge, 1996.

Costanza, R., d'Arge, R., De Groot, R., Farber, S., Grasso, M., Hannon, B., Limburg, K., Naeem, S., O'Neill, R. V., Paruelo, J., Raskin, R. G., Sutton, P. and van den Belt, M. "The Value of the World's Ecosystem Services and Natural Capital." *Nature* 387 (1997): 253–260.

Cudlin, P., Prokopova, M., Francirkova, T., Buresova, R., Smrz, T. and Boucnikova, E. "System NATURA 2000 utilization for purposes of biotope valuation." *Ekológia (Bratislava)* 24 (2005): 52-68.

Curtin, D. "Ecological Citizenship." In *Handbook of Citizenship Studies,* edited by E.F. Isin and B.S. Turner, 293-304. London: Sage, 2002.

Gough, S. and Scott, W. "Promoting environmental citizenship through Learning: Toward a theory of change." In *Environmental citizenship*, edited by A. Dobson and D. Bell, 263-285. Cambridge, MA: MIT Press, 2006.

de Groot, Rudolf S., Wilson, Matthew A. and Boumans, Roelof M. J., "A Typology for the Classification, Description and Valuation of

Ecosystem Functions, Goods and Services." *Ecological Economics* 41 (2002): 393-408.

Dean, H. "Green citizenship." *Social Policy & Administration* 35 (2001): 490-505.

Dobson, A. and Sáiz, A.V. "Introduction." *Environmental Politics* 14 (2005): 157–162.

Dobson, A. and Bell, D. *Environmental citizenship*. MIT Press, 2006.

Farber, S.C., Costanza, R. and Wilson, M.A. "Economic and Ecological concepts for valuing ecosystem services." *Ecol. Econ.* 41 (2002): 375-392.

Faustmann, M., "On the Determination of the Value Which Forest Land and Immature Stands Possess for Forestry." In *Martin Faustmann and the Evolution of Discounted Cash Flow*, edited by M. Gane. Oxford Institute, 1849.

Fisher, B., Turner, R.K. and Morling, P. "Defining and classifying ecosystem services for decision making." Conference Information: Valuing Wild Nature Workshop, MAR 12-16, Univ. Norwich, England. *Ecological Economics* 68 (2009): 643-653.

Folke, C., Holling, C.S. and Perrings, C. "Biological diversity, ecosystems, and the human scale." *Ecological applications* 6 (1996): 1018-1024.

Gough, S. and Scott W. "Education and sustainable development: a political analysis." *Educational Review* 58 (2006): 273-290.

Hailwood, S. "Environmental Citizenship as Reasonable Citizenship." *Environmental Politics* 14 (2005): 195–210.

Hartley, D. "Green citizenship." *Social policy and Administration* 35 (2001): 490-505.

Lange, G.M. *Manual for environmental and economic accounts for forestry: a tool for cross-sectoral policy analysis*. FAO Forestry Dept., 2004.

Latta, P. A. "Locating democratic politics in ecological citizenship." *Environmental Politics* 16 (2007): 377-393.

Loreau, M., Naeem, S., Inchausti, P., Bengtsson, J., Grime, J.P., Hector, A., Hooper, D.U., Huston, M.A., Raffaelli, D., Schmid, B., Tilman, D. and Wardle, D.A. "Ecology - Biodiversity and ecosystem functioning: Current knowledge and future challenges." *Science* 294 (2001): 804-808.

Luque, E. "Researching Environmental Citizenship and its Publics." *Environmental Politics* 14 (2005): 211–225.

Lyons, K.G., Brigham, C.A., Traut, B.H. and Schwartz, M.W. "Rare species and ecosystem functioning." *Conservation Biology* 19 (2005): 1019–1024.

MEA. *Ecosystems and Human Well-being: Synthesis. Millennium Ecosystem Assessment.* Washington DC: Island Press, 2005.

Melo-Escrihuela, C. "Promoting Ecological Citizenship: Rights, Duties and Political Agency." *Acme* 7/3 (2008):113-134.

Moran, B. and Bann, C. *The Valuation of Biological Diversity for National Biodiversity Action Plans and Strategies.* UNEP, March, 2000. Accessed September 28, 2010. www.undp.org/bpsp/thematic_links/enveco.html.

Nelson, G. C., Bennett, E., Berhe, A.A., Cassman, K., DeFries, R.. Dietz, T., Dobermann, A., Dobson, A., Janetos, A., Levy, M., Marco, D., Nakicenovic, N., O'Neill, B., Norgaard, R., Petschel-Held, G., Ojima, D., Pingali, P., Watson, R. and Zure, M. "Anthropogenic drivers of ecosystem change: An overview." *Ecology and society* 11 (2006): 29.

Odum, HT. *Environmental accounting - emergy environmental decision making*. New York: John Wiley & Sons, 1996.

Rees, W.E. "Understanding Urban Ecosystems: An Ecological Economics Perspective." In: *Understanding Urban Ecosystems*, edited by A. R. Berkowitz, Ch.H. Nilon and K.S. Hollweg, 115-136. New York: Springer-Verlag, 2003.

Ripl, W. "Water: the bloodstream of the biosphere." *Phil. Trans. R. Society Lond. B* 358 (2003). 1921-1934.

—. "Management of Water Cycle and energy flow for ecosystem control: the energy-transport-reaction (ETR) model." *Ecological Modelling* 78 (1995): 61-76.

Seják, J., Dejmal, I., Petříček, V., Cudlín, P., Míchal, I., Černý, K., Kučera, T., Vyskot, I., Strejček, J., Cudlínová, E., Cabrnoch, J., Šindlar, M., Prokopová, M., Kovář, J., Kupka, M., Ščasný, M., Šafařík, M., Roušarová, Š., Stejskal, V. and Zapletal, J. *Hodnocení a oceňování biotopů České republiky* (in Czech). Praha: Český ekologický ústav, *MŽP*, 2003.

Seják, J. and Cudlín, P. "On measuring the natural and environmental resource value and damages." *Studia Oecologica* 2 (2010): 53-68.

Seják, J., Cudlín, P., Pokorný, J., Zapletal, M., Petříček, V., Guth, J., Chuman, T., Romportl, D., Skořepová, I., Vacek, V., Vyskot, I., Černý, K., Hesslerová, P., Burešová, R., Prokopová, M., Plch, R., Engstová, B. and Stará, L. *Hodnocení funkcí a služeb ekosystémů České Republiky* (in Czech), Usti nad Labem: *FŽP UJEP*, 2010.

Seyfang, G. "Shopping for sustainability: can sustainable consumption promote ecological citizenship?" *Environmental Politics* 14 (2005): 290-306.

Smith, M. *Ecologism: Toward Ecological Citizenship*. Milton Keynes: Open University Press, 1998.

Turner, R.K., Paavola, J., Cooper, P., Farber, S., Jessamy, V. and Georgiou, S. "Valuing nature: lessons learned and future research directions." *Ecol. Econ*. 46 (2003): 493-510.

Turner, R.K., Pearce, D. and Bateman, I. *Environmental Economics, An elementary introduction*. London: Harvester Wheatsheaf, 1994.

Wackernagel, M. and Rees, W.E. *Our ecological footprint – Reducing Human Impact on the Earth*. Gabriola Island: New Society Publishers, 1996.

CHAPTER THREE

LIVING GREENFULLY, EATING GREENFULLY

ROBERT GOODLAND[1]

Introduction

The term "Green" is much used and abused these days. Catchphrases such as "Live Green" and "Eat Green" are used to define and market all sorts of products and lifestyles, including many that are not in fact environmentally sustainable. This article will explain what is necessary for fully "Green" life on earth – what this article will call "Living Greenfully" and "Eating Greenfully."

In Copenhagen in December 2009 and almost a year later in Cancun, policymakers have convened for two major conferences to agree on a new international treaty to reverse climate change. Both times they have largely failed to agree on their stated objectives. Yet between the two events, temperatures hit new highs in seventeen countries. The previous record was in 2007, with new highs in fourteen countries. For example, last August outdoor temperatures in Moscow rose above 100 degrees for the first time, making people wear face masks to perform daily activities.

Even were policymakers to have reached an agreement in Copenhagen or Cancun, it would likely have reflected their stated objectives, which focused principally on the development of renewable energy infrastructure to reduce atmospheric carbon.

Yet the common prescription of renewable energy infrastructure to keep temperatures from rising more than two degrees centigrade is estimated to cost eighteen trillion dollars and take decades to install[2] – long past the several years that most climate experts estimate remain before the tipping point for climate catastrophe.

Still, the prescription for renewable energy must be filled to keep emissions down over the long term. Meanwhile, something must be done about excess carbon dioxide in the atmosphere today that will take at least a century to dissolve, without new capacity to absorb it. There is only one known way to create such new capacity in time: It is to allow some land now set aside for livestock and feed production to regenerate forest – by replacing a sizeable amount of today's livestock products with substitute products that are tasty, easy to prepare, and inexpensive. It involves no waiting for a new international treaty, and no trillions of dollars for new infrastructure.

The Livestock-Climate Connection

In 2006, the FAO published a 388-page report entitled *Livestock's Long Shadow,* assessing for the first time in a major publication the greenhouse gas (GHG) emissions attributable to livestock's supply chain – from forests cleared to supermarkets.[3]

Three years later, Jeff Anhang and I prepared an article for *World Watch,*[4] in which we consider whether any sources of GHG emissions might have been missed in FAO's *Livestock's Long Shadow.* The key ones that we found missing are in the land set aside for both livestock and for feed production, along with several other significant sources. So our article suggests that livestock's shadow is not only long but colossal, responsible for at least 51% of human-caused GHG emissions.

Our analysis has been widely cited, including on the Government of China's official climate change website.[5] Chris Mentzel, the CEO of Clean Energy Maui LLC, has written that our analysis persuaded him that a one percent reduction in meat would have the same environmental effect as $3 trillion of solar energy financing.[6]

We developed our case using our environmental assessment experience at the World Bank Group. In my 23 years there, I wrote most of the Bank's environmental policies and assessed environmental risk in numerous industries, countries, and projects. For my co-author's name to be on our article, it had to be cleared by World Bank Group management, which required extensive peer review.

Our case is the only pragmatic one available for reducing global warming quickly. Most climatologists estimate that few years remain before climate change becomes irreversible. Yet the common idea of replacing fossil fuel infrastructure with renewable energy requires decades to implement. It must still be pursued to keep emissions down over the long term. Meanwhile, alternatives to livestock products can be scaled up and can significantly reduce emissions quickly and inexpensively.

Some people will have trouble believing that eating meat can cause any change in climate, let alone imperil humanity. However, they may not know that the world's population of land-based livestock has grown sixfold since 1960 – so more than 50 billion animals will be raised in 2010 alone.

One quarter of land worldwide is now used for grazing livestock, and a third of all arable land is used for growing feed. Twenty percent of the Amazon forest has been destroyed, fivefold the area of England, mainly for livestock and feed production. When tropical forest is burned, not only are greenhouse gases emitted, but the world's largest carbon sink is steadily shrunk.

Lower estimates than ours of livestock-related emissions don't count all of livestock's indirect inputs. They also don't count the direct impact of livestock respiration – or its reflection in foregone carbon absorption in land set aside for livestock and feed production.

Our analysis proposes counting either carbon in respiration or foregone carbon absorption attributable to livestock. That's because reality no longer reflects the old model of the carbon cycle, which proposes that photosynthesis perfectly offsets respiration. That model assumed roughly constant levels of respiration and photosynthesis on Earth. But respiration has increased exponentially with livestock – while these animals have caused a dramatic decline in the Earth's photosynthetic capacity, along with large and accelerating increases in soil carbon volatilization.

Assessing Assessments

Greenhouse gas emissions and climate change are transboundary – while animal feed, animal products, and alternatives to animal products are often sourced in the form of commodities flown, shipped, and trucked all over the world, often with weak capacity for tracing their source. For example, livestock in many parts of the world consume marine feed whose source is usually difficult to trace but probably distant; and fish is often mislabeled as wild from a nearby source when it is actually farm-raised from a distant source, where it was fed grain whose source may be difficult to trace but is probably distant. Such facts suggest that the climate risk of animal agriculture should be assessed primarily at the global level.

Some experts sometimes argue for changing the subject to overall human consumption, and argue for reducing it and/or reducing human population. These may be worthy goals. However, efforts to fulfill these goals have been made for many years – while growth in human population and in human consumption has increased. It may always be difficult to control human urges; while livestock's urges are already almost completely controlled.

For fully assessing the environmental impacts of any product and its alternatives – such as livestock products and their alternatives – a tool of choice for environmental professionals is often Lifecycle Assessment. Within Lifecycle Assessment, the most widely-used protocol for assessing GHGs is the "GHG Protocol."[7] This protocol sets out three levels of greenhouse gases (GHGs) to be assessed:

- LCA-1, which is to count all GHGs directly attributable to activity of the industry or project under assessment;
- LCA-2, which is to count all GHGs not directly attributable to activity of the industry or project under assessment, but attributable to activity related to products or services purchased by the industry or project, or otherwise incorporated into the activity under assessment (such as materials, fuel, and/or electricity purchased by the sponsor); and
- LCA-3, which is to count all GHGs attributable to the lifecycle and supply chain (upstream and downstream) of the project or industry under assessment that are not counted under LCA-1 or LCA-2.

A number of media reports have cited lifecycle assessment of livestock-related GHGs by Frank Mitloehner, a professor at the University of California at Davis, who has acknowledged receiving funding from the livestock industry.[8] Dr. Mitloehner has made numerous misstatements about the lifecycle assessment of livestock-related GHGs. Some media outlets have followed up with critiques of both Professor Mitloehner's claims and some initial media coverage.[9] For example, in one of his presentations:[10]

- There is a misstatement that the use of fertilizer, pesticide, irrigation, etc. comes under LCA-3 – when it actually comes under LCA-2.
- The figure of 21 for the Global Warming Potential (GWP) of methane is used – without updating it to the now commonly-used figure of 25 for a 100-year timeframe – and without explaining why a 100-year timeframe would be more appropriate than a 20-year timeframe, by which the GWP of methane would be over 70.
- There is a misstatement in the Q&A that a figure of 3% of U.S. greenhouse gas emissions represents LCA-3 (i.e., "true LCA") – when 3% is actually US EPA's estimate of LCA-1 and LCA-2 at

most. There is also omission of any appropriate reference to recent scholarly criticism of US EPA's estimate.[11]

- There is omission of any assessment of LCA-3 – assessment of which is good practice according to the most widely-used international protocol for GHG lifecycle assessment.

The Worldwatch Institute recently posted a description of the World Bank Group's approval of livestock-climate analysis by Jeff Anhang and me.[12] In the first comment under the initial posting, one can see a major promoter of grass-fed livestock endorsing our analysis as applicable to intensively-raised livestock. Now that even some livestock promoters endorse our analysis, it is surprising to see environmentalists not do so. Our analysis is not only three years more recent than *Livestock's Long Shadow*, but Jeff Anhang and I are environmental assessment (EA) specialists[13] – while none of the authors of *Livestock's Long Shadow* is an EA specialist.

According to the EA benchmark most used by financial institutions, EA of activity with major risks should be guided by independent EA experts.[14] According to international good practice, an independent expert is defined in part as someone who does not collaborate with the industry to be reviewed. Independent experts guiding EA are expected to advise on the terms of reference for any EA; key issues and methods for preparing the EA; recommendations and findings of the EA; and implementation of the EA's recommendations.

Within the practice of EA, Analysis of Alternatives is the tool of choice for environmental professionals assessing the environmental impacts of alternatives to the activity proposed as the primary choice, such as animal agriculture. Analysis by Jeff Anhang and me incorporates Analysis of Alternatives, whereas *Livestock's Long Shadow* does not.

Sectoral environmental assessment (SEA) is the tool of choice for environmental professionals assessing environmental risks in any industrial or economic sector, such as animal agriculture. Jeff Anhang and I have previously collaborated in developing SEA of the livestock sector.[15]

Good practice in EA is for environmental risks and impacts never to be considered offset by economic benefits. For example, on both environmental and economic grounds, the World Bank's 2001 livestock strategy statement recommends that financial institutions "avoid funding large-scale commercial, grain-fed feedlot systems and industrial milk, pork, and poultry production."[16] Strangely, the lead author of that statement was Cornelius De Haan – who later became one of the co-authors of *Livestock's Long Shadow*. In assessing that report's objectivity, it seems relevant to note that it not only omits any recommendation for avoiding funding any livestock projects, but instead promotes a doubling of livestock production by 2050.

Jeff Anhang and I were subject to a World Bank Group rule requiring prior approval by management of any article by staff. So to gain approval for our "Livestock and Climate Change," it had to undergo extensive peer review within the World Bank Group by specialists on climate change, forests and soils, agribusiness, and development and climate change. Conversely, we are unaware of any peer review undergone by *Livestock's Long Shadow*.

FAO analysis is often cited as UN analysis – but the FAO is only one of 19 UN specialized agencies. They are all autonomous; none is governed by any of the 6 main UN bodies. While the FAO is one UN specialized agency, the World Bank and IFC – longtime home base for Jeff Anhang and me – are two UN specialized agencies.

Another UN agency has promoted our analysis: "Goodland and Anhang… [showed] at least 51% of all GHG emissions are attributable to livestock production... This represents an enormous shift in perspective... what may be a large-scale paradigm shift in the approaches to mitigating climate change." [17] Our analysis has also been promoted by the UN World Business Council for Sustainable Development.[18] Though our article critiques the FAO, it has actually invited us to deliver presentations in Rome and Berlin.[19]

Bad Breath

The two most cited analyses of the livestock-climate connection are *Livestock's Long Shadow* and analysis by Jeff Anhang and me. The most significant difference between the two is that the latter counts carbon dioxide in the respiration of the estimated 56 billion head of land-based livestock raised in 2009, while the former does not.

The second biggest discrepancy between the two analyses concerns methane attributable to livestock – which is important due to methane's relatively short half-life of 8 to 12 years as compared with carbon dioxide's 100 years. *Livestock's Long Shadow* uses the outdated Global Warming Potential (GWP) for methane of 23 instead of instead of our more appropriate GWP of 72, which is measured over the more appropriate timeframe of 20 years instead of *Livestock's Long Shadow's* 100 years.

When it comes to respiration, the specific metabolic rate of the main species raised in animal agriculture is broadly consistent, with each animal dissipating about two watts per kilogram to stay alive.[20] To reflect that, a roughly constant amount of carbon dioxide per unit of weight of livestock is respired, regardless of species.

Transportation fuels accounted for about 6 billion tons of carbon dioxide in 2009. In comparison, carbon dioxide from the breath of livestock raised in 2009 accounted for about 10 billion tons of carbon dioxide. Not only are the approximately 10 billion tons of carbon dioxide from the breath of livestock in 2009 invariable regardless of animal type, but there are also billions more tons of GHGs that are more or less invariable from these aspects of livestock production:

- GHGs from transporting each kg of livestock product;
- GHGs from solid and liquid waste from livestock and their products;
- GHGs attributable to livestock by-products such as leather and feathers;
- GHGs from the substantially higher amount of refrigerating, cooking, and packaging of meat versus alternatives; and
- GHGs attributable to carbon-intensive medical treatment of millions of cases worldwide each year of zoonotic illnesses (such as swine flu) and chronic degenerative illnesses (such as coronary heart disease, cancers, diabetes, and hypertension leading to strokes) linked to the consumption livestock products – but not to alternatives.[21]

Accordingly, close to half the GHGs attributable to the lifecycle and supply chain of livestock products are more or less invariable, regardless of the type of livestock produced. Therefore, while significantly more GHGs are attributable to beef than to other meats because of cattle's grazing, feed, enteric fermentation, and manure management, that significance is much smaller than commonly thought – and no particular meat product is likely to have a GHG footprint more than 25 percent lower than any other. As a result, replacing beef with chicken and pork would not result in any appreciable slowing of climate change.

Yet an article written by two authors of *Livestock's Long Shadow* proposes gradual replacement of beef by chicken and pork as a sustainable way of doubling meat consumption by 2050.[22]

Some experts have questioned why carbon dioxide in human respiration wouldn't be counted if it is considered legitimate to count carbon dioxide in livestock respiration. However, the world's livestock mass is about double human mass. So offsetting human respiration requires about half as much photosynthesis as does livestock respiration.

Also, photosynthesis might justifiably be considered to offset human respiration if meat and dairy substitutes can be produced with minimal change in land use, and minimal soil carbon releases. In comparison, livestock production now releases high levels of soil carbon worldwide, while more soil carbon releases occur because grain production worldwide must be maximized in order to produce the amount of meat and dairy consumed today. Meanwhile, emissions attributable to human exhalations and excretions have been seen as socially and politically justifiable choices of GHGs to remain uncounted.

Some experts have proposed that reducing animal agriculture might stimulate an increase in wild animals that would respire as much carbon dioxide as do livestock. However, studies have consistently shown that emissions attributable to wild animals are minimal relative to livestock's.[23]

Whether the true figure for GHGs attributable to animal agriculture is closer to 18 percent or to 51 percent, it is clear that GHGs attributable to livestock are inordinately high, have been increasing fast, and need to be reduced, as they pose the most urgent of risks to all life on earth (humans, animals, and plants).

Opportunity Knocks

For at least the past twenty years, the dominant global strategy for stopping climate change – indeed, practically the only strategy – has been for advocates to engage with politicians to reduce the use of fossil fuels. Every year this strategy has been implemented, the use of fossil fuels has increased. This has happened despite the fact that power utilities are among the most regulated businesses in the world.

Worldwide, the food industry is much less regulated than the power industry. So even if one could develop and implement a strategy to engage politicians to reduce the number of livestock, the chances of success from such a strategy are small.

It seems unlikely that the U.S. Congress will pass laws that would promote any significant climate change strategy at least through 2012, now that Republicans are on the ascent in Congress. Even before this development, American legislators have never dictated terms to the food industry. For many years, the truth has been the opposite, as the food industry has dictated terms on Capitol Hill.

So it seems up to the food industry and the general public to move forward with "Eating Greenfully." Yet from within the general public and the food industry, a number of questionable claims are often made.

Some claim that pasture-raised livestock make meat climate-friendly. But only about 8 percent of meat is produced from entirely pasture-raised livestock, and little land is available to increase this amount without further deforestation. These animals emit up to triple the methane released by factory-farmed livestock, and contribute greatly to soil carbon vaporization.

Some claim that technology can adequately mitigate livestock-related emissions. But available technology can mitigate emissions by only a trivial amount, and is impractical for most livestock.

Some argue that many poor people have no alternative to raising livestock for their livelihoods. But tens of millions of poor people's livestock have died recently due to climate catastrophes. Replacing these animals would risk a similar fate for the new livestock. Meanwhile, supporting new livelihoods for those whose livestock die in climate disasters would be less risky. Microfinance, off-grid electricity, computers, and mobile technology have generated dramatic growth in many poor rural communities.

The concept of "sustainable livestock" has recently been much promoted, in the British Parliament and elsewhere. However, while 8% of meat derives from wholly pasture-raised livestock, the figure is closer to 50% of livestock that get pasture-raised for some or all of their lives (according to the FAO). Whether the figure is closer to 8% or to 50%, most of the land used worldwide for livestock and feed production can regenerate forest.

In woody vegetation and the soil beneath, much more carbon can be sequestered than in grasslands now set aside for grazing and feed – as much as the increase in atmospheric carbon since 1980, according to James Hansen, the U.S. government's top climate scientist. Lord Stern and others say that saving or regenerating forest is the most cost-effective way of stopping climate change.

More demand for the same land seems bound to grow as production of biofuels increases due to peak oil. So there seems very little room left for romantic visions of grazing cows.

Nevertheless, some prominent British authors promote extensification of livestock as sustainable.[24] Such promotion often aims to have the cages opened of intensively-raised livestock – rather than to have people stop consuming intensively-produced livestock products.

Probably the only way all of today's livestock products could be extensively produced would be by having all remaining forest in the world destroyed. It should be obvious that this is a very unsustainable proposition. Not only is the promotion of "sustainable livestock" likely to increase deforestation; it also makes people more comfortable with eating all types of livestock products. This happens when intensive producers often falsely advertise their products as "free-range," projecting images of extensive open fields – though their production is actually intensive. Consumers often don't investigate before buying products based on such greenwashing. Media outlets advance such greenwashing when they blaze misleading headlines such as "How Eating Meat Can Save the Planet."[25]

Some assert that promoting locally-produced food would be a key way to minimize GHGs. However, research on the concept of "food miles" has shown that buying locally produced food may not minimize GHGs, but rather increase them. For example, one study concludes that it takes four times as much energy to produce British lamb locally than to import it from New Zealand (due to grain feeding of British sheep and heating of their sheds in winter).[26]

Some suggest that reducing food waste should be a key priority. However, food waste has been increasing gradually over the years, and involves problems that have been widely yet fairly evenly dispersed across the world. So it is a rather intractable problem on the one hand; on the other hand, there is no obvious pragmatic way to solve it.

In fact, among all categories of food, livestock products likely generate by far the most waste, as a large proportion of these products – along with extra packaging to protect them from their extreme perishability – are discarded precisely because of their perishability. In addition, some of their bone, fat, and gristle is used efficiently by slaughterhouses – but much gets discarded by butchers and by consumers.

While no clear new opportunity exists to reduce food waste generally, replacing livestock products with alternatives is a pragmatic business opportunity that can manage what is mainly a problem in the scale of livestock production that has ballooned in just the past half century – and which is helpfully concentrated at the production stage around the livestock industry, and at the consumption stage in urban areas.

Ethical Aims and Claims

Large animal welfare groups often pursue extensification of livestock as a key way to resolve ethical issues in animal production. Those groups have allies in some promoters of free-range animal agriculture who assert that extensification provides environmental benefits.

But according to the FAO and many environmental experts, extensification would significantly increase adverse environmental impacts, especially deforestation, and by extension GHGs. They assert that increased intensification will reduce GHGs.

Such debate over intensification *versus* extensification of livestock has been waged for many years, with no resolution in sight. Rather than attempt to resolve the environmental impacts of animal agriculture by focusing on animal welfare, a surer path to resolve the environmental impacts of animal agriculture would be through focusing on them directly.

Animal welfare groups might also usefully consider that the best strategy to promote "Eating Greenfully" will likely involve advocating to policymakers, industry, and individuals that they do something that fits their existing priorities, rather than something that requires them to change their priorities. Animal welfare ranks nowhere on most policymaker's lists of priorities. Conversely, climate change is among the very top few priorities for many policymakers, industries, and individuals across the world. Activists can implement a quick, global strategy of educating these audiences on how climate change endangers life on earth in the near term, but can be reversed through better alternatives to meat and dairy products.

Surveys consistently show that consumers buy food items primarily on the basis of price and quality. Quality includes ease and speed of preparation and delivery, cleanliness of retail outlets, and good taste. Only a small percentage of consumers have ever chosen food on the basis of animal welfare. Activists have a long track record of agitating for change in diets based on animal welfare and the healthfulness of meatless diets that has coincided with more meat-eating, not less, year after year.

Conversely, climate-friendliness has a reasonable chance of being accepted as part of the quality of food products, in part because there is no practical alternative for consumers to reverse climate change. In fact, animal welfare activists are increasingly accepting and even promoting the notion that allowing climate change to continue unabated is unethical.

Even if activists were to make no explicit ethical claim around a climate strategy, that wouldn't mean that a climate strategy might not be the most ethical strategy to win a world of more "Eating Greenfully." According to an overwhelming consensus of thousands of climate experts, the world has only 5-10 years to stop irreversible, catastrophic climate disruption. It would seem that allowing climate change to continue unabated is unethical.

When animal activists aim for "culture change" in the age of climate change – only 5-10 years before irreversible climate disruption – it would seem ethically sound for them to recognize that culture change is normally generational, taking typically 50-100 years to achieve, under the best of circumstances, with reversals not uncommon (e.g., consider the "Dark

Ages"). So working toward culture change could be profoundly unjust to animals, if activists could better succeed with a climate change strategy. A strategy of culture change might not only extend the misuse of animals for decades longer than under a climate change strategy, but mean that activists are resigning themselves to the possibility of extinction of most life on earth.

The legal status of animals must be decided upon legislature by legislature. Yet a new legislature sometimes reverses laws passed by a previous legislature. In any event, given the urgency of climate risk, there is no time to lobby all the world's legislatures. An emergency often requires suspending society's normal focus on justice, until the emergency is quelled. Yet justice for animals can actually be achieved through a climate strategy advocating "Eating Greenfully."

Some argue that animal activists should partner with livestock producers to improve animal welfare, and create an "enriched environment" for livestock. But if today's level of livestock production is unsustainable, then a different strategy would seem to be called for. Consider, for example, that when Ralph Nader advocated successfully against the Corvair and various unsafe practices by American carmakers in his 1965 book *"Unsafe at any Speed,"* neither he nor any Japanese carmaker ever negotiated with General Motors or the U.S. government officials for an "enriched environment" for the Corvair.[27] More recently, nobody has gotten anywhere advocating for an "enriched environment" that would allow tube TVs or analog phones to survive in the marketplace.

Eating Greenfully

It should now be understood that "Living Greenfully" involves most critically "Eating Greenfully." Yet habits form around people's choices of food – and these choices are induced by fiscal measures and marketing, which have strongly promoted livestock products.

However, fiscal measure and marketing can promote alternatives instead. Meat and dairy products can be replaced with substitutes such as seitan-based "chicken," soy-based "beef," nut-based milks, and coconut-based ice cream. Trying tasty new foods is commonly considered desirable.

Meat and dairy substitutes can be promoted like digital technology. Within a decade, manufacturers have switched almost entirely from analog televisions and telephones to digital versions, propelled by savings in materials and energy use, along with other improvements. Like digital technology, meat and dairy substitutes can deliver better quality at lower cost, while fulfilling the world's priority of reversing climate disruption.

Agriculture is outdoors to a unique degree, exposing it to greater risk from livestock-related emissions than any other industry's risk from the same emissions. So the food industry has a compelling commercial incentive to reduce these emissions. Meat and dairy substitutes require no subsidies or offsets. Consumers can buy more of them tomorrow.

In other words, "Eating Greenfully" requires replacing livestock products with substitutes. This is the only way for industry and the public collaboratively to take a single, powerful action to reduce climate change quickly.

References

Brainard, 2010, "Meat vs. Miles" in the *Columbia Journalism Review* at http://www.cjr.org/the_observatory/meat_vs_miles.php.

Calverd, Alan, July 2005, "A radical approach to Kyoto," in *Physics World*, 56.

Fairlie, Simon, 2010, *Meat, A Benign Extravagance*, East Meon, U.K., Permanent Publishers.

Goodland, Robert, 1999, "Livestock sector environmental assessment" in *Nachhaltigkeit in der Landwirtschaft: Landwirtschaft in Spannungsfeld zwischen Okologie, Okonomie und Sozialwisenschaften*, Hardtlein, M., et al. (eds), Berlin: E. Schmidt Verlag [for] Umwelt Stiftung: Deutsche Bundesstiftung Umwelt.

Goodland, Robert and Jeff Anhang, 2009, "Livestock and Climate Change: What if the Key Actors in Climate Change are Cows, Pigs and Chickens?" in *World Watch* 22(6): 10-19 at www.worldwatch.org/node/6294. Sources and resources are at www.worldwatch.org/ww/livestock; responses to readers are at http://awellfedworld.org/sites/awellfedworld.org/files/pdf/WWMLivestock-ClimateResponses.pdf.

Goodland, Robert, 2010, "Replace 25% of Livestock Products to Reduce Global Warming" at http://www.weblogtheworld.com/countries/africa-africa/replace-25-of-livestock-products-to-reduce-global-warming.

IEA, 2010, presentation by Nobuo Tanaka, Executive Director, at www.iea.org/speech/2010/Tanaka/IEA_press_conf_cop16.pdf.

Lory, J. A., R. E. Massey, and J. M. Zulovich, 2010, "An Evaluation of the USEPA Calculations of Greenhouse Gas Emissions from Anaerobic Lagoons" in *Journal of Environmental Quality*, 2010; 39 (3): 776 DOI: 10.2134/jeq2009.0319.

McWilliams, James, 2010, "Carnivorous Climate Skeptics in the Media" in *The Atlantic* at www.theatlantic.com/food/archive/2010/04/carnivorous-climate-skeptics-in-the-media/39177/.

Mentzel, Chris, 2009, "The Copenhagen Fools" at www.mauiweekly.com/page/content.detail/id/500866/The-Copenhagen-Fools.html.

Monbiot, George, 2010, "I was wrong about veganism. Let them eat meat – but farm it properly" in *The Guardian* at www.guardian.co.uk/commentisfree/2010/sep/06/meat-production-veganism-deforestation.

Nader, Ralph, 1965, *Unsafe at Any Speed*, Massachusetts, Knightsbridge Publishing.

Pitesky, M., K. Stackhouse, and F.M. Mitloehner. 2009, "Clearing the Air: Livestock's Contribution to Climate Change" in *Advances in Agronomy*, 103: 1-40.

Russell, Geoff, 2010a, "Monbiot renders me speechless," Part 1 at http://animalliberation.org.Au/blog/83-monbiot-sides-with-muddle-headed-romanticism.html and Part 2 at http://animalliberation.org.au/blog/81-monbiot-sides-with-muddle-headed-romanticism.html.

—. 2010b, "Of brains, biceps and baloney" at http://bravenewclimate.com/2010/10/31/of-brains-biceps-and-baloney/.

Saunders, Caroline and Andrew Barber, 2007, *Comparative Energy and Greenhouse Gas Emissions of New Zealand's and the UK's Dairy Industry,* Research Report No. 297, Christchurch, New Zealand, Lincoln University, at http://researcharchive.lincoln.ac.nz/dspace/bitstream/10182/144/1/aeru_rr_297.pdf.

Steinfeld, Henning, Gerber P., Wassenaar, T., Castel, V., Rosales, M. and Haan, C., 2006, *Livestock's Long Shadow: Environmental Issues and Options*, Rome, U.N. Food and Agriculture Organization.

Steinfeld, Henning and Pierre Gerber, 2010, "Livestock production and the global environment:Consume less or produce better?" in *Proceedings of the National Academy of Sciences* at http://www.pnas.org/content/107/43/18237.

Subak, Susan, 1994, "Methane From The House Of Tudor And The Ming Dynasty: Anthropogenic Emissions In The Sixteenth Century" at http://www.uea.ac.uk/env/cserge/pub/wp/gec/gec_1994_06.pdf.

UNESCO, 2010, *Energy Flow, Environment and Ethical Implications for Meat Production* at

http://unesdoc.unesco.org/images/0018/001897/189774e.pdf.
World Bank, 2001, *Livestock Development* at www-wds.worldbank.org/external/default/WDSContentServer/WDSP/IB/2001/12/11/000094946_01112104010387/Rendered/PDF/multi0page.pdf.
York, Helene, 2010, "The Myth of Green Beef" in *The Atlantic* at www.theatlantic.com/food/archive/2010/04/the-myth-of-green-beef/38810.

Notes

[1] Robert Goodland is grateful to Jeff Anhang for assistance with this article.
[2] IEA, 2010.
[3] Steinfeld et al., 2006.
[4] Goodland, R. and Anhang, J., 2009.
[5] From personal communication: "I am very excited to tell you that your article has been posted on the Chinese Government's official Climate Change Website (www.ccchina.gov.cn/cn/NewsInfo.asp?NewsId=20278)... This is a huge honor, because only high-level Chinese policymakers' articles or speeches are allowed on the website. The website attracts millions of eyeballs every day."
[6] Mentzel, 2009.
[7] The Greenhouse Gas Protocol (at http://www.GHGProtocol.org) is the most widely used international accounting tool for government and business leaders to understand, quantify, and manage greenhouse gas emissions.
[8] Pitesky, Stackhouse, and Mitloehner, 2009.
[9] E.g., see York, 2010, Brainard, 2010, and McWilliams, 2010.
[10] See www.extension.org/pages/Animal_Ag%E2%80%99s_Role_in_Greenhouse_Gas_Production#Clearing_the_Air:_Livestock.E2.80.99s_Contributions_to_Climate_Change.
[11] Lory et al, 2010.
[12] Goodland, 2010.
[13] I have served as President of the International Association of Impact Assessment.
[14] See www.equator-principles.com/documents/Equator_Principles.pdf.
[15] Goodland, 1999.
[16] See World Bank, 2001, p. 65.
[17] See UNESCO, 2010, p. 10.
[18] See www.wbcsd.org/plugins/DocSearch/details.asp?type=DocDet&ObjectId=MzYyMjU.
[19] These can be seen at http://awellfedworld.org/sites/awellfedworld.org/files/pdf/FAOConsult12-09.pdf and http://awellfedworld.org/sites/awellfedworld.org/files/pdf/GoodlandFoodIndustryBerlinJan2010.pdf
[20] Calverd, 2005.

---

[21] Russell, 2010b provides a thorough explanation of why livestock products are unneeded for good human health.
[22] Steinfeld and Gerber, 2010.
[23] E.g., Subak, 1994.
[24] E.g., Fairlie, 2010 and Monbiot, 2010, rebutted in Russell, 2010a.
[25] E.g., see *Time Magazine* at www.time.com/time/health/article /0,8599,2024133,00.html
[26] Saunders and Barber, 2007.
[27] Nader, 1965.

CHAPTER FOUR

VEGETARIANISM AND FOOD GOVERNANCE: SUSTAINABILITY AND ECOLOGICAL JUSTICE

HEATHER MCLEOD-KILMURRAY

Many people are overwhelmed with stories of environmental and economic doom and gloom. Climate change and environmental destruction seem insurmountable problems, or at least ones whose solution requires overwhelming sacrifices and huge costs, from individuals to major corporations, from the local to the international level. However, recent evidence[1] suggests that by doing one thing - reducing the consumption and production of meat, particularly meat produced by intensive industrial farming – it is possible to achieve a win-win-win situation: improved environmental sustainability, improved human health, and greater social justice. So why are laws, policies and individual actions not moving in this direction? How does the structure of food governance as it currently exists encourage and structure a mode of food production that places the dietary desires of wealthier individuals and nations, as well as the agricultural and trade priorities of a handful of countries and multinational corporations, above the food, economic and environmental survival needs of the current and future generations of the rest of the world? This paper will address some of these questions.

The Context

Intensive livestock production and the consumption of its products (meat, eggs, dairy) has grown exponentially in the last 50 years, primarily at the demand of wealthy countries, and wealthier individuals in all countries.[2] Yet the economic, social and environmental impacts of animal-based diets, and particularly intensive livestock production, have been all but ignored in the law and policy contexts.

While the relationship between meat eating and climate change has been very well-documented by Goodland and Anang,[3] among others,[4] I want to particularly emphasize the harms of industrial livestock production. Intensive Livestock Operations in Canada (ILOs), or Concentrated Animal Feeding Operations in the United States (CAFOs), gather very large numbers of cows, pigs, chickens or other animals into industrial warehouses. The scope of these ILOs is astounding. "Each year in the United States, 10 billion land animals are raised and killed for meat, eggs, and milk. Statistically, farm animals comprise 98 percent of all animals in the country with whom we interact directly, and that staggering percentage does not even include the estimated 10 billion aquatic animals killed annually for human consumption."[5] The purpose was to make livestock production faster, less labour-intensive[6] and less economically costly for producers. Yet feeding these animals grains they were not designed to eat, gathering their waste into large storage facilities, and using antibiotics to treat their increased diseases and health problems resulting from confinement, unnatural conditions and close proximity, have all made this mode of livestock production far more costly for the planet, in economic, environmental and social ways.[7] Eliminating or reducing the consumption of animal products would eliminate most of these harms. More traditional approaches to farming and animal husbandry would at least significantly reduce them.[8]

It is interesting to note that a large motivation for the growth of these ILOs and CAFOs can be found in the history of corn. While the full details are available elsewhere,[9] the main idea is that subsidies and increased centralization of corngrowing created a need to find uses for this surplus grain, and feeding it to livestock was one solution. It also meant that industrial farmers could feed livestock more cheaply with corn and soy than small scale farmers with their own grazing land, creating strong economic incentives for centralization and industrialization of livestock production. So neither human health and animal welfare, nor environmental sustainability or even human food preferences, have been the driving force behind industrialization of livestock production.

One of the greatest environmental impacts of intensive livestock production is the emission of greenhouse gases (GHGs). According to the Intergovernmental Panel on Climate Change (IPCC),[10] livestock production creates approximately 14% of global anthropogenic GHG emissions.[11] Goodland and Anhang put the figure much higher: "total GHGs attributable to livestock products is actually closer to 32,564 million tons annual CO_2e, or 51% of the worldwide total",[12] because they include the full "life cycle" effects of livestock production,[13] which can include deforestation,[14] desertification, fertilizer production and use, housing, slaughtering and processing, transport costs, etc.[15]

Livestock GHG emissions include CO_2 but also methane produced by the enteric fermentation[16] of ruminants and the decomposition of their waste gathered and stored at the factories. Methane is more atmospherically harmful per volume than CO_2,[17] but "its half-life in the atmosphere is only about 8 years, versus at least 100 years for CO_2,"[18] so tackling methane emissions is a very rapid and cost-effective way of slowing the pace of climate change.[19] ILOs also produce nitrogen which is a GHG when it enters the air, but it also harms the soil and groundwater.[20] Excess amounts are put into the environment through the extensive use of fertilizer required to grow the feedcrops for industrially farmed animals. It is also a product of manure. ILOs and CAFOs turn manure - what was once a natural, organic and beneficial fertilizer - into a significant source of waste and emissions by accumulating large amounts of it in holding tanks or spreading it in large quantities on soils.[21] In contrast, "small scale, mixed and diversified at the farm level production systems improve the fertility of the soil and can thus sequester more carbon dioxide, while ... preserving and improving genetic biodiversity...making [these farms] more adaptive to climate change."[22]

Intensive livestock production causes other environmental harms such as "habitat loss and degradation, erosion, water depletion, pollution and salinization, agrochemical contamination, [and] animal waste and air pollution,"[23] and water pollution.[24] It creates a vicious circle, because through its use of land and emissions it causes climate change, which in turn reduces agricultural productivity.[25] It also reduces biodiversity among domesticated animals[26] but also wildlife, as the land used for feedgrain and pasture reduces the ecosystems of other species.

Finally, and perhaps most persuasively, it is an extremely inefficient use of global resources to turn grain into meat, due to "the feed-conversion ratio". "One kilogramme of edible boneless beef requires around 20 kg of animal feed and 15,500 litres of water to produce ... One calorie of food energy obtained from beef requires inputs of 9 calories of food energy from plants and 40 calories of fossil fuel energy."[27] In other words, "the ratio of energy input to output for industrially produced meat can reach as high as 35:1."[28] This wastefulness means that the American diet would only feed 2.5 billion people globally.[29] Building a global food governance system that encourages this way of eating and producing food can be argued to be a form of cultural domination. This injustice to people, animals and ecosystems requires the attention of law and policy makers.

Environmental Sustainability and Ecological Justice

Reducing the harms from ILOs and CAFOs involves science, economics, and other fields. This paper examines what role law can play by integrating sustainability into the regulatory and policy frameworks that affect ILOs and CAFOs in order to achieve the overarching goals of fairness and justice. To date, law and policy have been part of the problem in structuring, enabling and encouraging this mode of food production.

While sustainable development has long been the catchword in environmental law,[30] its vagueness has reduced its potential as a guiding principle. Greater specificity of priorities and components can be found in "sustainability", and the even more precise "environmental sustainability", particularly where it has been defined clearly. The Australian *Environment Protection and Biodiversity Conservation Act 1999* does so in detail:

> 3A Principles of Ecologically Sustainable Development
> The following principles are *principles of ecologically sustainable development:*
> (a) decision-making processes should effectively integrate both long-term and short-term economic, environmental, social and equitable considerations;
> (b) if there are threats of serious or irreversible environmental damage, lack of full scientific certainty should not be used as a reason for postponing measures to prevent environmental degradation;
> (c) the principle of inter-generational equity – that the present generation should ensure that the health, diversity and productivity of the environment is maintained or enhanced for the benefit of future generations;
> (d) the conservation of biological diversity and ecological integrity should be a fundamental consideration in decision-making;
> (e) improved valuation, pricing and incentive mechanisms should be promoted.

Robert Gibson argues that sustainability should include "socio-ecological system integrity, livelihood sufficiency and opportunity, intergenerational equity, intragenerational equity, resource maintenance and efficiency, socio-ecological civility and democratic governance, precaution and adaptation, and immediate and long term integration."[31] This idea of ecological sustainability is something that should be taken into account by all those undertaking calculations of the profitability and feasibility of future food production. Without this, the true costs and benefits of livestock production will be impossible to calculate.[32]

Further, Klaus Bosselmann[33] argues that in the environmental context, the law should be a tool for achieving not only sustainability but also "ecological justice" which he defines to include (i) *intragenerational justice* or "concern for the poor";[34] (ii) *intergenerational justice*, or "concern for future generations";[35] and (iii) *interspecies justice*.[36] Clearly, laws and policies which structure a system that damages the environment as well as the social and economic well-being of many for the benefit of a few are unjust. Reversing the trend toward increased industrialization and increased production and consumption of meat can achieve these three kinds of justice through a single course of action.

Law and Policy of Food Governance

What is the law doing to stop, or to promote, industrial livestock production at both the domestic and international levels? The WTO Agreement on Agriculture (AoA) has arguably helped to facilitate intensive livestock production. Penelope Simons explains that "[s]ecuring the interests of big business was one of the goals of the negotiations on agriculture. …Ritchie and Dawkins allege that the US and the EU used their political clout in the Uruguay Round negotiations to ensure that, among other things, the AoA rules would serve the interests of their major corporate actors by facilitating an increase in trade volume of agricultural products. This was accomplished by requiring member states to reduce domestic prices to that of global prices, reducing subsidies to farmers to remove tariff and non-tariff barriers while protecting export subsidies."[37] The provisions of the AoA supported industrial farming, allowing the U.S., the E.U. and other OECD countries to continue their subsidy programmes, benefitting multinational corporations and large agribusinesses.

The behaviour of these transnational corporations is not regulated by the WTO, so that agreement has not prevented the massive concentration of power in the agricultural industry. "An UNCTAD study on market concentration in the agricultural input industry found that six major agrochemicals companies control approximately 77 per cent of the global market (with the top three companies controlling 50 per cent), and the top four seed companies control an estimated 30 per cent of the global market for seeds, while five corporate groups control a significant percentage of the agricultural technology market."[38] This vertical and horizontal integration of the market[39] means that this handful of players controls the food chain. "For example, ...[i]n terms of meat production they might produce animal feed, raise and slaughter the animal, and process the meat. ... Farmers have little negotiating power in a system where they have increasingly fewer choices among sellers of inputs and buyers of outputs (who may even be the same company or family of companies)."[40] Apart from small farmers everywhere, this has significantly harmed developing countries because free trade means they now import more food but export less, "leading to unemployment, and hence displacement among farmers and increases in large landholdings."[41] This has not changed significantly as a result of the Doha or Uruguay Rounds of WTO negotiations.[42] The FAO argues that the very act of globalizing livestock production and trade creates increased environmental, economic and social risks: "while globalization can increase national income, create employment and improve nutrition, it also 'globalizes' potential threats to livelihoods, human health and the environment. ... [It] entails massive flows of capital investment, information and technology, rapidly changing food preferences and 'dietary convergence', the adoption of increasingly demanding international standards and moves to greater market integration and concentration of ownership."[43] This global trade structure has helped to facilitate and even encourage intensive livestock farming and its harms.

In addition to trade rules, there are other mechanisms that can concentrate power and reduce the freedom and welfare of small-scale farmers and consumers. The TRIPs[44] Agreement under the WTO governs trade related private intellectual property rights. As genetically modified grain and livestock, such as the Enviropig,[45] become larger factors in the livestock industry, it may be possible for large players to gain even greater power over the global food chain.[46] Further, the creation of national public standards in domestic markets can lead to harmonization of rules which can effectively reduce the power of local actors. For example, "U.S. regulations controlling water pollution from feedlots have accelerated the concentration of the industry because the mandated control techniques are cheaper to use in large facilities."[47] Private standards, such as those created by multinational corporations, can have similar effects.[48]

Domestic laws and policies are also relevant. In the U.S., subsidies to farmers, as well as the de-regulation of agricultural markets since the mid-1990s,[49] have had significant impacts. Wise[50] argues that removing subsidies to grain growers and "bringing feed costs closer to their costs of production"[51] would make IFOs costlier and family farmers more competitive, domestically and internationally. Durning and Brough claim:

> "[i]n Western industrialized nations the degree of government support for livestock and feed producers is in the billions of dollars. Meat, dairy and feed producers benefit from a variety of measures such as guaranteed minimum prices, government storage of surpluses, feed subsidies, import levies, and product insurance. .... [These programs] are designed to

promote self-sufficiency in food production, to raise farmers' incomes, and to aid rural communities. They have the unintended side effect, however, of boosting intensive forms of livestock production."[52]

Also, in the U.S. intensive livestock production is "woefully under-regulated".[53] Cassuto argues this is because of "a nearly unbroken series of [Congressional] decisions to exclude farms and farming from the burdens of federal environmental law, with states mainly following suit. The combination of a powerful agricultural lobby, the family farm's hold on the collective national imagination, and the short-term profits of industrial agriculture have proved too potent a mix for any would-be regulators."[54]

Canada, is similarly under-regulated. Federal law does little, and most provinces have left regulation to municipalities, who require facility operating permits, and regulation of nutrient management plans (NMPs) (primarily to protect groundwater); setback or minimum distance separation requirements; public information and public notice requirements; requirements for professional certification; and requirements for financial guarantees and technical assistance.[55] Right-to-farm legislation is also found in Canada, as well as the United States, which "protects farming activities from nuisance actions, provided the operation conforms with 'normally acceptable agricultural practices.'"[56] There has been discussion of regulatory reform in Canada's food industry, but almost none relates to concerns about environmental sustainability, human health or animal welfare.[57] The message is that "[a] framework is needed that could bring about a more flexible regulatory system for the sector, one that supports innovation and competitiveness."[58] The focus is on efficiency and reduced government regulatory burdens, not environmental or social sustainability.

In analyzing the robustness of Canadian standards for ILOs against U.S. and Mexican approaches, Speir, Bowden et al. conclude that Canada's standards vary across jurisdictions, focus primarily on water pollution, and could be improved. They argue that environmental rather than agriculture ministries should have primary responsibility for ILOs because agricultural agencies have always been promoters of agriculture, while environmental regulators are more used to an enforcement role.[59] They also argue that full lifecycle management of manure and wastewater should be undertaken, and new technologies created, to reduce environmental harm. They note that none of the regulations deal with "worker health, antibiotic, hormone and specific pathogen issues" and that these are crucial to environmental and human health protection.

Having outlined the general structure of the legal system pertaining to industrial livestock, we turn now to assess how this system measures up to an analysis of its sustainability and ecological justice.

The Sustainability and Ecological Justice of Intensive Livestock Production

Assessing the law and policy of food governance and of ILOs in particular, it becomes clear that this mode of food production is neither sustainable, nor ecologically just.

Sustainability

An undertaking that produces up to 51% of global GHGs is a major cause of climate change and all of its related harms. Refusing to reduce these emissions, which are easier to reduce and will lead to significant and immediate impacts on climate change, is not sustainable. The short term "cost savings" of efficient, centralized livestock production is at the expense of the long-term health of the planet, contrary to the first principle of ecologically sustainable development.

Beyond climate change, these food factories cause air, water, and soil pollution and deforestation. There is no component of the ecosystem that they do not significantly harm, in contrast to the greater ecological efficiency and sustainability of vegetarian diets, or smaller-scale animal husbandry and farming.[60] The mono-cropping and forced surplus grain production of a limited number of grains have a significantly negative impact on biodiversity, as does the focus on a limited number of livestock breeds in intensive facilities. Reducing meat and animal product consumption, and moving away from the increased industrialization of livestock production, are effective ways of reducing the environmental footprint of food production and achieving food sustainability.

From an economic sustainability perspective, the main driving force for industrializing livestock production is economic efficiency and profit – it costs producers less, they can produce more "animal units" more quickly, and they can pass on these savings to consumers, who are increasing in numbers globally. However, these economic "savings" are illusory. First, reducing the price of animal products in itself encourages increased consumption. Secondly, any economic savings are short term and focused on only a small percentage of the true economic costs of this mode of production. "[T]he price of meat might double or triple if the full ecological costs – including fossil-fuel combustion, groundwater depletion, agricultural-chemical pollution, and methane and ammonia emission – were included in the bill."[61] It has been suggested that the global adoption of a low-meat diet for the next 20 years "would reduce by 50% the expected costs of mitigating climate change up to 2050."[62]

Wasting grain to produce meat is inefficient. Other costs include lost livelihoods of small-scale farmers and the ripple effect across rural communities, where tourism, reduced real estate values and other economic effects result from establishing an intensive livestock operation.[63] Also, "a diet rich in animal protein …is [not] a wise development strategy. It creates dependence on imports for food and can widen the gap between rich and poor…. Higher meat consumption among the affluent frequently creates problems for the poor, as the share of farmland devoted to feed cultivation expands, reducing production of food staples."[64]

The global food governance regime has also reduced food security. The cost of food has risen dramatically due to the global food structure, with its subsidies, trade barriers and other instruments of the "free" market which alter the costs of food, unrelated to the true environmental, social and economic costs of food production. This system encourages the demand for meat and animal products, increases food prices and food insecurity, and the pace and extent of climate change These are all economic as well as environmental and human rights costs. Finally, monopolies and reduced competition are not generally good for the economy. "[T]oday just 20 feedlots feed half of the cattle in the US and these are directly connected to 4 processing firms that control 81% of the beef processing."[65] This can create vast market distortions and reduce economic efficiency and certainly fairness. Even from the solely economic perspective, the economic benefits of ILOs do not outweigh their costs.

In terms of social sustainability, many of these environmental and economic impacts bring significant social harms with them. The impact on the small scale farming community has been vast, in both the developed and developing worlds. "Between 1994 and 2001 …the number of farms raising livestock in the U.S. declined by over 60 percent,"[66] and "[t]he exploitation of contract farmers by corporate animal factories is a growing problem."[67] This system which puts so much control in the hands of a small number of multinational corporations and agribusinesses results in significantly reduced power and choice for individual farmers, local communities and developing countries to determine their own priorities and preferences for what to eat, and how to produce it. The system creates a food chain alien to ecologically sound ways of sustaining life and alien to the food, sustenance, and lifestyle preferences of large groups of people. The physical concentration of ILO and CAFO operations creates a significant physical and psychological disconnection between producers and consumers, and between different types of farmers. The food is less natural, less proximate, less "human", than food produced locally.

An animal-based diet rich is also detrimental to human health. The increased fat content, bacteria, obesity, risk of cancers and heart disease are all results of a more meat-heavy diet.[68] The risk of transmission of disease from animals to human consumers, as well as effects of air emissions from ILOs and CAFOs on neighbouring residents, are also of concern.[69] Clearly, industrial livestock production is not sustainable. It also has negative impacts on all three branches of ecological justice.

Ecological Justice

Intragenerational justice involves fairness between members of the present generation. This focuses on "resource egalitarianism" meaning that the welfare of one group of people or nations should not be at the expense of others. However, the industrial livestock system produces greater food dependence, higher costs, and the privileging of the food preferences of wealthier people and nations. It also increases the inequality between small and large-scale farmers, and between developed and developing countries. Using grain to feed animals for human consumption is a waste of resources, and causes significant changes to resources use patterns in countries tied to the international market economy. Water, air and soil pollution reduce the availability and utility of these other resources for all, but particularly in those countries where deforestation and desertification are undertaken to supply pasture for livestock or the grain for meat consumption in other countries.

The impacts of climate change create both intragenerational and intergenerational injustices. As discussed above, the significant GHG emissions of this mode of production cause climate change, which it is widely known will affect those in developing countries more severely.[70] Regions where food is already scarce and extreme environmental events are increasing are already suffering the impacts. If emissions are not significantly reduced, the impacts on future generations will be even more acute. Additionally, as the Stern Report[71] has so clearly indicated, postponing mitigation measures imposes greater costs on future generations. Reducing the natural capital future generations can enjoy is an injustice that the law and policy of food governance needs to address.

Finally, what is perhaps most obvious but least discussed in relation to intensive livestock operations is its gross violation of interspecies injustice. Putting living creatures capable of feeling pain and distress into overcrowded, unnatural and often painful environments where they are artificially made to grow more quickly, are prevented from engaging in most of their natural behaviours, are put at risk of greater illnesses and experience very high levels of stress, fear, pain and early death,[72] is unjust. Such treatment of human beings would clearly be a violation of numerous domestic and international laws, and any conception of justice. Finally, as discussed above, apart from these animals themselves, this mode of production also reduces biodiversity, among farmed animals but also among wildlife whose ecosystems are reduced or destroyed to meet the needs of expanding livestock production.

We turn now to some suggestions for how law can contribute to reversing this trend and improving the sustainability and ecological justice of the food system.

Climate Change Negotiations and Agriculture

Since our focus is the environmental impact of ILOs and CAFOs and particularly their GHG footprint, it is useful to examine their treatment by the international laws dealing with climate change. Article 2 of the United Nations Framework Convention on Climate Change (UNFCCC)[73] provides that the treaty aims to stabilize GHGs "at a level that would prevent dangerous anthropogenic interference with the climate system … within a time-frame sufficient to allow ecosystems to adapt naturally to climate change, to ensure that food production is not threatened and to enable economic development to proceed in a sustainable manner." Article 4(1)(c) provides that state parties have an obligation to "[p]romote and cooperate in the development, application and diffusion, including transfer, of technologies, practices and processes that control, reduce or prevent anthropogenic emissions of greenhouse gases not controlled by the Montreal Protocol in all relevant sectors, including the energy, transport, industry, *agriculture*, forestry and waste management sectors." So while agriculture is less emphasized in climate change debates than other sectors, the UNFCCC explicitly includes agricultural emissions and envisages achieving mitigation at least in part through this sector.

In the Kyoto Protocol[74] itself, Article 2(1) provides that Annex I countries "shall: (a) Implement and/or further elaborate policies and measures in accordance with its national circumstances, such as: …(iii) Promotion of sustainable forms of agriculture in light of climate change considerations." Article 10 states that "[a]ll Parties, taking into account … their specific national and regional development priorities, [and] circumstances, … shall: …(b) Formulate, … regional programmes containing measures to mitigate climate change and measures to facilitate adequate adaptation to climate change" and that "(i) [s]uch programmes would, *inter alia*, concern the energy, transport and industry sectors as well as *agriculture*, forestry and waste management." Finally, Annex A explicitly states that sources of GHG emissions include "*Agriculture*: Enteric fermentation, Manure management, Rice cultivation, Agricultural soils, Prescribed burning of savannas, Field burning of agricultural residues, Other." The FAO has called for bringing projects that reduce the impact of livestock, including CAFOs, within the scope of the Clean Development Mechanism (CDM).[75] This is a strong recognition that CAFOs are significant contributors to emissions and that treaty negotiators agree that reducing their impact is a significant way to mitigate global warming. However "the inclusion of emissions by sources and reductions by agricultural sinks is optional under article 3.4 of the Protocol and only a few countries have elected to do so. Currently within the AWG-KP, aspects of land use, land use change and forestry accounting are under consideration, as is the possibility of expanding the scope of the CDM to include, inter alia, cropland management, grazing land management, wetland management and soil carbon management in agriculture."[76]

In the various discussions to arrive at a binding agreement to replace Kyoto, the Bali Action Plan did permit work on "cooperative sectoral approaches", and this contained a specific section on agricultural sector. While Copenhagen itself was silent on this issue, since then the Subsidiary Body on Scientific and Technological Advice (SBSTA) was created to specifically study the "scientific, technological and methodological aspects of agricultural mitigation."[77] In addition, some countries who have voluntarily declared their emissions targets under Copenhagen have explicitly referred to cuts in agricultural emissions.[78] This indicates there may be hope that international law may be on a path to focusing on binding targets to reduce the harms of agriculture, and since ILOs and CAFOs are among the largest emitters, this would impact them. It would be ideal if negotiators would specifically address intensive livestock production as key sub-issue in agriculture in the climate change regime

Other Recommendations

Guiding Principles

In order for law and policy to reduce the harms of ILOs, law and policy reform should proceed by putting ecological sustainability and justice, rather than economic efficiency and short term profit, as the first priority in agricultural policy. Guiding principles of sustainability, precaution and justice for future generations have been included in a wide variety of domestic and international laws, and should also be included in any environmental, agricultural or trade law agreements, laws and policies which directly regulate, or have an indirect role in shaping, the context of industrial livestock production.

At the international level, a treaty on animal rights, or including animals within existing rights treaties or covenants, would be an immense contribution to ending the harms intensive livestock production. Since that seems a distant dream, other immediate steps could be taken.

Significant reforms to WTO rules in relation to agriculture have long been called for, and are necessary to stop law from being a significant cause of the problems presented by ILOs, including the paradigms of constant growth, centralization and commodification. Globalization itself in the context of agriculture needs to be revisited, as do development programmes affecting agriculture and livestock production. "International development agencies have contributed to the number of destructive livestock projects in the Third World.... Lending ventures tried to graft [Western] ranching practices onto [Latin America and Africa], with little understanding of either the environment or the people who operated in it."[79] Programmes have often also encouraged privatization of land, which has itself led to greater concentrations of power and wealth and a reduction in resources available for others. However, many development institutions have moved away from encouraging livestock production because of the social and economic harms that can result.[80] In addition, subsidies, trade barriers, technical requirements and standards, all of these tools which necessitate concentration of power and industrialized, centralized processes, must be addressed. Finally. laws which directly address corporate concentration in agricultural markets that gives them unprecedented market power are also needed as another tool for reducing the prevalence of industrial livestock production.

In addition to climate change treaties with caps on GHG emissions, perhaps we need enforceable targets for livestock consumption. "[T]he European Union and other high-income, developed countries should reduce production and consumption of meat and milk to one third below current levels over the next decade (by 2020) and to at least 60% below current levels by 2050."[81] While these goals could perhaps be met by including the full GHG emissions of animal products in countries' GHG calculations, it may raise awareness of the direct impact of meat eating, and industrial meat production, further to have specific, separate targets for these sources of emissions.

At the domestic level, again specific animal rights legislation would have the strongest impact in reducing ILOs but this is not likely in the current political context. In the meantime, legislation is needed specifically to address ILOs at the provincial level, to ensure uniformity in practices. Adequate resources for enforcement by environmental authorities are also necessary. Right to farm legislation must also be amended or judicially interpreted to ensure that intensive production of livestock is not viewed as "normal farming practices". Yet the Ontario statute, for example, specifically states in its preamble that "[a]gricultural activities may include intensive operations that may cause discomfort and inconveniences to those on adjacent lands,"[82] meaning these could be interpreted as "normal farming practices" which are protected by the act. Such legislation should not be permitted to use the logic of protecting farms to shield massive industrial livestock operations from liability for harms caused by emissions or pollutants.

In addition, domestic subsidies must be adjusted.[83]There have also been suggestions to create tax schemes to discourage both intensive livestock production and also meat consumption. A carbon tax may be sufficient to impact ILOs and CAFOs due to their status as large industrial emitters, but a specific tax on meat or animal products[84] may also be necessary to encourage individual consumers to turn to alternatives that are less costly to sustainability goals by increasing public awareness of the climate effects of livestock. This would also signal to individuals their power to reduce these harms through changes in food choices. Other tools to achieve these goals are labeling and full cost pricing of intensively produced animal products. Governments could go even further by better subsidizing organic and sustainable farm practices. These legal measures would be strengthened by extra-legal means such as broader public education on the costs of meat-intensive diets, and the benefits of local, organic, sustainable farming.

There are many reform options available. What is not an option is to leave unchallenged a global structure of food production that is so costly to sustainability and ecological justice. The law can and must play an important role in the achievement of the triple benefits – environmental, economic and social - that reducing meat consumption and industrial livestock production would achieve.

Notes

[1] Robert Goodland and Jeff Anang, 2009, "Livestock and Climate Change: What if the Key Actors in Climate Change are Cows, Pigs and Chickens?" in *World Watch* 22(6): 10-19 at www.worldwatch.org/node/6294.
[2] "National meat-consumption figures generally reflect income levels, with rich countries eating more meat than poor ones. …[G]global meat production has …nearly quadruple[d] since 1950." Alan B. Durning and Holly Brough *Taking Stock: Animal Farming and the Environment* (Paper 103, Worldwatch Institute, July 1991) at 9 and 11.
[3] Goodland and Anhang, *supra* note 1.
[4] Jeremy Rifkin *Beyond Beef: The Rise and Fall of the Cattle Culture* (New York: Dutton, 1992).
[5] The Humane Society of the U. S. "The Welfare of Animals in the Meat, Egg and Dairy Industries" online www.petfinder.com/docs/library-articles/animal-welfare-overview.pdf at 1.
[6] Joan Dye Gussow "Ecology and Vegetarian Considerations: Does Environmental responsibility demand the Elimination of Livestock?" (1994) 59 (Suppl) Amer. J. of Clinical Nutrition 1110S at 1112S.
[7] Compassion in World Farming *Beyond Factory Farming: Sustainable Solutions for Animals, People and the Planet* (2009) at 9 www.ciwf.org.uk/includes/documents/cm_docs/2010/b/beyond_factory_farming_report_2009_exec_main_final.pdf.
[8] Karl-Heinz Erb et al. *Eating the Planet: Feeding and Fuelling the World Sustainably, Fairly and Humanely* (Nov. 2009, Social Ecology Vienna) www.ciwf.org.uk/includes/documents/cm_docs/2009/e/eating_the_planet_full_report_nov_2009.pdf.
at 8. Their study analyzed the biomass potential of the earth and its ability to satisfy a range of particular diets. Some argue that eliminating livestock would not be sustainable - Gussow *supra* note 6 at 1114S -1115S.
[9] David N. Cassuto "The CAFO Hothouse: Climate Change, Industrial Agriculture and the Law" (Animals and Society Institute Policy Paper, 2010) www.animalsandsociety.org/resources/details.php?id=821, citing among others Michael Pollan *The Omnivore's Dilemma: A Natural History of Four Meals* (2006, The Penguin Press) at 48. See also Rifkin, *supra* note 4.
[10] IPCC *Climate Change 2007: Synthesis Report, Fourth Assessment Report*, "Summary for Policymakers" www.ipcc.ch/pdf/assessment-report/ar4/syr/ar4_syr_spm.pdf.
[11] The FAO puts this estimate at 18% "in carbon dioxide equivalents": Food and Agriculture Organization of the United Nations (FAO) *Livestock's Long Shadow* (Rome, 2006) www.fao.org/docrep/010/a0701e/a0701e00.HTM. at xxi.
[12] The FAO's estimate is 11.8%, *ibid*. See Goodland and Anhang, *supra* note 1.
[13] See Goodland and Anhang, *supra* note 1. Se also Compassion in World Farming (CIWF) *Global Warning: Climate Change and Farm Animal Welfare – A Report by Compassion in World Farming – 2008*(Revised 2009)

www.ciwf.org.uk/includes/documents/cm_docs/2010/g/global_warning_full_report_revised_2009.pdf.

[14] Gussow,*supra* note 6 at 1114S.

[15] Canada's agriculture GHG inventory does not include these sources either: Beyond Factory Farming "Fact Sheet: Climate Change and Intensive Livestock Production" (June 2009) http://beyondfactoryfarming.org/files/climatechange2.pdf. See also CIWF *Global Warning*, *supra* note 13.

[16] For details of enteric fermentation see Christian N Jardine et al. *Methane UK* (Environmental Change Institute, University of Oxford). www.eci.ox.ac.uk/research/energy/downloads/methaneuk/methaneukreport.pdf.

[17] J"A ton of methane represent twenty-one times the warming potential of a ton of carbon dioxide." John C. Dernbach and Seema Kakade "Climate Change Law: An Introduction" (2008) 28(1) Energy L J 1 at 6.

[18] Goodland and Anhang *supra* note 1 at 13.

[19] Jardine et al., *supra* note 17 at 6.

[20] Cassuto *supra* note 9 at 10. See also Durning and Brough, *supra* note 2 at 18.

[21] CIWF *Global Warning supra* note 13 at 13; Dernbach and Kakade, *supra* note 18 at 6. "In Hungary … [f]rom 1960 to 1986, as the number of animals slaughtered in the country doubled, the use of manure as a fertilizer fell by one-fifth. Farmers opted for easy-to-spread chemical fertilizers instead of trucking manure from centralized feeding facilities." Durning and Brough *supra* note 18 at 18.

[22] *Ibid.* The UNCTAD/WTO International Trade Centre's report "Organic Farming and Climate Change" (2007) www.ifoam.org/growing_organic/1_arguments_for_oa/environmental_benefits/pdfs/FiBLStudyOrganic_Farming_and_Climate_Change.pdf concludes that organic farming contributes to climate change mitigation, and is better for adaptation. See also Beyond Factory Farming: "Fact Sheet: Climate Change and Livestock Production" *supra* note 16.

[23] Cassuto *supra* note 9 at 8.

[24] *Ibid.* at 9.

[25] McMichael et al. "Food, livestock production, energy, climate change, and health" (2007) Lancet 370: "The IPCC's Fourth Assessment Report concludes that, by 2020, crop yields could increase by 20% in east and southeast Asia, but decrease by up to 30% in central and south Asia, and that rain-fed agricultural output could drop by 50% in some African countries."

[26] "At least one livestock breed a month has become extinct over the past seven years, and around 20 percent of the world's livestock breeds are at risk of extinction." International Centre for Trade and Sustainable Development "Conference Agrees Steps to Safeguard Farm Animal Diversity" (2007) http://ictsd.org/i/news/biores/9342/. See also *Global Warning supra* note 13 at 24.

[27] *Beyond Factory Farming supra* note 7 at 19.

[28] Cassuto *supra* note 9 at 7, citing The Pew Commission on Industrial Farm Animal Production *Putting Meat on the Table: Industrial Farm Animal Production in America* (2008) www.pewtrusts.org/our_work_report_detail.aspx?id=38442. "Half of the soy produced in the United States, and more than half of the corn, is fed to domestic livestock. Vertically integrated multinational agribusinesses export

most of the rest of U.S. corn and soy production to be fed to industrial livestock facilities in other countries": Dennis Olson "Below-Cost Feed Crops: An Indirect Subsidy for Industrial Animal Factories" (June 2006, Institute for Agriculture and Trade Policy: Trade and Global Governance Program) www.agobservatory.org/library.cfm?refid=88122 at 2.

[29] *Beyond Factory Farming supra* note 7 at 6.

[30] G. Brundtland, ed. *Our Common Future: The World Commission on Environment and Development* (Oxford: Oxford University Press, 1897).

[31] Robert B. Gibson has itemized how the sustainability approach can be practically applied: "Sustainability-based assessment criteria and associated frameworks for evaluations and decisions: theory, practice and implications for the Mackenzie Gas Project Review" (2006) www.ngps.nt.ca/Upload/Joint%20Review%20Panel/Specialist%20Advisors/Dr.%20Robert%20Gibson/sust_asmt_MGP_fnl.pdf.

[32] Erb et al. *supra* note 8 at 24.

[33] "Ecological Justice and Law" in B. Richardson and S. Wood, eds., *Environmental Law and Sustainability* (Hart, 2006) 129.

[34] What Wil Kymlicka calls "resource egalitarianism" in "Indigenous Rights and Environmental Justice" in Wil Kymlicka *Politics in the Vernacular: Nationalism, Multiculturalism and Citizenship* (Oxford Scholarship Online, 2003) ch.7.

[35] As authoritatively expounded by Edith Brown Weiss in *In Fairness to Future Generations: International Law, Common Patrimony and Intergenerational Equity* (Tokyo: United Nations University, 1989).

[36] "The justice of the relationship between humans and the rest of the natural world", Bosselman *supra* note 34 at 129.

[37] Penelope Simons "Binding the Hand that Feeds Them: The Agreement on Agriculture, Transnational Corporations and the Right to Adequate Food in Developing Countries" in W Shan, P Simons, D Singh (eds) *Redefining Sovereignty in International Economic Law* (Hart Publishing, Oxford2008), 399 at 414. See also "Agriculture and the WTO: Does Membership Make a Difference?" CATPRN (Canadian Agricultural Trade Policy and Competitiveness Research Network)Working Paper 2010-03 May 2010; and Allan N Rae "Trade in Livestock Products and the WTO Millennium Round: Projections to 2005 and Problems with TRQs" (Meeting of the International Agricultural Trade Research Consortium (IATRC), Auckland, New Zealand, Jan. 2001).

[38] Simons *ibid.* at 414, citing UNCTAD, 'Tracking the Trend Towards Market Concentration: The Case of the Agricultural Input Industry' (20 Apr 2006) UNCTAD/DITC/COM/2005/16 at 3, 8 and 25ff.

[39] See also Jerry Spear, Marie-Ann Bowden, et al. "Comparative Standards for Intensive Livestock Operations in Canada, Mexico, and the United States" (Paper prepared for the Commission for Environmental Cooperation, 4 February 2003) www.cec.org/Storage/49/4168_Speir-etal_en.pdf at vi: "One aspect of this concentration is the dominance of a small number of companies, often meat processors, who increasingly own the animals, feed mills and brand-names of the finished product through a process called 'vertical integration'."

[40] Simons, *supra* note 38 at 416-17.
[41] Joyeeta Gupta "Global Sustainable Food Governance and Hunger: Traps and Tragedies" (2004) 106 Br. Food J. 406 at 410. See also Durning and Brough *supra* note 2 at 39.
[42] Jennifer Clapp "Demystifying Doha: Making Sense of the WTO Agricultural Trade Talks" (Sept. 5, 2006), Harvard International review web reports http://hir.harvard.edu/demystifying-doha.
[43] FAO, Agriculture and Consumer Protection Department "Globalization and Livestock" (2005) www.fao.org/ag/magazine/0504sp1.htm.
[44] TRIPS: Agreement on Trade-Related Aspects of Intellectual Property Rights, Apr. 15, 1994, Marrakesh Agreement Establishing the World Trade Organization, Annex 1C, THE LEGAL TEXTS: THE RESULTS OF THE URUGUAY ROUND OF MULTILATERAL TRADE NEGOTIATIONS 320 (1999), 1869 U.N.T.S. 299, 33 I.L.M. 1197 (1994).
[45] Wency Leung and Jessica Leeder, "The World's Most Controversial Pig" (Globe and Mail, Nov. 25, 2010) http://v1.theglobeandmail.com/servlet/story/LAC.20101125.BIG8FOODPIG1125 ATL/TPStory/?query=Biotechnology&pageRequested=all&print=true.
[46] Gupta *supra* note 42.
[47] Durning and Brough *supra* note 8 at 43.
[48] FAO "Globalization and Livestock" *supra* note 44 at 2. See also Jason Konefal et al. "Governance in the agro-food system: Backlighting the role of transnational supermarket chains" (2005) 22 Agriculture and Human Values 291.
[49] Olson *supra* note 29 at 1.
[50] Timothy A. Wise "Identifying the Real Winners from U.S. Agricultural Policies" (Global Devel. and Envir. Institute (Tufts University), Working Paper No. 05-07, Dec. 2005) http://ideas.repec.org/p/dae/daepap/05-07.html.
[51] *Ibid.* at 6.
[52] Durning and Brough *supra* note 2 at 35:"Centrally planned states have subsidized animal products almost as vigorously as market-oriented nations."
[53] Cassuto *supra* note 9 at 7.
[54] *Ibid.* at 8, citing J.B. Ruhl, "Farms, Their Environmental Harms, and Environmental Law" (2000) 27 Ecol LQ 263 at 267.
[55] See Speir, Bowden et al, *supra* note 40 at xi-xii and 110-111.
[56] *Ibid.* at 55. See, e.g., Ontario *Farming and Food Production Protection Act*, SO 1998, c 1 and Saskatchewan's *Agricultural Operations Act*, SS 1995, c A-12.1.
[57] The welfare of livestock animals in Canada is governed by 3 federal statutes and some provincial laws. The federal statutes are the *Criminal Code* RSC 1985, c C-46, s. 446; the *Health of Animals Act,* SC 1990, c 21 and the *Meat Inspection Act* RSC 1985, c 25 (1st Supp). See Jackie Wepruk "A Report on Animal Welfare Law in Canada" (for Alberta Farm Animal Care, June 2004) at 1. www.afac.ab.ca/lawsregs/awlcanada.pdf.
[58] *Regulatory Reform in Canada's Agri-Food Sector"* (2009) www.capi-icpa.ca/pdfs/CAPI_Regulatory%20Framework%20_March%204%202009_.pdfTh e Report supports this approach by citing the OECD's own review of Canada's

system, yet this OECD report had nothing to do with agriculture, but with approaches to regulation in general: "OECD Praises Canada's Regulatory Reforms and Encourages Sustained Momentum," October 29, 2002, www.oecd.org/document/24/0,3343,en_2649_34141_1835224_1_1_1_1,00.html.

[59] Speir, Bowden et al *supra* note 40 at 114.

[60] Durning and Brough *supra* note 2 at 41.

[61] *Ibid* at 42.

[62] CIWF *Beyond Factory Farming supra* note 7 at 11.

[63] Sierra Club of Canada "Intensive Livestock Operations: An Introduction" www.sierraclub.ca/national/programs/health-environment/food-agriculture/ilo-introduction.shtml.

[64] Durning and Brough *supra* note 2 at 29-32.

[65] Lang "Food Industrialisation and Food Power: Implications for Food Governance" (2003) 21 Devel. Pol. Rev. 555.

[66] Olson *supra* note 29 at 2.

[67] *Ibid.*

[68] CIWF *Beyond Factory Farming supra* note 7 at 10.

[69] Pew Commission News Release "Pew Commission Says Industrial Scale Farm Animal Production Poses 'Unacceptable' Risks to Public Health, Environment" (April 2008) www.pewtrusts.org/news_room_detail.aspx?id=38438.

[70] *Livestock's Long Shadow supra* note 11 at 81; Benjamin J. Richardson et al., eds. *Climate Law and Developing Countries: Legal and Policy Challenges for the World Economy* (Edward Elgar 2009) at 1, citing Sir Nicholas Stern *The Economics of Climate Change* (Cambridge University Press, 2007) 92-93.

[71] Stern, *ibid.*

[72] Peter Singer and Jim Mason *The Way We Eat: Why Our Food Choices Matter* (Rodal Books, 2006).

[73] 1771 UNTS 107; S. Treaty Doc No. 102-38; U.N. Doc. A/AC.237/18 (Part II)/Add.1; 31 ILM 849 (1992).

[74] UN Doc FCCC/CP/1997/7/Add.1, Dec. 10, 1997; 37 ILM 22 (1998).

[75] *Livestock's Long Shadow supra* note 11, Executive Summary at 238. The CDM is a tool under Kyoto permitting Annex I countries to get credit toward their emission reductions by funding GHG reduction projects in other jurisdictions.

[76] FAO "Towards a Work Programme on Agriculture: A Submission to the AWG-LCA by the Food and Agriculture organization of the United Nations" (2010) http://unfccc.int/resource/docs/2010/smsn/igo/081.pdf.

[77] The Ad Hoc Working Group on Long-term Cooperative Action under the Convention (AWG-LCA) does some "policy-focused work on agriculture" as well. See FAO "Towards a Work Programme on Agriculture".

[78] See "Agriculture and Climate Change: An Agenda for Negotiation in Copenhagen" (2020 Vision for Food, Agriculture and the Environment, May 2009) www.africaclimatesolution.org/features/Focus16_01_-_20090415.pdf and "Agriculture, Food Security and Climate Change in Post-Copenhagen Processes" (An FAO Information Note)

http://foris.fao.org/static/data/nrc/InfoNote_PostCOP15_FAO.pdf; and "Towards a Work Programme on Agriculture: A Submission to the AWG-LCA by the FAO" http://unfccc.int/resource/docs/2010/smsn/igo/081.pdf.

[79] *Ibid* 38.

[80] *Ibid.* at 39.

[81] CIWF *Global Warning supra* note 13 at 6.

[82] *Farming and Food Production Protection Act, 1998*, *supra* note 58.

[83] Cassuto *supra* note 9 at 14-15. Olson agrees *supra* note 29 at 3.

[84] Noam Mohr "A New Global Warming Strategy: How Environmentalists are Overlooking Vegetarianism as the Most Effective Tool Against Climate Change in Our Lifetimes" (An EarthSave International Report, August 2005) www.earthsave.org/globalwarming.htm.

PART II

ETHICAL ISSUES AND ECOLOGICAL INTEGRITY

INTRODUCTION

KLAUS BOSSELMANN

Ecological Integrity is based not only on scientific evidence, as shown in part 1 of this book, but also on ethical reasoning. In fact, scientific and ethical discourses are not disconnected, but related and complementary as will be shown in the following part 2 of this book. The integrative and interdisciplinary nature makes ecological integrity the powerful reference that it has become for the development of truly sustainable policies, laws and economics.

Each of this part's four chapters are reflections of the close nexus between science and ethics. As we explore the various manifestations of the global crisis, we become aware that they are all manifestations of a crisis in perception. Human beliefs and perceptions are fundamental to how we see the world and how we assess the potential for change. The dominant corporate and governmental view of the world is deeply distorted as it continues to ignore ecological realities. Understanding and acknowledging ecological interdependencies is, therefore, the essential first step in appreciating the connections between sustaining the human enterprise and sustaining all other forms of life.

Perhaps the most striking manifestation of humanity's crisis is the ongoing failure of achieving a global solution to climate change. In his analysis of the Copenhagen Accord, Donald Brown shows how nations, some more than others, have occupied themselves with narrow, short-term economic self-interest virtually devoid of recognizing global responsibilities. This preoccupation has presented itself over a period of nearly twenty years with Copenhagen – and recently Cancun – highlighting what fundamentally is a crisis of state-centered governance. The self-interest of powerful nations and corporations consistently prevents the international community from doing "the right thing", both ecologically and socially. While Brown has his focus on climate change ethics, he also shows how ethical failure is coupled with ignorance of what is actually at stake. Both the Copenhagen Accord and the Cancun Agreement lack not only urgency and commitment, but also a very basic recognition of ethics and justice.

This basic failure is also the starting-point for Peter Brown's essay on politics in an ecological age. Insisting on ecology as a collection of mature sciences, Brown asserts the emergence of a coherent worldview that is far superior to the reductionist assumptions of current environmental and economic policies. As we enter the Anthropocene, we need to rethink the basics of environmental and economic governance. With reference to six different areas – from anthropological concepts of the person to the politics for an ecological age - Brown shows how a coherent and scientifically-based worldview can be expressed in clear psychological, ethical, political and economic terms. He makes the case for a different, albeit not entirely new agenda for the human prospect.

To a degree, Sheila Collins builds upon this new agenda and asks how it relates to the various other agendas or scenarios currently being discussed. Her chapter starts from the premise of a current "breakdown" of the world's economy, architecture of capital accumulation and political power to then examine alternative scenarios. There are five future scenarios that can be discerned from the present juncture. Four of these scenarios, however, appear as compromising, unsustainable or plainly suicidal; only the fifth scenario offers a sustainable perspective. Importantly, Collins presents the various scenarios as discourse-related, not as matters of inherent objectivity. This is particularly true for the fifth, i.e. the "commonwealth of life" scenario. The "big question" of how likely an ethically advanced new order is, can be answered in quite tangible terms. According to Collins, it will be realistic in so far we are able to learn from "myriad examples around the world of places where pieces of such a new order have already been assembled".

The final chapter by Helmut Burkhardt can be read as a commentary on the scenarios promoted by Donald Brown, Peter Brown and Sheila Collins. The conditions for sustainability of a civilization are complex, but can be described as a set of sustainability imperatives. Each of these imperatives is centred around the notion of integrity; Integrity of the individual, the community, the ecosystem and of the cosmic environment is the essence of sustainability. Imperatives derived from this essence include the achievement of personal integrity, the commitment to societal integrity, the preservation of ecological integrity and a sense of cosmic integrity. Burkhardt maintains that even though cosmic conditions are beyond human influence, they need to be remembered. The cosmic equilibrium, largely provided through solar activity, is threatened by self-inflicted climate change.

In concluding Part 2, we aim to close the circle. Anthropogenic climate change appears to be the tip of the iceberg. It cannot be successfully dealt with in isolation from human beliefs and perceptions. Will those in public office or with private influence assume responsibility in recognising the iceberg and how profoundly we all must change?

CHAPTER FIVE

A COMPREHENSIVE ETHICAL ANALYSIS OF THE COPENHAGEN ACCORD

DONALD A. BROWN

Introduction

If climate change must be understood as a civilization challenging ethical problem, what can be said about the positions taken by governments and results achieved at the December 2009 Copenhagen conference.

To evaluate what happened in Copenhagen one must understand that the Copenhagen meeting was only the last in almost two decades of meetings that have failed to achieve a global solution to climate change. Copenhagen was the 19th meeting of governments from around the world that have been meeting every year since 1990 to forge a comprehensive climate change regime. Copenhagen was also the 15th conference of the parties (COP-15) since the United Nations Framework Convention on Climate Change (UNFCCC) came into effect in 1994 (UN 1992).

For more than twenty years some nations have been taking positions on climate change that raise serious ethical concerns. Copenhagen meeting was no exception. However, there were two issues that arose with a new force in Copenhagen. They were the intensity and frequency of calls for: (a) global justice, and (b) increased funding for adaptation programs in vulnerable developing countries (Brown 2009a).

Yet, at the conclusion of the Copenhagen conference, as we shall see, little was accomplished in response to these issues or the other climate change disputes that have now plagued climate negotiations for almost two decades. Although, as we shall see, some have pointed to a few positive Copenhagen outcomes, most observers have judged COP-15 to be a disaster.

The Path to the Copenhagen Accord

The Copenhagen conference took place from December 7-19, 2009. Copenhagen was intended to be the culmination of a two-year negotiating process that was agreed to in Bali, Indonesia, in December 2007.

In 1990 negotiations began that led in 1992 to opening for signature and ratification of the UNFCCC. This treaty itself does not contain binding greenhouse gas (GHG) emissions limitations for countries but nevertheless includes numerous other binding national climate change obligations.

To understand the significance of what happened in Copenhagen, it is necessary to understand the goals and objectives for an international climate regime that were originally set out in the UNFCCC. Among other things, for instance, the parties to the UNFCCC agreed that:

(a) They would adopt policies and measures to prevent dangerous anthropogenic interference with the climate system,
(b) Developed countries should take the first steps to do this,
(c) Nations have common but differentiated responsibilities to prevent climate change,
(d) Nations may not use scientific uncertainty as an excuse for not taking action, and
(e) Nations should reduce their GHG emissions based upon "equity" (UN 1992).

Some national proposals in Copenhagen, seventeen years after the UNFCCC was agreed upon, failed to abide by many promises made by governments in the UNFCCC.

As of December 2009, the UNFCCC had 192 parties, a number that includes almost all countries in the world including the United States which ratified the UNFCCC in 1994.

The UNFCC is a "framework" convention because it has always been expected that additional requirements would be added to the framework in updates that are known as "protocols" or in annual decisions of the conferences of the parties.

The first major addition to the UNFCCC was the Kyoto Protocol which was negotiated in 1997 because the international community had been convinced by emerging climate change science that developed nations needed to be bound by numerical emissions reductions targets. The Kyoto Protocol entered into force on February 16, 2005 and currently has 190 parties. The United States is the only developed country that never ratified the Kyoto Protocol.

Under the Kyoto, Protocol, the developed countries agreed to reduce their overall emissions of six greenhouse gases by an average of 5.2%

below 1990 levels between 2008-2012. The developing countries had no binding emissions reductions obligations under Kyoto.

The Copenhagen negotiations were necessary because the emissions reductions obligations of developed countries set out in the Kyoto Protocol expire in 2012.

At climate negotiations at COP-13 in Bali, Indonesia in 2007, parties to the UNFCCC agreed to replace the Kyoto Protocol with an agreement that would create a second commitment period under the UNFCCC and would include binding emissions reductions for developed countries and new programs on adaptation for developing countries, deforestation, finance, technology transfer, and capacity building. This agreement is referred to as the Bali Roadmap which also called for articulating a "shared vision for long-term cooperative action," including a long-term global goal for emission reductions.

The Bali decision also recognized that developing countries could make contributions to solving the climate change through the development of Nationally Appropriate Mitigation Actions (NAMAs), that is climate change strategies for developing countries. The NAMAs, however, would not constitute binding emissions reduction requirements for developing countries in contrast to the binding obligations of developed countries in the Kyoto Protocol that would be further developed in Copenhagen.

At Bali the parties also agreed on a two-year negotiating process to achieve the objectives of the Bali Roadmap. Under this action plan, nations would proceed on two negotiation tracks. One under the UNFCCC and the other under the Kyoto Protocol.

Intense negotiations in preparation for Copenhagen took place during the two years between Bali and Copenhagen including four separate meetings in 2009 alone. In these deliberations, many contentious issues surfaced. Among other things, these disputes included particularly strong disagreements about the magnitude of developed country emissions reduction commitments and institutional arrangements and funding amounts for financing developing country needs for technology cooperation, adaptation, reducing emissions from deforestation, and capacity building.

Although some progress was made on a few issues in the two year lead-up to Copenhagen, little progress was made on the major issues and particularly on commitments for GHG emissions reductions and funding for adaptation, deforestation programs, and technology transfer.

As Copenhagen approached, optimism about a Copenhagen deal faded although there was a short spurt of renewed hope several weeks before the conference started as the US, China, and a few other nations publicly made commitments on emissions reductions.

The Copenhagen meeting was different than previous COPs for the noticeable increase in attention paid to the ethics and justice dimensions of climate change and adaptation needs (Brown 2009a). Yet, as we shall see, none of this appeared to motivate developed nations to make commitments to reduce GHG emissions consistent with what science claims necessary to protect those most vulnerable to climate change.

During the Copenhagen conference representatives from poor vulnerable nations begged developed countries to: (a) commit to reduce GHG emissions to levels necessary to prevent dangerous climate change. and (b) to fund adaptation programs in developing countries that are necessary to protect the most vulnerable from climate change impacts that could be avoided and compensate for damages that could not be avoided.

Despite these pleas, not much happened during the conference to resolve the most contentious issues until US President Obama appeared on the morning of the last day, Friday, December 18, 2009. For much of that day, President Obama negotiated with Chinese premier Wen Jiabao, Brazilian president Luiz Inacio Lula da Silva, Indian Prime Minister Manmohan Singh and South African President Jacob Zuma (Lerer 2009). Yet, a large part of this time of these negotiations was focused on a dispute between the United States and China on whether China would agree to monitoring and verification of Chinese climate change commitments.

President Obama could not commit to anything in Copenhagen that he could not get through the US congress. Because a climate change bill that had passed the US House was very weak compared to what science says was necessary to protect the world's poorest people, the United States took a position in the lead-up to Copenhagen that continued to be the weakest of all the developed countries on emissions reductions.

Because none of the developed countries were willing to make emissions reduction commitments congruent with what science was saying was necessary to protect them, some of the most vulnerable developing countries saw the developed countries' positions as ominous, perhaps a death sentence.

The United States was not alone among developed countries in refusing to make emissions reductions commitments consistent with what science was saying was necessary to prevent dangerous climate change. In fact, no developed country made such a commitment in the lead-up to Copenhagen (Brown 2009b).

A notable example of the acrimony between developed and developing countries during the pre-Copenhagen period were a provision that had been inserted in the negotiating text that claimed that developed countries were responsible for "ecological debt." As we shall see, this assertion was strongly rejected by Todd Stern, lead US climate negotiator, who said: "I actually completely reject the notion of a debt or reparations or anything of the like" (Roberts 2009). In addition, US negotiator Jonathan Pershing further said, "The donor countries have only so much largesse." This statement appeared to be premised on the idea that the United States has no obligations to developing countries for past pollution and that payments to developing countries for adaptation are simply a mater of charity.

In response to this denial of historical responsibility for climate change, Vandana Shiva, a prominent social activist for developing countries, said: "I think it is time for the United States to stop seeing itself as a donor and recognizing itself as a polluter, a polluter who must pay for its pollution and its ecological debt. This is not about charity. This is about justice" (Roberts 2009).

President Obama personally negotiated the Copenhagen Accord during last hours of the conference. Yet, to get this deal, President Obama had to ignore many of the positions of the most vulnerable nations that were unresolved in the two negotiating documents that had been created in the lead-up to Copenhagen.

Although the United States was unwilling to acknowledge financial responsibility for climate change, President Obama was at least willing to recognize that the United States should make commitments on emissions reductions and funding, a profound change from the prior US administration. Most likely because of a potential future, but yet unrealized, new US direction, President Obama managed to obtain support from many nations including some prominent developing countries on the Copenhagen Accord, the non-binding, political statement that emerged from the Obama-led negotiations.

Despite growing scientific evidence that the world is running out of time to protect the poorest people in the world from the harsh impacts of climate change, the best the international community could do in Copenhagen was to agree on the weak, compared to what was hoped for, non-binding document, the Copenhagen Accord.

Politically President Obama's hands were tied in the negotiations because of domestic political constraints.

For domestic political reasons, the US President also wanted agreement from China and other large developing countries on transparent procedures for verifying their non-binding emissions reduction commitments.

Those opposing climate change legislation in the United States often have argued that it would be unfair to the United States if it was bound to reduce GHG emissions and China was not required to do the same. In fact, a decade earlier, when the Kyoto Accord was under consideration in the United States, opponents of the Kyoto deal frequently ran TV commercials that argued that the Kyoto Protocol was unfair to the United States because China was excluded from emissions limitations. This argument was often made without visible critical comment in the United States even though the United States had committed itself to take the first steps to reduce emissions along with other developed countries under the UNFCCC. For instance, the Heritage Foundation argued that the Kyoto Protocol should not be ratified because the Protocol excluded developing countries from binding emissions reductions including China (Coon 2001).

When China became the world's largest emitter in terms of total tons of GHGs in the fall of 2008, this fact was widely publicized in the United States as if China had moved into first place in terms of global responsibility. The US press frequently announced that China had become the largest polluter, a fact that ignored that the United States emits more than four times the amount of GHG than China on a per capita basis.

Although President Obama originally negotiated the Accord with just four other countries, in the last few hours of the Copenhagen conference the United States successfully convinced most large emitting countries to support the Accord.

As a result, in the wee hours of Saturday mourning, the Copenhagen parties agreed to reference the Copenhagen Accord in a COP decision that "took note" of the Copenhagen. The COP "took note" of the Copenhagen Accord rather than adopting it because a consensus was not possible and decisions by a COP require a consensus. At least 25 countries that were asked by the COP President to participate in a high level meeting of the "friends of the President" eventually accepted the Accord, at least four Parties spoke out against it (Tuvalu, Sudan, Bolivia, and Venezuela) (Werksman 2009).

The Copenhagen Accord will not solve the climate change crisis as even President Obama acknowledged because it does not contain commitments necessary to avoid dangerous climate change nor deal adequately with other issues including GHG gases coming from deforestation and adaptation issues.

The Copenhagen Accords is a three page document whose most significant elements are:

- **Long-Term Goals:** The parties agreed that deep cuts in global emissions are required according to science and as documented by the IPCC Fourth Assessment Report, with a view to reducing global emissions in order to limit the increase in global temperature to below 2°C. The agreement also calls for an assessment of the implementation of the Accord to be completed by 2015 including examining whether the long-term goal should be a temperature rise of 1.5°C.

- **Adaptation**: The document states that adaptation to the adverse effects of climate change and the potential impacts of response measures is a challenge faced by all countries, and that enhanced action and international cooperation on adaptation are urgently required in developing countries, especially in the least developed countries, the small island states, and Africa. The parties to the Accord also agree that developed countries shall provide adequate, predictable and sustainable financial resources, technology and capacity building to support adaptation.
- **Financing For Poor Nations**: The document provides that developed countries shall set a goal of mobilizing jointly $100 billion a year by 2020 to address the needs of developing countries and that the funds will come from a wide variety of sources, public and private, bilateral and multilateral. An annex carries the following short-term financing pledges from developed countries for 2010-2012: EU - $10.6 billion. Japan - $11 billion. United States - $3.6 billion
- **Emissions Reductions**: The document provides for countries to voluntarily commit to GHG mitigation plans in two separate annexes, one for developed country targets and the other for the voluntary pledges of major developing countries. The developed countries committed to implement individually or jointly quantified emissions reductions targets for 2020 to be submitted by 31 January 2010. The developing countries may identify commitments that will be included in the annex to the Copenhagen Accord. Neither developed nor developing country commitments under the Accord are legally binding.
- **Verification of Climate Change Promises.** A sticking point for a deal, largely because China refused to accept international controls, the section on monitoring of developing nation pledges is one of the agreement's most significant provisions. The document provides that emerging economies must monitor their efforts and report the results to the United Nations every two years, with some international checks to meet Western transparency concerns but "to ensure that national sovereignty is respected."

- **Forest Protection.** The accord "recognizes the importance of reducing emission from deforestation and forest degradation and the need to enhance removals or greenhouse gas emission by forests," and agrees to provide "positive incentives" to fund such action with financial resources from the developed world.

Arguments for and against Conclusions that Copenhagen was a Disaster or a Positive Step Forward

The Copenhagen Accord has triggered serious disputes among world leaders, academicians, journalists, and participants as to whether it was an unmitigated disaster or a positive step toward finally dealing with climate change.

For instance, sharply negative assessment of the Copenhagen has been made by John Sauven, executive director of Greenpeace UK, who said at the conclusion of Copenhagen: "The city of Copenhagen is a crime scene tonight, with the guilty men and women fleeing to the airport (BBC 2009).

James Curran, the director of science at the one of the Scottish government's top advisors on climate change said "It really couldn't be any more serious," and "It's staggeringly frightening and deeply disappointing that Copenhagen has failed. This is extremely dangerous for Scotland and the world" (Herald Scotland 2009).

Yet, even though almost all commentators that have drawn positive conclusions about the Accord have also expressed some reservations about this agreement, some have concluded that on balance it was a positive step.

And so some environmental NGOs are split about whether the Copenhagen Accord was a disaster or a positive step forward.

Ethical Analysis of Copenhagen Accord

Any proposed post-Kyoto regime must as a matter of at a minimum:

- Require sufficient greenhouse emissions reductions to assure that the international community is on a greenhouse gas emissions reduction pathway that will prevent dangerous climate change harm. This is sometimes referred to as the environmental sufficiency criteria.
- Begin to base differences among national allocations on the basis of equity and justice. This is sometimes referred to as the equity criteria.

In addition to these criteria previously identified to guide ethical analysis of a post-Kyoto regime, the rise in importance in the adaptation agenda and the almost certain fact that it is too late to protect poor countries from the need to cope with adaptation issues, there is clearly a need to add additional ethical criteria for evaluating second commitment period regimes under the UNFCCC. That is, any proposed post-Kyoto regime must:

- Assure that those responsible for climate change provide adequate, predictable adaptation funding to enable developing countries and in particular the most vulnerable developing countries to do what is necessary to avoid climate change damages in cases where it is possible to take action and to prevent damages, or be compensated for climate change damages in cases where it is impossible to take protective action. We will refer to this as the just adaptation criteria

These three criteria, that is environmental sufficiency, equity, and just adaptation constitute the minimum ethical considerations that any climate regime must satisfy, they don't capture all ethical questions raised by climate change.

As the world approached the Copenhagen negotiations in late summer of 2009, some nations were taking positions on the Copenhagen agenda as if GHG emissions reduction commitments can be guided by national interest alone and not responses to international ethical duties and obligations. For this reason, the major emitters were approaching Copenhagen in a way that appeared to be avoiding their ethical obligations and acting as if narrow national economic interest is a valid justification for their climate change positions. Yet, ethics would require these nations to act in accordance with duties and responsibilities to other vulnerable nations, not self-interest alone.

The Copenhagen Accord and the way it was negotiated does nothing to change the conclusion that most nations failed to live up to their ethical obligations in approaching Copenhagen. In fact, as we shall see, many of the considerations that seemed to be the driving forces behind the content of the Copenhagen Accord are based upon ethically problematic assumptions and premises.

a. Environmental Sufficiency Criteria

Not only does the Copenhagen Accord fail to require sufficient GHG emissions reductions to assure that the international community is on a GHG emissions reduction pathway that will prevent dangerous climate change harm, the emissions reductions commitments that have been identified under the Accord almost guarantee that millions of poor people, plants, animals, an ecosystems will be harmed by climate change if the consensus science position turns out to be correct.

Although it is still possible that nations in the next few years will revise upward their GHG emissions reductions commitments to levels that will protect the most vulnerable people and countries, very recent science has concluded that the world is running out of time to do this.

As we have seen, the Copenhagen Accord does set a target that nations should work together to limit human caused additional heating to 2°C. There are, however, several ethical problems with this target. They include the following:

- Any additional warming from current levels is ethically problematic because current temperatures are already dangerous for some vulnerable people around the world and an additional 1 °C temperature rise is already locked in by prior emissions. Because any additional warming from current levels could have serious consequences to those most vulnerable to climate change, those who are most vulnerable should have as a matter of procedural justice rights to consent to put at risk by the additional 2°C goal adopted in the Accord. Yet, some of the most vulnerable nations in Copenhagen wanted the Copenhagen outcome to adopt a 1.5 °C goal. Although the Copenhagen Accord specifies that the 1.5 °C can be reconsidered in five years, many scientists would argue that since global peaking of GHG emissions must take place in the next few years and therefore waiting five years virtually guarantees that the 1. 5°C is unachievable.
- The Copenhagen Accord appeared to reject an idea that was under consideration in the negotiating text that world needed to set not just a temperature limit goal but needed to commit to an atmospheric GHG concentration level that would give any temperature goal practical significance. To give the warming target practical meaning it is necessary to identify an atmospheric GHG concentration goal because there is scientific uncertainty about how much warming will be produced by different levels of atmospheric

GHG concentrations, a scientific issue usually referred to as the "climate sensitivity" problem. Therefore, if the international community is serious about protecting vulnerable people it must take a position on climate sensitivity by adopting a atmospheric GHG concentration goal. The negotiating text under consideration before the Copenhagen Accord contained unresolved alternative atmospheric GHG concentration goals such as 350 and 450 parts per million of CO2. Only by adopting an atmospheric GHG concentration target, can anyone analyze the sufficiency of national commitments. And so the 2°C warming limit adopted by the Accord is virtually worthless in determining whether national emissions reductions commitments are adequate to protect human health and the environment from climate change.

- There is substantial scientific evidence that even the 1.5 °C temperature limit would not be sufficient to protect those most vulnerable to climate change. For instance, a recent paper by Jim Hansen and seven other authors concluded that additional warming should be limited to 1°C warming and to do this existing atmospheric concentrations of CO2 must not only not be allowed to rise the small amount to 450 ppm of carbon equivalent but must be reduced from existing levels of 385 ppm to 350 ppm CO2. (Hansen et al. 2008) According to this paper, the world has likely already shot past the level of atmospheric concentrations that will lead to dangerous climate change for many. Under this view, the world has already used up all of the assimilative capacity of the atmosphere and biosphere that has been available to buffer against dangerous climate change. Given this, a strong ethical argument can be made that all nations have a duty to try to prevent additional warming of almost any amount, while the Copenhagen Accord legitimizes an additional 2°C warming. Given that the Copenhagen Accord can also be understood to legitimize any national GHG emissions target that is proposed voluntarily, even if it is insufficient to achieve the 2°C temperature limit goal adopted by the Accord, let alone the duty to try and prevent any additional warming, the Accord is ethically problematic. Given that a case can be made that current levels atmospheric GHG concentrations are already harming or putting people and ecosystems at risk, it is difficult to make an ethically acceptable case that atmospheric GHG concentration targets higher than current levels are justified unless consent is given by those who are already being harmed by warming. Yet the Copenhagen Accord assumes that an additional 2°C warming is acceptable. The national GHG emissions reduction commitments that have been identified to be included in the Annex to the Copenhagen Accord are not sufficient to even achieve the 2°C temperature limit goal. As one commentator has explained:

1. As a group, the Copenhagen commitments for the biggest emitters, if confirmed, would imply a 28% increase of emissions above the 1990 level.
2. Compared with the business-as-usual scenario for those countries, emissions would be reduced by 21%.
3. Assuming that the rest of the world continues on a business-as-usual path, global emissions would increase to about 48 gigatons of carbon dioxide equivalent (GT CO2-eq) by 2020. This represents a 29% increase with respect to 1990, a 5% increase with respect to 2005 and a 16% reduction with respect to business-as-usual
4. The Copenhagen declarations are clearly insufficient to control global warming below 2°C - even if they are substantial when compared with the business-as-usual scenario. What at first glance seemed like good news - the emission-reduction declarations - turns out to be bad news. The declarations are inconsistent with the 2°C temperature target, even though the target is reiterated in the Copenhagen Accord itself (Carraro and Massetti 2010).

For this reason, the Copenhagen Accord is ethically problematic because it appears to legitimize emissions reductions commitments that are inconsistent with its temperature limit goal.

We have also seen that the United States approached negotiations of the Copenhagen Accord as if the United States need not make emissions reductions commitments unless it could secure commitments to reduce GHG emissions from high-emitting developing nations. No nation may deny its duty to reduce its emissions to its fair share of safe global emissions on the basis that others who are contributing to the harm have failed to cease harmful behavior (Brown 2009c). This is so because no nation or person has a right to continue destructive behavior on the basis that others who are contributing to the harm have not ceased their destructive behavior.

And so, if some nations are not willing to reduce their emissions to levels consistent with what justice requires of them, no nation, including the United States, can refuse to reduce its emissions to its fair share of safe global emissions levels on the basis that others won't act.

Although the United States is well within its rights to obtain promises of other nations to contribute to solving the climate change problem, it may not as a matter of ethics condition its willingness to reduce its emissions to levels required by justice on other nations' behavior. That is, although it may be in everyone's interest if the United States encourages others to make GHG emissions reductions commitments, the United States may not refuse to reduce its emissions to its fair share of safe global emissions on the basis that others have not acted. The United States could ethically link non-obligatory climate change actions on other's participation in climate change solutions but must agree to do what ethics requires of it in reducing emissions without regard to the actions of others.

Therefore, from the standpoint of the environmental sufficiency goal, the Accord fails to satisfy the requirement that any post-Kyoto regime must assure that the international community is on a greenhouse gas emissions reduction pathway that will prevent dangerous climate change harm.

b. Equity Criteria

The second minimum ethical criteria that all Copenhagen proposals must meet is the requirement that emissions reduction proposals must be consistent with what "equity" and "justice" demands of them Equity requires that each nation reduce its emissions to its fair share of safe global emissions. And so, each nation's emissions reduction levels should be based upon what distributive and retributive justice demands, not on national self-interest. Although there are different theories of distributive justice that lead to different national allocations, many justifications for national GHG emissions allocations fail to satisfy any ethical scrutiny. In other words, it is not necessary to know what perfect justice requires to conclude that some proposals for national GHG allocations are unjust. One such common approach to national GHG emissions reductions commitments that fails to satisfy any ethical scrutiny is the claim that all nations must reduce emissions by the same amount without regard to whether a nation is a large or small contributor to the climate change problem, an approach often referred to as 'grandfathering' or equal reductions from existing emissions levels. It would appear that some of the national commitments that will be included in the Accord are based upon grandfathering emissions reductions from existing levels.

Since most nations entered the Copenhagen negotiations as if national interest rather than global responsibility to others was an adequate basis for national climate change policies, the Copenhagen Accord has failed to satisfy equity criteria. In fact, in the lead-up to Copenhagen, most of the justifications for national commitments that had been announced by countries to reduce emissions were exclusively focused on whether they met global goals to reduce GHG emissions unadjusted by equity considerations.

There have been several proposals discussed by the international community about second commitment period frameworks that would expressly incorporate equity into future GHG emissions reductions pathways. Two such frameworks are known as "Contraction and Convergence" (C&C 2009) and "Greenhouse Development Rights" (GDR) (Bear and Athanasiou 2009) frameworks. In the lead-up to Copenhagen, all major GHG emitting nations ignored the C&C or GDR frameworks or any other comprehensive framework that took equity into account. In fact, the Copenhagen Accord allows each nation to identify its emissions reduction commitment based upon voluntary national considerations without regard to equity.

Therefore, the Copenhagen Accord is a complete failure in satisfying equity criteria.

c. Just Adaptation Criteria

The third minimum ethical criteria for judging any second commitment period under the UNFCCC is that it must provide adequate funding to support adaptation programs in developing countries given that some developing countries have done nothing to cause climate change and must take steps to avoid harsh impacts.

In Copenhagen there was a huge blossoming of interest in adaptation to climate change and the linking of the adaptation agenda to everything else. All of a sudden the world has awoken to specific adaptation questions such as: (a) Who is going to pay for climate change damages?, (b) How should these monies be administered?, (c) To whom should they go?, (d) How to set priorities among adaptation needs, and (e) How much money will be made available for growing adaptation needs? From the standpoint of the poorest developing countries, adaptation issues have been a high priority for some time but now these issues are high on the negotiating agenda for all because every day it is becoming more difficult to prevent climate change damages. In fact, for some parts of the world adverse climate change impacts are not a future threat but a present reality.

Over the last several years, developing countries have pushed the adaptation agenda to center stage in international climate change negotiations. In Bali, a decision was made to create a new adaptation fund as part of the UNFCCC climate change architecture and to put adaptation on an equal footing with mitigation. For the last few years, adaptation has been growing in importance in international climate change negotiations, yet the energy, force, manner, and specificity with which the adaptation agenda is being negotiated is new.

For instance, for the first time some developing countries succeeded in getting the words "ecological debt" into the negotiating text. One proposed paragraph provides:

The guiding principles of the Convention should support subparagraphs (b) and (c) above, in terms of common but differentiated responsibilities and respective capabilities, historical responsibilities in greenhouse gas emissions and the related historical ecological debt generated by the cumulative greenhouse gas emissions since 1750 and the most recent scientific information (UNFCCC 1992: 17).

The negotiating text leading up to Copenhagen called adaptation "urgent" and provides that the "polluters shall pay" through "mandatory" contributions for "new, predictable, and additional sources of funding" for:

- vulnerability assessments,
- adaptation planning,
- adaptation implementation,
- short-term shocks and long-term climate shifts,
- special funding amounts for particularly vulnerable countries including small island states and African countries affected by desertification, droughts and floods,
- disaster readiness and relief,
- creation of a new international body under the UNFCCC to monitor and determine adaptation needs, assess capacity building needs of developing countries, oversee the creation of adaptation funds, and creation of national adaptation funding,
- creation of regional adaptation centers, and
- creation of new international adaptation committees that collect, analyze, and disseminate information on adaptation.

Yet, it was the amount of the adaptation funding agenda identified in the negotiating text that has profound historical significance for developed countries. According to one provision in the proposed negotiation text, funding for adaptation should be in the range of $70 to $140 billion per year until 2020 and then updated after. Another proposal calls for mandatory adaptation in the amount of 0.5 % of GDP for developed countries.

The Accord acknowledges that enhanced action and international cooperation on adaptation are urgently required in developing countries, especially in the least developed countries, the small island states, and Africa. The Accord also states that that developed countries shall provide adequate, predictable and sustainable financial resources, technology and capacity building to support adaptation. Yet, as we have also seen, the United States expressly denied that this funding is a duty or responsibility but a matter of "largess."

Although short- and long-term funding amounts for a variety of developing country climate change needs are also recognized in the Accord, this document does not acknowledge that this funding is in fulfillment of national responsibilities and in fact may be satisfied from 'public and private' sources.

In the Accord, the developed nations only promised to take steps to mobilize funding from a variety of sources, not provide dedicated funding sources.

As we have seen above, the assertion that the United States has no duties for adaptation was made by US chief negotiator Todd Stern who said:

> I actually completely reject the notion of a debt or reparations or anything of the like. For most of the 200 years since the Industrial Revolution, people were blissfully ignorant of the fact that emissions caused a greenhouse effect. It's a relatively recent phenomenon (Revkin and Zeller 2009).

And so, the United States' denial of responsibility for adaptation appears to be based upon lack of knowledge of harm. Although there are some interesting ethical questions about when this responsibility should start, the international community could have been put on notice of potential climate change harms on the following dates:

- In 1824, Joseph Fourier described the greenhouse effect for the first time
- In 1894, Swedish chemist Svante Arrhenius calculated that a doubling of CO_2 would increase global temperatures by 5 0 C.

- In the late 1950s, direct measurements of atmospheric CO2 demonstrated rising levels in direct proportion to fossil fuel use around the world,
- In 1965, President Johnson's Scientific Advisory Committee warned that climate change was a "real concern."
- In the 1970s computer model based studies were warning the world that climate change would create adverse climate change impacts.
- In 1981, Albert Gore organized a climate change hearing on Capitol Hill.
- In 1985, scientists at Villach, Austria conference reached consensus that global warming was happening and international treaties were needed to curb emissions.
- In 1988, the UN-led Intergovernmental Panel on Climate Change (IPCC) was established to assess state of knowledge on climate change.
- 1990, IPCC issued its first report concluding that climate change was a serious problem.

And so the United States cannot deny that it was on notice about the harms of climate change for many decades and certainly at the very latest by 1990. Therefore, if responsibility for harms starts upon notice of potential harm, a strong claim can be made that all developed countries have been on notice for many decades.

Because the Copenhagen Accord made no binding commitments on the part of developed countries for adaptation and provides no reliable adaptation funding mechanism, The Copenhagen Accord must be understood to fail the just adaptation criteria.

Conclusions-Climate Change Ethics after Copenhagen

The Copenhagen Accord is ethically problematic for reasons stated in this post among other reasons. In summary, the commitments made by nations in the Accord are not environmentally sufficient, distributively just, nor provide for just adaptation responses for vulnerable developing countries.

The next climate change negotiations took place Cancun in December 2010. Although it may still be possible for developed nations to take more ethically responsible positions on an urgently needed global climate change solution, the world is running out of time to do this according to the consensus scientific view.

The longer the world waits to reduce its GHG emissions, the more important and expensive will be the adaptation agenda and the steeper emissions reductions commitments will need to be to protect vulnerable developing countries. Each failure to develop a global solution to a climate change regime will make it more difficult to forge a just climate change regime.

It would appear that US domestic action on climate change is a key to success at the international level. In the months ahead there are sure to be many arguments made against US climate change legislation in the United States. If these arguments follow past patterns, they will predominantly be claims based on narrow, short-term national economic self-interest devoid of any recognition of duties, responsibilities, and obligations to others to prevent climate change If this is the case, those interested in an ethically justifiable global solution to climate change need to turn up the volume on the ethical problems with US climate change responses.

The US press has utterly failed to emphasize the ethical dimensions of climate change and as a result most Americans see little problem with judging national climate change policies on the basis of national economic interest. As a result, ethics and justice are the crucial missing elements in the US national conversation about climate change policies.

References

Baer, P., and T. Athansiou, Greenhouse Development Rights Framework (GDR), July 3, 2009, http://gdrights.org/publications/, last visited.

BBC, Key Powers Reach Compromise At Climate Summit, Dec. 19, 2009, http://news.bbc.co.uk/2/hi/8421935.stm.

Brown, D, (2009a) *Two Climate Change Matters Move To Center Stage In Copenhagen With Profound Implications for Developed Nations: Ethics and Adaptation*, ClimateEthics, http://climateethics.org/?p=331.

—. (2009c), *Ethical Issues Raised By US Blue Dog Democratic Senators' Opposition to Climate Legislation - When May a Nation Make Domestic GHG Reduction Commitments Contingent on Other Nations' Actions*, http://climateethics.org/?p=206.

—. (2009b), Ethical Failures of National GHG Emissions Reduction Proposals Approaching Copenhagen, http://climateethics.org/?p=147.

Carraro, C., E, Massetti, Two Good News From Copenhagen? www.voxeu.org/index.php?q=node/4490, Jan. 15, 2010.

Coon, Charli, Why President Bush Is Right to Abandon the Kyoto Protocol, Heritage Foundation, May 11, 2001, www.heritage.org/Research/EnergyandEnvironment/BG1437.cfm.

Global Commons Institute. Contraction and Convergence (C&C), July 14, 2009, www.gci.org.uk/contconv/cc.html, last assessed.

Hansen, J. et al, Target CO2 : Where Should Humanity Aim?, 2008 www.columbia.edu/~jeh1/2008/TargetCO2_20080407.pdf.

Herald Scotland, Copenhagen climate deal 'a disaster for the planet', Dec. 20, 2009, www.heraldscotland.com/news/transport-environment/copenhagen-climate-deal-a-disaster-for-the-planet-1.993374.

Lerer, Lisa, President Obama's Dramatic Climate Meeting, Politico, www.politico.com/news/stories/1209/30801.html, Dec. 18, 2009.

Revkin, Andrew, and Tom Zeller, U.S. Negotiator Dismisses Reparations for Climate, New York Times, Dec. 9, 2009, www.nytimes.com/2009/12/10/science/earth/10climate.html.

Roberts, David, Is The 'Climate Debt' Discussion Helpful?, Grist, www.grist.org/article/2009-12-17-is-the-climate-debt-discussion-helpful/, Dec. 2009.

United Nations, United Nations Framework Convention on Climate Change, (UNFCCC), 1992, U.N. Doc. A/AC.237/18, reprinted in 31 I.L.M. 849 (1992).

Werksman, Jacob, World Resources Institute (WRI), "Taking Note" of the Copenhagen Accord: What It Means, Dec. 20, 2009, www.wri.org/stories/2009/12/taking-note-copenhagen-accord-what-it-means.

CHAPTER SIX

ENVIRONMENTAL POLICY AND THE ANTHROPOCENE

PETER G. BROWN

The current disarray in environmental policy is, in part, a result of a failure to connect it to a coherent and scientifically-based worldview with a related ethics. This failure contributed to slipping from the relatively safe Holocene of the last 10,000 years into the unstable and much more hazardous Anthropocene where humans have profound effects on Earth's systems. Many of the assumptions on which our inadequate environmental policies are based derive from political liberalism, which is itself being destabilized. . I suggest six areas where the implications of a scientifically based worldview need to be developed so that an ecological political economy, respectful of the Earth and the life on it, can be envisioned and brought into being. Nothing short of a complete rethink is essential as we enter the Anthropocene.

I. Introduction

In the movie *The Magic Christian,* Peter Sellers plays the richest man in the world. In one magnificent scene he is conducting a board meeting in the parlor car of a train. It is apparent that his enterprises are being mismanaged and his assets are in steep decline. He tells his managers that they are all fired. The train stops and they are all handed maps as they disembark. But none of the maps are maps of where they are.

To me this metaphor sheds considerable light on our circumstances. Many of our conceptual maps of the world do not track with current scientific understandings of reality. This is particularly true of our normative perspectives. Neo-classical economics, many of our ethical systems, much of philosophy and theology, law, finance, and politics are a-scientific. Environmental policy in our time is made up of an amalgamation of the ideas from these systems; and these have achieved modest successes in areas as diverse as air quality and protected areas. But each of them individually and all of them together have failed to stem, or in most cases even to call into serious question, the well-documented and escalating mayhem of our time. The result has been moral and conceptual bewilderment and the enthusiastic dismemberment of the Earth's life support systems.

We have been able to muddle along without really understanding where we are in part because, for millennia, the human project has been small compared with the overall size of the Earth's biophysical systems. But it is well known that this game is up and that there are now steep and, in the context of historical time, irreversible, human-caused declines in Earth's life support systems. The emergence and maturation of the sciences of ecology, biogeophysics, along with many other disciplines, has done much to document the overshoot of the human project relative to these systems. Faced with the choice between very substantial correction or collapse, however, humans in general are taking the road to collapse as is evidenced by destabilizing the climate system, acidifying the oceans, massive biodiversity loss, etc. .

The news is not *all* grim. The science of ecology, systems and quantum theory, far from equilibrium thermodynamics, new discoveries about resilience, and many related disciplines offer the opportunity for the construction of a new, scientifically-grounded worldview on which to build a different future for ourselves and the rest of life with which we share heritage and destiny. If developed this could replace the discordant and disconnected perspectives that liberalism offers. A broad scientific perspective both undermines these liberal residuals and has the potential to replace them with new and much more fruitful understandings of our relationship with life and the world.

II. Why we have no effective environmental framework(s)

The conceptual tools most commonly used in justifying and formulating environmental policy in liberal political economies are simply not up to the task at hand. They are: the idea of protecting the vulnerable; sustainability, including the precautionary principle; adaptive management and resilience theory; and a cluster of ideas grounded in neo-classical economics: cost/benefit analysis, ecosystem services, and the perceived need for economic growth.

II. A. Liberalism's Portfolio

Protecting the Vulnerable. An ancient idea that underlies much of environmental policy is that we should protect the weak. Protecting the vulnerable suggests that the goal of air quality legislation, for instance, should protect those with impaired coronary and/or pulmonary function, as well as the health and welfare of the general population. [1] Ironically, many of these policies are formulated in order to protect the weak from the byproducts of the liberal project itself; such as the side effects of chemically engineered compounds. [2]

Sustainability. The idea that all generations should behave in a way so as not to compromise the needs of future generations of persons became a touchstone for environmental policy after the publication of *Our Common Future* in 1987.

The Precautionary Principle. This idea has recently been formulated to ensure that the absence of conclusive scientific evidence is not used as a means to delay taking decisive action with respect to protecting the environment.

Adaptive Management and Resilience Theory. Adaptive management emphasizes the importance of experimentation and flexibility. It is often coupled with a kind of "discourse ethics" that offers all "stakeholders" a place at the table. "Resilience theory" stresses the need to manage natural (and sometimes social) systems so as to ensure that they retain their flexibility to return to a previous "healthy" or functional state if perturbed.[3]

Neo-classical Derivatives

Cost/benefit analysis. The most influential framework in setting environmental policy is cost/benefit analysis. The core idea is very simple. Environmental policies should not be pursued unless the benefits exceed the costs; if there is more than one policy option the one yielding the largest net benefits should be chosen. Any net increase in human welfare is a benefit, and any decrease a cost. Money is typically used as a surrogate for happiness in this model.

Growth. While cost/benefit analysis finds its foundations in microeconomics, recently there have been attempts to ground environmental policy in macroeconomics. For example, the Stern Review argues that we should seek to stabilize climate so as not to undercut the prospects of future economic growth.[4]

Ecosystem Services. There are now a number of attempts to place monetary values on what nature provides to humanity. A number of rationales are offered. First, in a culture centered around money, this is the only way that we can expect policy makers and the public generally to recognize what nature gives us. Second, once we recognize these values we will be more likely to try to preserve nature to ensure that these services are continued. Third, once we understand the monetary value of nature we will be able to assess the "true" cost of its destruction. And that it will cost more to replace the ecological service than protect it [5]

Exceptions

Liberal environmental policy does not completely ignore the well-being of all other life forms. Laws such as the Species at Risk Act in Canada and the Endangered Species Act in the United States are designed to protect other species from going irretrievably extinct. There are other laws and treaties focused on conservation of biodiversity, e.g. Migratory Bird Convention Treaty, Convention on International Trade in Endangered Species, laws on the management of public lands in the United States, protected area legislation, and the like. However, the constraints imposed by these laws can be relatively easily overridden, and enforcement is often unfunded and weak, and none of them deal with the fact that the human project is too big for the biosphere.[6]

II B. The Scientific Challenge to the Liberal Portfolio

The science of ecology is, of course, but one of the scientific developments that have challenged mainstream Western cultures assumptions about the place of humans on the Earth and in the Universe. Developments in evolutionary, organism and molecular biology, equilibrium and far from equilibrium thermodynamics, quantum theory, astrophysics, cosmology, neuroscience, and certain branches of theology are just some of the elements of a new understanding of who we are, where we came from, and where we may be headed. Yet, the conceptual, moral and theological underpinnings of economic and political liberalism have not been systematically altered to reflect our enormous advances in scientific understanding. As a consequence, the frameworks available for environmental policy, which are derivatives of liberalism, have not been able to ground an even a minimally adequate response to life's declining prospects.

Of course, there is a large literature pro and con on each of the standards in the liberal portfolio. It is not my purpose to review that literature here. Rather I want to show that the science of ecology, and its cognate disciplines mentioned above, brings a perspective that casts doubt on them all. This is so for at least five reasons.

Large-scale system collapse. A major shortcoming is that all these frameworks are designed for decisions at the margin. What are the costs and benefits of a particular policy with respect to greenhouse gases? The vulnerability standard assumes that the population as a whole is healthy and resourceful and can provide the resources to protect the weak. Adaptive management assumes that we can control the course of events through incremental experimentation and social learning or that we should prepare for surprise more than control because systems are recognized to be non-linear. In both approaches it is assumed that we can avoid or at least understand positive feedback loops that are loaded with momentum, resistant to incremental solutions, and therefore can spiral out of control. What none of the standards help with is how to respond to radical, nonlinear changes in whole systems such as is currently underway in the oceans and the climate system.

Ecology and its cognate disciplines bring a distinctive and urgent additional consideration to the fore: the fact that the human enterprise has gotten too big. This is expressed by a number of measures including the idea of "footprint," safe planetary boundaries, overshooting carrying capacity and the like. These measures all contain the same basic message: that there are limits to the Earth's life support systems and we are at the very least approaching them or have in many cases passed the ability of the Earth's life support systems for self-renewal.

Rethinking the Idea of Community. Each of the standards in the liberal portfolio assumes that the moral community is limited to individual human persons. The lessons from the sciences, that human beings are related to all Earthly life and that we have co-evolved with the Earth itself, are simply not recognized by mainstream political discourse and its stepchild, environmental policy. We are, as Aldo Leopold said, plain members of the natural world, not its lords and masters.[7] Yet the bulk of the standards in the liberal portfolio assume, typically without any argument at all, that the moral community is limited to humans. Many discussions of "sustainability," for example in the Bruntland Report, regard the objective to be securing the well-being of future persons, and are compatible with turning all of nature into the means for satisfying this objective.

From Resources to Respect. Once we see ourselves as members of the natural community, the idea that the Earth and the life on it are resources intended only for humanity becomes absurd. The world is not a collection of sources for satisfying human desires or a place where we can dispose of the waste stream inevitably created by that satisfaction. Rather, it should be seen as a commonwealth where all species interact with each other and the planet's biophysical systems in such a manner that all life thrives, both ours and all the others, so that life as a whole can continue on its nearly four billion year old journey of travail and splendor.

What do we seek? Most standards used in grounding and assessing environmental policy are attempts to avoid any serious encounter with systematic ethics, metaphysics or theology. As a result they lack a theory of the "good" that is grounded in a scientific understanding of whichever thing they are allegedly designed to protect.

This fact is well exhibited in sustainability, adaptive management, resilience theory, protecting the vulnerable, and the precautionary principle. The standard of sustainability as formulated in the Bruntland Report is compatible with subordinating the entire natural world to the satisfaction of human needs. Adaptive management assiduously avoids saying what management is for; though it does not rule out concern with the well being of natural systems.

Resilience theory seems to say that it is only the good for man that counts; that the point of management using this framework is utilize natural systems as fully as possible; but not to jeopardize its long term capacities to supply services to humanity.[8] The idea of protecting the vulnerable, so well-grounded in human-centered Judeo-Christian thought, is unable to say what to do about the vulnerability of nature. The precautionary principle, allegedly a standard to protect nature and ourselves in the face of scientific uncertainty, is regularly turned on its head by arguments that we need to be careful not to impede economic growth, or reduce employment opportunities. Without a connection to a worldview that helps us decide what we should aim for these standards will continue to be manipulated in ways that illustrate the current human failure to come to grips with our place in, and duties toward, the world in which we live.

This is, in part at least, a byproduct of liberal societies' strong tendency to avoid questions of theology and metaphysics in the name of preventing the discord and outright conflict that such discussions have historically provoked. This trend was regrettably reinforced by logical positivism in the 20th century—which in certain versions denied that metaphysics was even possible, let alone desirable. The consequence has been that most modern societies neither contain nor can construct a coherent, grounded narrative to inform and justify their fundamental purpose. Thus, by default, money has become the common language we are all encouraged, even required, to speak.

Lacking the Courage of their Metaphysics. Some standards that are developing in environmental policy are, or at least seem to be, partially grounded in an ecological/scientific worldview. Resilience theory draws many of its insights from ecology and related sciences. Yet, as noted above, the main direction of the literature is to identify thresholds, so that humans can extract the most that they can from the world without causing a cataclysm, an irreversible decline in the systems that provide things that are valuable to us.

Perhaps the most regrettable step backwards comes, paradoxically, from ecological economics itself, the discipline that issued a clarion call for grounding economics in a scientific understanding of the world. For it is a branch of this discipline that has been the leader in the valuing-ecosystem services craze now sweeping the world. The underlying idea that the natural world is here to serve humanity and that that service can be quantified by the "true" neoliberal "good," money, harkens back to the very metaphysics that have been overturned by the last two centuries of science. This is not say, of course, that nature does not provide humans with enormous benefits. But it is highly dangerous to value them monetarily, as if that's the only way to know their true value. This is a problem of incommensurable values and not knowing where to end commodification and monetization in order to focus on other expressions of value.

III. Exploring the Ecological and Systems Worldviews to Construct an Ecological Political Economy to Ground Environmental Policy

With this background in mind, here are six questions that are essential to answer if we are to construct a much saner and safer worldview, one that is capable of grounding environmental policies so that they truly enhance and support life's prospects. At the same time, these questions help move the discussion away from reliance on the ineffective liberal portfolio and toward an ecological political economy. 1) What is the nature of the person? 2) What do we know about what we know? 3) What should we do and not do? 4) What would economics look like? 5) How should we rethink how we think and what we measure? 6) What are the most promising politics for an ecological age? The answers to these questions are intertwined and as they emerge will constitute the tissue of a new understanding to inform our understanding of our relationship with life and the world.

III. A. 1 What is the nature of the person? The "rational person," who coolly seeks to maximize their own interests and assumes everyone else is doing the same, is a cornerstone of neo-classical economics. This mythological figure has been repeatedly challenged over the years, most recently by behavioral economics and psychology, although without much success in the eyes of the mainstream. This conception of the person is a partial mixture of rationality as conceived during the Enlightenment and the hedonism of thinkers like Bentham and Mill. It contains a narrowly individualistic notion of "the good," from which compassion and empathy, as well as community and connections, have mostly been stripped away.

The continued findings from evolutionary biology, cognitive science, quantum physics and systems theory are helping to find very different answers to the question of who humans are and what they ought to be. Quantum physics, for example, provides a very different perspective on the human self than that contained in neo-classical economics. Because events at the quantum level cannot be directly perceived by the human senses, we are not normally aware that every aspect of physical reality emerges through the interaction of fields and quanta. But from the perspective of our most advanced scientific knowledge, this is the ground for our existence in physical reality. As Robert Nadeau puts it: "The part we call 'self' emerges from and is embedded in a seamless web of activity that is the entire cosmos. Any sense we have of being separate or disconnected from this ground of being… is an illusion that is not in accord with the actual character of [physical] reality."[9]

Systems theory tells us there are no individuals as the concept is normally understood, and that human beings live in complex, interlocking environments with other lifeforms that include one another, all the species in their particular ecosystem, hitchhikers and symbiots like bacteria and viruses and much more, with most of this activity well below the level of consciousness. These systems influence and sometimes actually dictate much of our behavior. Seen from the perspective of contemporary systems science, the human self is highly sensitive to initial conditions, subject to multiple feedback loops and given to wide variation in subjective/ behavioral outcomes. For example, a huge variety of factors—the weather, a sudden collision with another life form, an indisposition—will change what we do even after a course of action or thinking has been undertaken. We are relational and permeable with respect to energy and matter, and live in a world of shared semiotic meanings. Conscious reasoning is not the primary motivator of our actions—and a great deal of what we think is our knowledge is tacit and creaturely. The self is emerged in and entangled by the brain/body/environment/culture/cosmos.[10] If we understood from "the beginning" that the human self is fully embedded our policies with respect to toxins, for example, would likely be very different. We would not regard the world as something "out there" to be exploited; but as part of who we are.

III. A. 2. How the Market Manufactures the Person. Once we recognize the inherently embedded character of the human self, it should come as no surprise that this self is also very much shaped by the human-evolved institutions and assumptions that surround us. As Steven Marglin puts it: "Markets organize the production of goods and services, but at the same time markets produce people: they shape our values, beliefs and ways of understanding in line with what makes for success in the market. Markets thus exist in a kind of symbiosis with the discipline of economics, shaping people to fit the assumptions of the discipline even as economists shape the world in the textbook image of the self-regulating market. A new economy will need a new economics, which goes beyond the calculating, self-interested, individual to take account of community, compassion, and cosmos."[11]

The brain is a complex, adaptive system, and this is what makes it malleable. The behavioral characteristics of any person, whatever the source of their reinforcement, become possible through the establishment of neurological pathways that then become ingrained. The more the pathways are used, the more they will be used. The more they are associated with the pleasure centers of the brain, the stronger the incentives are to increase their use—something long understood, at least at some levels, by advertisers. What humans and other animals do—how much time we spend on the computer, for example—actually changes how our brain is constructed.

III. B. What do we know about what we know? From the point of view of contemporary science, the idea of the world as certain and predictable is, at best, an approximation of reality that is held by human societies from time to time. This view, a legacy of the scientific revolution and the Enlightenment of the 17th and 18th centuries, that the world is made up of quantifiable and stable parts, has been replaced or modified by the more recent science of the 19th and 20th centuries, by a view that emphasizes relations and systems. Systems theory is now an established foundation of a scientific understanding of the Universe.

III. B. 1. The Importance of Uncertainty. The systems that make up the universe have multiple, interactive feedback loops, as well as both fragile and robust initial conditions. Knowledge of such systems is typically incomplete, because all knowledge can only be an approximation of reality; so surprises should not surprise us. The universe itself is a complex adaptive system "ever advancing into novelty," to use Whitehead's phrase. In that universe, equilibria or static, predictable states—the centerpiece of the neo-classical model—are rare, perhaps even delusional. The overall reality, as Heraclitus wrote, is *change.* In a world of complex systems successfully seeking to maximize a variable like GDP is sure to bring chaos and instability in its wake, since it will perturb the other, constantly changing and evolving systems of which it is a part. For example, the worldwide commitment to economic growth is destabilizing the climate system, even though the latter is certainly not part of the mental construct typically used to think about the economy.

III. B. 2. All Knowledge is Approximate and Provisional. The Human ability for abstract reasoning is one of the great adaptive capacities that we have, but it also has many shortcomings. Consider the example of a map of Quebec. A map should tell us as much as possible about the land and water in the Province and it will do this better the more it approaches the size and complexity of what it is trying to represent. That means that the ideal map would be the size of the Province itself; but then it would not be very useful. So we leave lots out on maps and typically make them very small. In the same manner, our abstractions leave out most details, even though much of what we *need* to know is lost in achieving the benefits of abstraction.

III. C. What should we do and not do? For ethics the implications of moving to a worldview compatible with contemporary science are both deep and wide, though often, not new. Here are four interrelated examples:

III. C. 1. Uncertainty and unpredictability should ground *epistemological humility.* The scientific fact that all human knowledge is partial and provisional has profound implications for action. It should lead us to treat the urge to manage complex systems with enormous caution, while at the same time recognizing that, at the present level of overshoot of ecological capacity, some sort of orderly pullback is essential. In *Water Ethics,* my co-author Jeremy Schmidt and I have called for *compassionate retreat*—a concerted effort to reduce the human impact on Earth's life and its life support systems.[12]

III. C. 2. For far too long people have embraced a form of "exemptionalism"; the idea that human beings are special, in some miraculous way not a part of nature, and are therefore not subject to its sanctions, controls and limitations. This has led to absurd ideas, such as thinking we can control "pests" with compounds that will affect them—but not us. Recognition that we share heritage and destiny with all the other life on this planet, and the dependence of life on physical and chemical evolution, should lead us to *expand the moral community.* The attitude of human dominion of the world must be replaced with *respect and reciprocity* toward *all that is.*

III. C. 3. Once we recognize that humans, like any other native lifeform, are in a reciprocal relationship with the Earth, the duty to *help restore* the massive damage to the Earth's living system caused by our species comes into clear and central focus. An ethics of atonement for our lack of respect and responsibility in the past must inform the every action of the children of a new Enlightenment. Yet, we must not fall back into the trap of trying to manage and force complex systems—much of our current trouble is a result of this attitude. Rather we must enable the reconstruction of Nature and stand aside (often in awe) as she flourishes afresh. Atonement, in this case, is more like being a midwife, than a surgeon.

III. C. 4. Energy broadly defined is a fundamental good, that underlies all other "goods," is energy. It enables the far from equilibrium, autocatalytic living organisms. This repositions the eminent philosopher John Rawls' concept of "primary goods" such as income, wealth, and opportunity, to a secondary status. They all depend on energy. From the perspective of an ecological political economy wasting that which makes life itself possible is a very fundamental moral wrong.[13] And so it is, in the moral codes of most traditional and aboriginal peoples.

III. **D**. ***What would economics look like if it were embedded in a scientific understanding of life and the world?*** How would economics and its goals have to be rethought? Here are some of the ways:

III. D. 1. Rather than a system with little connection to a scientific understanding of the world, the human economy must be conceived of as an open system, whose processes can and must be described in terms of the expenditure of energy and matter.

III. D. 2 The ability of the economy to grow in terms of throughput is inherently limited. The ecosphere stays in far from equilibrium conditions maintained by sunlight with slow rates of growth. The current economic growth imperative *far exceeds* the maintenance and growth capacity of these systems. Proposals to decouple growth from material and energy throughput must be subject to rigorous critique, and if deemed untenable, as seems likely given the record to date, acknowledged as necessary but also insufficient.

III. D. 3. “Production” dismembers the complexity of these far from equilibrium systems, and often produces wastes that are toxic to life, whereas an economics for the Anthropocene must work hand in hand with thermodynamic and other processes and produce digestibles in both quantity and quality.

III. D. 4. While many economists claim that their discipline is value free (while being freely prescriptive), certain ethical questions must be central to a new economics. These include: What is 1. Well-being of life and its support systems? 2. Poverty? How can it be vigorously addressed? 3. A fair share of Earth’s life support capacity for humanity, in relation to other species? 4. A fair share for countries ranging from Canada to Bangladesh?[14]

III. D. 5. A fundamental principle of macroeconomics, seen from the perspective of the limits of our knowledge and our planet, has to be *precaution*. The overriding objective of macroeconomic policy is stability, and as a consequence the “good” humans have to learn to be cautious about is the health of the Earth’s life support systems. This good is often pursued by traditional and aboriginal groups whose watchword was often restraint and self-imposed limits, as with harvests. These moral traits were often denigrated by Europeans for not favouring the energetic pursuit of individual gain, but they show that whole cultures can be very conversant with honoring this type of attitude toward life and the world.

III. D. 6. A key element in an ethically-grounded, macroeconomic system has to be an understanding that the right relationship between the economy and the Earth must be a mutually enhancing one. While this will seem perhaps hopelessly utopian, we need to point out that perpetual growth is also a utopian (and completely absurd) vision. In thinking about economics, insights from other disciplines, periods or cultures are often obscured or marginalized.

III. E. ***Rethinking how we think and what we measure*.**

III. E. 1. Humanity’s symbol systems shape our understanding of reality. Let’s think about just a subset of these systems, words, and one language, English. Basic underlying metaphysical categories of English are subject and object; nouns and verbs. This metaphysics does not track well with an ecological worldview. The human is not the only subject that must be considered in a value system. And subjectivity has lots of forms—that of the hawk being different than that of the hockey goalie. The world is not divided into that which is acted upon and that which acts—rather everything is in a state of current existence and future becoming. In other languages and cultures, these symbols often work in the opposite direction, helping their users to experience greater kinship with other forms of being. In Okanagan, for example, their name for themselves, “we who share the same skin” does not mean just the people of the region, but includes animals, plants, fish, streams rocks—everything that occurs in that ecosystem.

III. E. 2. Keynes was able to redirect economics because he came up with *new concepts* (such as “aggregate demand” and “liquidity trap”), and because he rethought some elements of ethics (such as the role of the “virtue” of thrift). We cannot get “there” from here without a thorough re-grounding of economics thinking that does the same kind of thing. Key terms like production, wealth and money need to be *rethought and re-grounded* and key conceptions in neo-classical economics like “externalities” and “public goods” need to be exposed for their weaknesses and in fact, completely rethought.

III. E. 3. We need to measure the human impact on Earth’s life support systems, such as in the idea of “safe operating space”[15], ecological footprint and the like. These are the measures that must be treated as the true indicators of a functional and supportive economy.

III. E. 4. We also need to conceptualize and measure fair shares of this “safe operating space” for individuals, species, and countries. For example, once there is an agreement on carbon concentrations in the atmosphere there must be an allocation of emissions rights.

III. E. 5. A complex systems conception of the person allows us to reframe the indicators of economic performance. Here are some examples:

Once we recognize our osmotic, or permeable character, for example, we open the door to measuring the success *of an economy* by reference to the toxic load in our bodies, or the POP’s in breast milk, and the like.

And once we recognize the role of the market in determining character and social values, we can no longer avoid the question: is the market reinforcing the creation of the kind of people we want? Or not? More broadly, is it helping to move us toward what civilization is supposed to be for, that is, the cultivation and elevation of the mind and spirit of a human animal who lives respectfully on the Earth, and who can easily express reverence for life and the sources of its being?

III. F. ***Politics for an Ecological Age.***

III. F. 1. The enabling conditions of the "flowering" of political liberalism are in the process of unraveling. The evolutionary processes that gave rise to complexity on Earth set the stage for a unique period in human history, which can be benchmarked as the beginning of agriculture. This organic complexity, and the attendant energy stored in it, began to be both systematically and massively enhanced and harvested. As William Ophuls points out in *A Requiem for Modern Politics* in the subsequent evolution into industrial societies, this attempt to enhance and then harvest has turned into a riotous feast on every one of the world's natural systems.[16] It has resulted in (1) extensive settlement of "new" territory; (2) explosion of human population growth; (3) exponential growth in the (wasteful) use of matter and energy; (4) deliberate destruction of complex natural systems in favor of simplified systems; (5) intense and sometimes bellicose competition; (6) chronic boom and bust cycles; and (7) mass extinctions, possibly including our own.

We are now in what Nicholas Georgescu-Roegen called the flamboyant period—a time now drawing to a close and almost certainly not to be repeated as we enter the Antropocene. This period gave rise to liberal politics and economics that arose from the ashes of the Middle Ages and the religious wars that attended its demise. **Yet liberal ethics, economics, and politics itself is now on the chopping block of history, as its assumptions crumble and its enabling conditions unravel.**

III. F. 2. What kinds of institutions are implied by and required for assuring a flourishing Earth and a long and prosperous human presence here? Once an Earth systems point of view is adopted, does it become evident that the system of nation states is intrinsically incapable of addressing the problems of the Anthropocene? Is there sufficient (or any) evidence that democratic institutions are likely to achieve essential long term fiduciary goals? Is a democracy based on an idea of responsible citizenship more of an illusion than a reality in the electronic age of bread and circuses?

III. F. 3 From the point of view of a scientific understanding of the world many of the assumptions underlying liberal politics are problematic and have to be reviewed with new ends in mind. Here are some of them:

Property. The idea of ownership of the Earth is difficult to ground. The idea of political or other human-created boundaries is problematical. The morality of turning over complex and fragile portions of the Earth to individuals who are poorly informed, often outright ignorant, and who have no concept of fiduciary duties, cannot be justified. The question that must be confronted is how property is to be understood once humans are seen as members, not masters, of the community.

Human Rights. Often rights concepts are not connected to *responsibilities* and hence the relational dimension of our lives is understated, even ignored. But even when responsibility is included, this concept leads us away from understanding the essentially embedded nature of human beings in the natural world, as well as the chimera of any separate, isolated "self." Extending the idea of rights to nature itself, such as in the Ecuadorian Constitution, is a healthy attempt to rebalance our discourse.

Self-regarding acts. Perhaps the core idea of political liberalism is that each person is free to act as he or she wishes so long as that action does not harm other persons. Two important sources of this idea are John Locke's *A Letter Concerning Toleration* (1689) and John Stuart Mill's *On Liberty* (1859). Locke held that our religious beliefs are internal matters and hence matters of faith should be beyond the legitimate reach of the state, whose principal tasks are external—to secure "life, liberty and property." Mill held that the state has no right to interfere in what he called "purely self-regarding acts."—though interpreting this phrase has proved contentious, even for Mill himself. Yet despite the pedigree conferred by the quality of these two philosophers, the assumptions these ideas contain that now underlie political and economic neo-liberalism are becoming very problematical.

What has happened since Locke wrote is that his rather sensible idea that what one *thinks* is private, has been transformed into the idea that one can *live* however one wants. And, as Clive Hamilton and others have noted, in the 20th century how one lives, indeed how one defines oneself, has come to be understood widely (especially in the Anglophone economies) though not universally, by what one consumes or "owns." When we connect the foundational principles of political liberalism to the 1st law of thermodynamics, the implications are highly distressing. This cosmic law states that matter is neither created nor destroyed. This means that the carbon released when fuel is burned in a Toronto traffic jam directly affects the interests of people and ecosystems around the world. When we connect this to the question of reparation duties to environmental refugees, we must see that how we *live* is harmful to others. It also raises the question as to whether there are *any* purely self-regarding acts. **And whatever may have been the case when the human population was small and the consequences of our acts inconsequential, in the "full world" in which we live there are *none—no actions which affect us alone.*** We have no choice but to recognize that, as Thoreau says, "our whole life is startlingly moral. There is never an instant's truce between virtue and vice."[17]

The Tyranny of Tolerance. Because it grew out of centuries of religious wars, it is understandable that liberalism would make tolerance a central virtue. What is often not recognized is that the view that religious and metaphysical views have to be tolerated has had the effect of not requiring the development of a common narrative about reality, and enabled the tragic disarray exhibited by environmental policy.

The Perils of "Progress." For at least the last several centuries our expectations about the future have been centered around the idea that the future will be better in some way than the present. This is very different from the cyclic view of traditional communities still in existence around the world, and even in Europe until the late Middle Ages. It has allowed us not to develop any theory about our *duties to the future*. In fact, neo-classical philosophy assumes that the future may be discounted since, in part, we will all be better off then than we are today.

The downward spiral from the Holocene to the Antropocene no doubt began many centuries, likely millennia, ago. [18] But the Western idea of "progress," which arguably took root about 500 years ago gave new impetus to these trends. Perhaps the *coup de grace* was delivered by the Post-War assimilation of the idea of progress with economic growth and an expanding population. "Progress,"now understood as increased consumption for massive human populations is now in the process of devouring what makes its possible.

IV. Conclusion

We have entered the Anthropocene with conceptions that did not even work in the relatively placid Holocene; indeed these failed conceptions distracted us from the monumental changes we induced, and thus helped to pave the way for the sweeping changes now underway. What I have sketched here is a far reaching agenda to rethink our circumstances; and to build and act on a different, but not mainly new, agenda for the human prospect. It must be grounded in humility and penance for what we have wrought. The key to a flourishing future, such as it may be, is to be found more in controlling ourselves than the world around us---and in a return to the ancient wisdoms that everything that is, is Holy.

I am indebted to Julie Ames, Margaret Brown, Holly Dressel, Geoff Garver, Janice Harvey, Bruce Jennings, Suzanne Moore, Robert Nadeau, Jeremy Schmidt. Gus Speth, and Laura Westra for comments on various drafts of this paper; and to my colleagues Mark Goldberg and Tom Naylor at McGill; and to the students in the classes we have taught together. I also wish to acknowledge the support of the Social Sciences and Humanities Research Council of Canada.

Notes

---

[1] This idea is well developed in Robert Goodin's *Protecting the Vulnerable* (Chicago: University of Chicago Press, 1985)
[2] See, for instance, Ulrich Beck *Risk Society* (London: Sage Publications, 1992)
[3] See, for instance, Brian Walker and David Salt, *Resilience Thinking* (Washington, DC: Island Press, 2006)
[4] *Stern Review on the Economics of Climate Change,* UK Treasury, 2006.
[5] Robert Costanza, *et. al;* "The Value of the World's Ecosystem Services and Natural Capita," *Nature* 253-260. 15 May 2007.
[6] See David Boyd's *Unnatural Law* (Vancouver: UBC Press, 2003) for a discussion of the disarray in the conception and enforcement of environmental laws in Canada.
[7] Aldo Leopold, *A Sand County Almanac* (New York: Oxford University Press, 1966).
[8] See, for example, Malin Falkenmark's and Carl Folke's "Ecohydrosolidarity: A New Ethic for Stewardship of Value—Adding Rainfall," in *Water Ethics* edited by Peter G. Brown and Jeremy J Schmidt (Washington; DC: Island Press, 2009) pp. 247-264.
[9] Draft paper by Robert Nadeau (and contents added).
[10] This description of the self is taken, with modifications from Wendy Wheeler's *The Whole Creature,* and draws heavily on conversations and correspondence with Robert Nadeau.
[11] An unpublished article by Steven Marglin.
[12] Peter G. Brown and Jeremy J. Schmidt, " *An Ethic of Compassionate Retreat",* in *Water Ethics, Op. Cit.,* pp. 265-283.
[13] Howard. T. Odum, "Energetic Basis for Religion," in *Environment, Power and Society* (New York: Wiley, 1971)
[14] Some of the points in this section are derived from the work of Georgiana Galiussi.
[15] Rockstrom *et. al.,* "Planetary Boundaries," in *Ecology and Society,* 14 September 2009
[16] William Ophuls, *Requiem for Modern Politics* (Boulder, Colorado: Westview Press, 1997) See especially pages 1-28.
[17] Henry David Thoreau, *Walden* (New York: W.W. Norton and Co, 1992) p. 146
[18] See, for example, William F. Ruddiman's *Plows, Plagues and Petroleum: how humans took control of climate* (Princeton: N.J.: Princeton University Press, 2005) for an analysis how humans have likely disrupted the climate cycle beginning with agriculture

CHAPTER SEVEN

THE PAST IS PROLOGUE, THE NEW WORLD IS YET TO COME: FIVE SCENARIOS FOR THE GLOBAL COMMONS IN AN AGE OF TRANSITION

SHEILA D. COLLINS

Abstract

The divisions seen in the breakdown of the 2009 Copenhagen climate talks and in the tepid outcome of the 2010 Cancún talks, are a manifestation of the shifting configurations of power that signal the end of the fourth cycle of capitalist accumulation, as theorized by world systems theorists, Giovanni Arrighi and Prem Shakar Jha. The paradigm of political decision making and capital accumulation that created the world-historical system of nation states and the global capitalist economy is no longer functional in the 21st century. This paper argues that we are now looking at a chaotic period during which the capitalist system's ability to generate further accumulation and thus to revive itself is blocked by the threat of climate change, the biocapacity of the earth system and the proliferation of a nuclear armed world. We are entering a world without global rules or precedents to suggest the way forward. Thus, the outcome of this process is unpredictable. Two scenarios, one dire, the other hopeful can be discerned from present circumstances.

> The crisis consists precisely in the fact that the old is dying and the new cannot be born; in this interregnum a great variety of morbid symptoms appear.
> —Antonio Gramsci[1]

Tragically, despite new evidence that the earth is warming at a much faster rate than anticipated by climate scientists, the much heralded

Fifteenth summit on climate change (COP 15), held in Copenhagen in December 2009 ended, not in a binding agreement to reduce greenhouse gas emissions as many had hoped going into Copenhagen, but in a cacophony of voices, a dissonant chorus of recriminations, and some weak pledges from Annex I countries to voluntarily reduce their emissions. The sixteenth meeting of the parties (COP 16) held in Cancún in 2010, was barely a step above Copenhagen. It established a mechanism for administering a "Green Climate Fund" for assisting developing nations ($100 billion annually) starting in 2020, but with no designated source of funds, and an amount incommensurate with the amount that will be needed. The countries did agree to set up an external mechanism for measuring, reporting and verifying the individual pledges on emissions reductions. There was no agreement on what happens when the Kyoto Protocol expires in 2012, simply an agreement to try again a year later.

The result is that the combined and, as yet, unenforceable pledges made by individual countries will not be enough to meet even the lower range of emission reductions--25 to 40% below 1990 levels--required for stabilizing concentrations of CO_2 at 450 ppm, the concentration that the IPCC associates with a 26 to 78% risk of overshooting a 2°C goal. Climate experts agree that more than a 2°C rise over pre-industrial conditions could signal climate catastrophe (Levin and Bradley 2010). Bolivia, which had held out for a binding agreement on greenhouse gases denounced both meetings as resulting in an outcome that will be disastrous for humanity (Press Release Plurinational State of Bolivia 2010).

These two climate summits were fraught with more tension than any other previous comparable gatherings. The conflicts that are apparent in the global climate negotiations process are commonly portrayed as a dichotomy: developed versus developing nations, polluters versus victims, environmentalists versus climate change skeptics. This, however, is a perceptual illusion created by a media that is compelled to present reality in quickly digestible sound bytes and by political actors seeking to cast blame. In fact, when one focuses the lens more closely one sees that the divisions are much more numerous and present a puzzle that needs to be explained. What we see are new divisions between previously cooperative powers: the EU versus the US; China and India versus the rest of the G77 and other divisions within each of these groupings; divisions among both business corporations and environmental organizations; and new alliances between previously disassociated nations and groupings of people—the governments of Venezuela, Bolivia, Nicaragua, Cuba and other island nations aligning with civil society groups--peasant farmers from the South and urban dwellers from the industrialized North ("Copenhagen Climate

Summit: Cracks Appear" 2009). One also sees the emergence of new political actors like the world's indigenous peoples who are no longer willing to be "extras" in a drama orchestrated by the world's major economic powers and are now claiming space on center stage and threatening to rewrite the script.

The Breakdown

What was evident at Copenhagen and Cancún is the dissolution of the traditional geopolitical order that has governed the world, at least since the fall of the Soviet Union and for much of the previous half century. Much had happened between the Bali meeting and the Copenhagen summit that had already begun to signal the end of the political/economic architecture of the post World War II era (Sassen 2010). [2]

In 2008, the developing-world bloc, wielding newfound power, had brought the latest Doha round of trade talks to a halt.[3] The United States, having tried to extricate itself from a war in Iraq that was eating away at its domestic budget, internal morale and international reputation now found itself in another quagmire in Afghanistan that was quickly spreading to Pakistan and threatening to widen throughout the Middle East. Despite its new president's campaign vow to take a less arrogant posture toward the rest of the world, the United States found itself confronted by an increasingly belligerent Iran and a rebellion in Latin American, its traditional "back yard," a ballooning deficit and debt, and a rising tide of libertarian sentiment at home that threatened to undo the once-visionary promise of a government that would "enable" rather than thwart individual ambition and restore America as a land of promise. Moreover, it now saw itself being challenged geopolitically by the rising powers of China, India and Brazil.

To make matters worse, in 2008 much of the world's economy had come crashing down, and with it, the architecture of capital accumulation, political power and its underlying energy base that was cemented in the post-World War II arrangements at Bretton Woods and in the formation of the United Nations. These arrangements--that had given the United States government the hegemonic role in the management of global finance, trade, and the military protection needed to assure the regime's continuation--as well as a favored role as first among equals at the United Nations—had now become dysfunctional, not only for the continuation of the system of accumulation as it has existed for the past sixty years, but also for the management of conflict and the sustainability of the life systems of the earth.

The Prologue

World systems theorists, Giovanni Arrighi (2008, 2010) and Prem Shankar Jha (2006) help to shed light on the historical trajectory of capitalism that has led to the current conjecture. The collapse of the world economy in 2008, in Arrighi's view, was simply the end result of a several decades-long sequence of changes in the world geopolitical/economic system that was following a traditional historical pattern, albeit with its own unique mixture of ingredients. The modern world-system, which Arrighi defines as "capitalism,"--an integrated system of global power that orients the relations between nations over the long sweep of world history--developed historically through four great cycles of capital accumulation, or phases of continuous change along a single developmental path, until growth along that path had attained its limits. What ensues is a period of chaos (long periods of interstate struggle) during which the system restructures itself and begins a new cycle of accumulation, but this time encompassing a larger territorial playing field, more densely connected networks of power and accumulation (including military power), and a new set of players in the driver's seat (Arrighi 2010: 10-15).

Initially embedded in and subordinated to networks of state power, networks of capital accumulation as they evolved and expanded came to be intertwined with the political realm, or the realm of the state, so that state actors had to be leaders, not just in the process of state-making and war-making, but in processes of capital accumulation as well (Arrighi 2010, 87). Each cycle of accumulation is also an acceleration of the previous cycle so that the time from beginning to end of the cycle becomes shorter than the last. During each cycle, a different constellation of intertwined financial and political networks emerges as hegemonic.[4] It is the state that "controlled or came to control the most abundant sources of surplus capital" that also came "to acquire the organizational capabilities needed to promote, organize, and regulate a new phase of capitalist expansion of greater scale and scope than the preceding one" (Arrighi 2010, 15). As a world system, capitalism oscillates between the logics of power associated with the conquest of territory (where wealth and power are by-products of territorial expansion) versus the accumulation of surplus value and territorial expansion as a means or by-product of such accumulation. According to Arrighi, it is the contradiction between these two logics that is the basis for the periodic upheavals associated with world capitalist history that result in a temporary resolution of the contradiction through the reorganization of political-economic space by the leading capitalist state (Arrighi 2010, 34-37).

Arrighi and Jha (2006) follow Braudel in tracing the origins of this pattern to the emergence of the powerful northern Italian city-states of the 13th and 14th centuries rather than to the later sixteenth century development of the nation-state system. As "world system," the Italian city-states of Venice, Genoa, Milan and Florence exhibited the characteristics that were to play out in the ensuing cycles of accumulation in which the Dutch (16th to 17th centuries), the British (18th to 20th centuries) and the Americans (early 20th but achieving hegemony only after WW II) were dominant. Venice was the prototype of the quintessential capitalist state in which a merchant capitalist oligarchy firmly held the state in its grip and where territorial acquisitions were subject to cost-benefit analysis and usually only undertaken as the means for accumulation by the merchant capitalist oligarchy. A second characteristic of the "system" that emerged was the balance of power at a certain stage in the cycle among the major powers that enabled the system to develop as a "mode of rule" within feudal society without being prematurely destroyed by competition that would inevitably lead to disabling wars. In later cycles of accumulation, that balance would be maintained by a hegemonic state.

Third, turning war making into a "protection-producing industry" (a kind of primitive military Keynesianism) enabled the city-states to expand their markets by collecting rent from hired soldiers. Intercapitalist competition, however, eventually intensifies in each cycle, leading to a fall in the rate of profit, increasing state indebtedness, and the turn to financialization as corporations seek to keep a growing proportion of their incoming cash flow in liquid form and to find more lucrative sources of profit in speculative ventures (Arrighi 2010, 372). Jha explains this as the "scissors movement of liquid capital accumulation and investment activities" (Jha 2006, xviii-xix). When money is no longer used to finance productive enterprise it is redirected to the hostile takeover of competitors' markets and territories and to war making activities needed to protect this system. Thus, as Jha points out, "each cycle of capitalism has therefore given birth to long spells of violence, as the core cities and nations have sought to reorganize the periphery to augment and prolong the profitability of capital" (Jha 2006, xix).

The Interregnum

Arrigihi's detailed history of the rise of the United States to world hegemon following the Second World War, the flowering of the "golden age" of capitalism from the 1950s through the 1960s, the beginning of the decline (signaled by the Vietnam War and the collapse of the Bretton

Woods regime in the early 1970s), and the temporary resurgence of capitalist euphoria in the 1990s, which Arrighi compares to the *belle époque* preceding the fall of the British empire, appears to repeat the broad patterns of the three previous cycles of accumulation he has analyzed. The collapse of the world capitalist economy in 2008 would then appear to portend the end of American hegemony. The United States remains militarily superior, but its position in world geopolitics is one of dominance only by virtue of having a monopoly on the means of violence.

But even that dominance is now found wanting by the inability of the American warfare state to master the quagmire of Middle Eastern politics or the proliferating cells of non-state actors sprouting up all over. The shifting interstate relations seen at Copenhagen reflect this final breakdown in the order that has governed world politics at least since 1945, and an attempt by heretofore subaltern state and non-state players to sort out and position themselves on a new playing field without rules or umpires.

Where all this is leading of course, is an open question. If past cycles of accumulation are a key to future patterns, then we should expect to see a period of international turbulence to be succeeded by the emergence of a new cycle of accumulation with a new constellation of players in the hegemonic role. Since each past cycle of accumulation required a greater spacio-temporal box in order to absorb the surplus capital, the next one would have to be even larger than the last.

The World-Historic Turning Point

There are three major impediments to a repeat of this pattern, however. One is that the economic growth upon which each of the preceding cycles of accumulation was predicated is no longer possible. We are already outstripping the biocapacity of the planet to feed and support its growing population and are running up against ineluctable limits. Second, the territorial expansion upon which each cycle was built is no longer possible. Third, if each previous cycle ended in a period of international turbulence, the next period of turbulence, in a world of proliferating nuclear weapons, could mean annihilation, not reorganization. Thus, we are at a world-historic turning point in human affairs where the precedents and patterns of the past give us no guidance. There have, indeed, been many instances of past civilizations destroying themselves through either warfare or environmental collapse, but these were isolated instances that did not affect the entire globe (Diamond 2005).

The unprecedented wealth and conspicuous consumption that has signaled the fourth cycle has depended on the harnessing of the fossil fuel system, and especially of oil. Until now capitalism appeared to have freed itself from the environmental constraints that had destroyed previous civilizations (Diamond 2005) by drawing from an apparently infinite capacity to reproduce and reinvent itself through the exploitation of these vast reservoirs of fungible fossil fuels. There were, of course, local energy bottlenecks, but the system could get around these by expanding the size of its container through seeking energy and resources from greater and greater areas of the world (Debeir, et. al. 1991). Since so much of modern civilization is built upon oil, it is widely recognized that the transition to other energy sources will mean a revolutionary rupture with every aspect of production and consumption. But it is assumed that the continued exploitation of fossil fuels is necessary to give us the time and financial ability to move through that rupture into a new mode of living. That is certainly the outlook that seems to be driving all of the world's major powers, including the rising economic powerhouses of China, India and Brazil, upon whom many are casting their bets to lead the next cycle of global accumulation.

But can we assume that the rapid upward mobility that China's elite—and for that matter India's and Brazil's elites--are embarked upon is not markedly capitalist in character, a character that seems to lead to an unfettered desire for power and wealth (Barboza 2010;Yardley 2010, A1)? A telling statistic is the fact that China now has the second largest number of billionaires in the world (Miller and Kroll 2010). Marxists argue that the M-C-M circuit has a tendency to objectively condition all relations in those societies in which it becomes embedded (Gulick 2009, 244). For those who don't buy the materialist thesis, there is still the psychological one—the "animal spirits" that seem to infect anyone who rises to the top of the financial heap (Akerlof and Shiller 2009). While China appears to be attempting at some level to head off some of the danger of an overheated growth trajectory, it nevertheless has to walk a delicate line in how far it can go to curb its appetite for fear of unleashing the fury of its population which has now tasted some of the fruits of Western style development and for fear of plunging global growth, of which it is now the engine, into another recession or worse (Bradsher [a] 2010, A1).

Yet physics and chemistry militate against China's leading the world out of its present malaise. Despite China's spectacular growth and its recent efforts to modify conditions in the countryside, this growth has been accompanied by a stark increase in the Gini coefficient among people and between regions which has resulted in massive social unrest and environmental degradation. The Chinese government has responded by announcing new labor laws, arresting corrupt local officials, enacting funding to develop rural areas, initiating an over half trillion dollar stimulus package to offset the recession and embarking on a massive greening of the economy.

However, while China is emerging as the most powerful source of green technology for the world, it is also building more coal plants (albeit allegedly "cleaner" and more efficient ones) (Bradsher [b] 2010; Mouawad 2010), increasing the number of cars on the road, engaging in deals for oil exploration and development with oil rich countries in Africa, Asia, the Middle East and Latin America, investing in the development of the extremely polluting oil sands extraction in Canada, and buying up big tracts of land in Africa to feed its thirst for energy in the form of oil and biofuels.[5] No matter how much it greens its own economy, its thirst for natural resources creates more pollution in other parts of the world, thus negating the benefits at home. Moreover, its pledge to reduce the intensity of its greenhouse emissions in relation to economic growth represents anywhere from a 15% decrease to a 204% increase in emissions versus 1990 levels, thus the concept is extremely elastic and depends on continued economic growth.

Likewise, India and Brazil's growth trajectories are problematic. India's feverish industrial development is simultaneously accompanied by the displacement of millions of farmers from their land, the growth of unsustainable megacities, and the degradation of its environment. Its agricultural sector, which still employs half the country's workforce and should be the basis of any viable long-term economic development program, has been treated as a source of votes rather than an engine of redistributive growth. Consequently agriculture shrank by 2.8 percent last year even as India's population grew ("Crop Circles" 2010).

Brazil has taken some steps to green its economy, to provide social benefits to its poor, and has set aside 37 percent of the Amazon to be protected. But since clear cutting is still going on, the potential loss of sixty percent of the rainforest to the earth's carbon sink capacity is alarming. Moreover, Brazil's commitment to biofuel production, once seen as a panacea for reducing fossil fuel use, may undermine the potential emissions savings of transitioning from fossil fuels to biofuels. A recent study suggests that factoring in indirect land use—cattle ranching displaced to forest lands by cropland expansion for biofuel—dramatically extends the amount of time needed for emissions savings from biofuel production to compensate for emissions from deforestation: by 40 years for cane ethanol and 211 years for soy biodiesel (Alcamo 2010).

If the IPCC climate scientists are right, we may not have the time and we certainly don't have the biocapacity to accommodate a restructuring of the capitalist system as historically constituted. Ecological footprint analysis provides a rough estimate of the earth's capacity to accommodate its population. According to the latest figures we are now using up in one year the biocapacity of the equivalent of 1.5 planets to provide the resources we use and to absorb our wastes (Ewing, et. al. 2009; Global Footprint Network). While they are still far behind their rich nation counterparts in per capita emissions, China and India have already overshot their own biocapacity and are forced to drain the biocapacity of other parts of the world to sustain their growth trajectories.[6] Brazil's biocapacity is 150 percent of its ecological footprint so it still has a way to go to catch up. However, if it continues on its present growth trajectory it too will eventually be in overshoot.[7] If everyone were to adopt the lifestyle of the average person living in the United States we would need five planets to accommodate their needs and wastes.

Moreover, the role of international trade is central to the loss of biocapacity. The capitalist accumulation process, rather than creating the necessary social and infrastructural matrix necessary to absorb its surplus capital has driven more and more of the world into an entanglement of trade and investment that has resulted in a spiral of economic destitution and environmental degradation, not only in the so-called underdeveloped areas of the world, but in the wealthy countries of the north. Even the US, the world's greatest agricultural powerhouse, is increasingly dependent on fragile offshore biocapacity. The end result is that all nations are approaching or exceeding global carrying capacity simultaneously such that there may not be any fall-back position (Rees 2010, 482). In addition, global trade makes it increasingly difficult to affix responsibility for carbon dioxide emissions, as a great deal of global carbon emissions are transferred in traded goods. Over one-third of the carbon dioxide emissions developed countries release into the atmosphere result from goods and services produced outside their borders (Davis and Caldeira 2010), with China being the largest "exporter" of carbon dioxide emissions. With very few exceptions, the fourth cycle has succeeded in penetrating almost every major area of the world. Thus, a spatial fix for economic growth is highly dubious.

The New World is yet to come: Doomsday or the Commonwealth of Life?

When so much is fluid it is presumptuous to try to predict the future. It appears from the current round of climate talks in Cancun that the stark reality of what we are up against has not yet penetrated the decision makers in whose hands lies the fate of the planet. The lords of finance seem to have learned nothing from the near collapse of the global economy in 2008 and the structural adjustment regimes they are now imposing on the countries of the industrialized North—those most responsible for climate change—are making it even harder for them to wean themselves from fossil fuels and invest in greening their own economies, much less help the poorer nations. It is likely that the present scenario in which individual communities and regions take incremental steps to build sustainable economies will not happen fast enough to prevent catastrophic climate change. 2010 has already been the hottest year since records have been kept and climate-related disasters are on the rise. Scientists tell us that we may have reached the tipping point and that we are headed by the end of the century for a 4 degree Celsius temperature rise.

When increasing global inequality, peak oil, environmental degradation and biocapacity loss are taken into account the darker scenarios of analysts like Hardin (1968, 1993), Jha (2006), Klare (2001, 2008), Kunstler (2005), Dyer (2009) and the US military and intelligence establishment (Center for Naval Analysis 2010; US Department of Defense 2010) among others, become more plausible. All in varying ways foresee a future of climate-induced crises that could feed terrorist movements, create waves of environmental refugees touching off religious and tribal conflict, spread contagious diseases, cause mass starvation, do vast damage to infrastructure, destabilize entire regions, topple governments and call forth unprecedented repressive actions by military establishments. Indeed, in many parts of the world we are already seeing the beginnings of such a process. In such scenarios, the end result, if left to run its course, is either global chaos or a global authoritarian order that sweeps away whatever was left of the social harmonizing capacities of the nation-state. In Jha's view, the global entanglement in which the entire population of the metropolitan countries—but especially its financial elites--has a vested interest in controlling economic and political events in dependencies across the world, implies a world of severe repression to keep it from exploding.

It is not at all clear if the international order can be stabilized nor what constellation of forces would be able to do so. War and terrorist violence continue to proliferate. The governments that should have been taking us in bold new directions have become so enmeshed in the system of power and corruption that they appeared paralyzed to do what is necessary. While few dare to say so out loud, some economists are predicting that an even more devastating global economic collapse may be just around the corner. Europe, which once seemed a major economic player that was making important strides in moving its economies toward sustainability is now imploding under the impact of the financial collapse; and the United States, wracked with internal dissention, is unable to deliver even the kind of climate legislation, weak as it was, that Obama had promised in Copenhagen. When global political chaos is paired with environmental breakdown, economic collapse and the destructive power of modern weaponry the only outcome is the end of the world as we know it. Such a scenario is unthinkable, yet must be considered plausable should the world fail to work out a solution to both the economic and environmental crises we face.

An alternative scenario to doomsday is articulated by a growing body of civil society organizations, indigenous and peasant peoples and some developing and island nations across the world (The Earth Charter; Shiva 2005; Simms 2009; Brown and Garver 2009; Berry 1998, 1999; Bosselmann 2010; Engel, Westra and Bosselmann 2010; Korten 2006; Daly and Cobb, Jr. 1994; New Economics Foundation). It is a scenario that begins from an ethical value perspective—some would say from a spiritual grounding in a deep understanding of the earth system and a sense of reverence for and appreciation of its complexity and beauty-- rather than from the perspective of political economy or vested power. Its proponents use terms like "participatory democracy," "earth democracy," "justice," "wholeness," "right relationship," "ecological integrity," "stewardship," "intergenerational equity," "the commonwealth of life."

While it's starting point is ethics, such a scenario does not eschew science. In fact, its ethical convictions are impelled by what climate scientists and ecologists have discovered, as well as by the lived experiences of people facing the effects of climate change around the world. What is clear to those espousing this scenario is that if human life is to survive, an entirely new way of ordering our productive, exchange and political life will be required on a global scale. As the British think tank, New Economic Foundation puts it,

> 'Business as usual' is no longer an option. However, halting and reversing our consumption of more and more 'stuff' appears likely to trigger a massive depression with serious unemployment and poverty. This is certainly true if all we do is 'apply the brakes' without fundamentally

> redesigning the whole economic system. We are facing a series of interlinked systemic problems – consuming beyond our planetary limits; untenable inequality; growing economic instability and a breakdown in the relationship between 'more' and 'better'. The only way to overcome these systemic problems is through a set of solutions which themselves address the whole (New Economics Foundation 2009).

There have been many different iterations of this scenario but it is easy to discern the broad contours. We must abandon the fossil fuel system as quickly as possible and put our resources into renewable energy sources: wind, solar, geothermal, biofuels and others that may come along. The natural environment can no longer be treated as an "externality," but as the very basis for all economic life. Thus, environmental costs and energy efficiency ratios must be factored into the exchange value of products. An economics of endless growth must be replaced by what some have called a "steady-state" or "dynamic equilibrium" economy (Simms 2009, 274), meaning that the production of most goods, including food, and the generation of most energy should take place as close to the end use as possible.

Developments in science and technology will be guided by an ethics that is based on the concept of ecological integrity, not profit. Provision for future generations and anticipation of future problems employing the precautionary principle must be built into the generation of any new technologies; and problems that are discovered will have to be addressed immediately and not put aside until later, covered up or ignored. The GDP as a measure of growth—never an adequate measure of a country's well-being--must be abandoned and growth must be redefined to include, not the accumulation of profit, but the accumulation of quality of life, of beauty, of time for family, community, recreation, creativity, contemplation, of support for the development of one's unique potential (Sen 1999). Trade and investment decisions must be based on the principles of cooperation, reciprocity, complementarity, and solidarity rather than competition, comparative advantage and beggar-thy-neighbor,[8] requiring rich countries to pay their ecological debts to the developing world.[9] The costs of a transition to renewable energy could be taken from reductions in military budgets around the world—especially that of the US--and through taxes on carbon and international financial transactions (Baker, et. al 2009). The UK-based New Economics Foundation, using Britain as a model, has laid out a range of the practical measures in terms of housing, business, taxation, property and energy policy that would be needed for one economy to transition to a steady state economy (New Economics Foundation).

Governance of such a system would have to include both global institutions to regulate transboundary problems as well as local and regional governance systems based on the logic of ecosystems rather than the logic of arbitrary territorial boundaries established through conquest and war. A new global monetary system is required based on a different and stable store of value and means of exchange. Verhagen (2010) has suggested that it be a carbon-based system. Political decision making should be as democratic, transparent and accountable as possible, scaled to the level of the decisions that are to be made. Brown and Garver (2009) have suggested four such global governance institutions.

How likely is such a new order? That, of course is the question. Some would call it hopelessly utopian. Making such a radical change in everything will require a spiritual conversion, for what we are suffering from is only the outward product of a crisis of the human spirit, "as reflected in a culture of alienation, disorientation and loss of meaning and purpose" (Bosselmann 2010; Berry 1998, 1999). Poverty, in a world of obscene luxury, as we know, breeds violence, yet numerous studies have demonstrated that having more "things" does not make one happy and that the inequality within and between nations that we are witnessing is making us physically, socially and mentally sicker (Wilkinson and Pickett 2009). There is also evidence that a majority of people around the world are fed up with the current capitalist system and are ready for a change. A poll conducted by the BBC World Service in November 2009 found that only 11 percent of people sampled across 27 countries said that capitalism was working well and almost a quarter feel it is fatally flawed. Moreover a majority of the sample in 22 of the 27 countries backed a more even redistribution of wealth.[10]

There are myriad examples around the world of places where pieces of the Commonwealth of Life are already being implemented. They demonstrate that what some call utopian is, in reality, eminently practical. Common to all of them is the vision of those who began these projects that life can be different, the will to carry out that vision, and the commitment to a set of ethical principles that prevents the vision from being corrupted when put into practice. Each of them demonstrates that humans are more creative than we think we are. Indeed, as part of the marvelous evolutionary chain of emergence, we can choose to be adaptive or to self-destruct (Kauffman 2008).

Thus far, however, these experiments, largely local or regional in scope, remain unconnected to a larger strategic framework for global governance. But can such a vision work on a global scale? The government of Bolivia, the first indigenous government in world history, believes that it can. President Evo Morales, dismayed by the failure at Copenhagen, called for a World People's Conference on Climate Change and the Rights of Mother Earth that took place in Cochabama during the third week of April, 2010. NGO's who shared this vision, as well as governments, academics, and scientists were invited to participate in a democratic, participatory process to forge a "people's program" to address climate change. Some 35,000 people from 70 countries responded to the call by attending and many more participated in the internet discussions leading up to it. The World People's conference issued a "People's Agreement" (World People's Conference) that among other things, called for a non-market-based 50% reduction in greenhouse gas emissions under the Kyoto Protocol by 2017, the payment by rich countries of their carbon debts, and an International Climate and Environmental Justice Tribunal.

It was a remarkable demonstration of grassroots democracy. "The only thing that can save mankind from a [climate] tragedy is the exercise of global democracy," said Bolivia's United Nations Ambassador Pablo Solon, a key organizer of the summit.[11] Bolivia succeeded in getting major parts of the "People's Agreement" integrated into the negotiating text for the COP 16 meeting, but in the end got only empty gestures. Whether or not the people's conference succeeds in getting global leaders to meet these goals at subsequent global climate summits remains to be seen. Nevertheless, they have become a benchmark around which a global people's movement to reclaim the earth has formed. Such a movement may be the only solution to the crisis that confronts us in the 21st century.

References

Akerlof, George A. and Robert J. Shiller. *Animal Spirits: How Human Psychology Drives the Economy and Why it Matters for Global Capitalism*. Princeton: Princeton University Press, 2009.

Alcamo, Joseph "Indirect Land-use Changes Can Overcome Carbon Savings From Biofuels in Brazil," *Proceedings of the National Academy of Sciences*, 107, no. 8 (February 23, 2010): 3388-3393.

Arrighi, Giovanni. *The Long Twentieth Century: Money, Power and the Origins of our Times*. New and updated version. London: Verso, 2010.

Baker, Dean, Robert Pollin, Travis McArthur and Matt Sherman, "The Potential Revenue from Financial Transaction Taxes," Working Paper 212, December 21, 2009. Available from Political Economy Research Institute www.peri.umass.edu/236/hash/7cd9f46fbb/publication/393/ (accessed June 9, 2010).

David Barboza, "Market Defies Fear of Real Estate Bubble in China," *New York Times*, March 5, 2010 www.nytimes.com/2010/03/05/business /global/05yuan.html.

Berry Thomas. *The Great Work: Our Way into the Future*. New York: Bell Tower 1999.

—. *The Dream of the Earth*. San Francisco: Sierra Club Books, 1998.

Bosselmann, Klaus. "Earth Democracy—a Position Paper," unpublished paper delivered to the Global Ecological Integrity Group, University of British Columbia, July 26, 2010.

Bradsher, Keith [a.] "China's Central Bank Hits Brake on Hot Economy," *New York Times*, February 12, 2010, A1.

—. [b] "China Outpaces US in Cleaner Coal-Fired Plants, *New York Times*, May 10, 2009 www.nytimes.com/2009/05/11/world/asia/11coal.html (accessed March 14, 2010).

Brown, Peter G. and Geoffrey Garver, *Right Relationship: Building a Whole Earth Economy*. San Franciso: Berrett-Koehler Publishers, 2009.

Center for Naval Analysis, *National Security and the Threat of Climate Change* http://securityandclimate.cna.org (accessed March 16, 2010).

"Copenhagen Climate Summit: Cracks Appear in Consensus of Developing Nation Bloc." *Telegraph*, December 9, 2009. www.telegraph.co.uk/earth/copenhagen-climate-change-confe/6774311/Copenhagen-climate-summit-Cracks-appear-in-consensus-of-developing-nation-bloc.html (accessed March 15, 2010).

"Copenhagen Accord" available at www.unfccc.int.

"Crop Circles." *The Economist*, March 13, 2010, 15.

Daly, Herman E. and John B. Cobb, Jr., *For the Common Good: Redirecting the Economy Toward Community, the Environment and a Sustainable Future,* updated and expanded. Boston: Beacon Press, 1994.

Davis, Steven J. and Ken Caldeira. "Consumption-Based Accounting of CO_2 Emissions," *Proceedings of the National Academy of Sciences*, March 8, 2010 DOI: 10.1073/pnas.0906974107 (accessed May 15, 2010).

Debeir, Jean-Claude, Jean-Paul Deléage and Daniel Hémery, *In the Servitude of Power: Energy and Civilization Through the Ages*, trans. John Barzman. London: Zed Books, 1991.

Diamond, Jared. *Collapse: How Societies Choose to Fail or Succeed*. New York: Viking/Penguin, 2005.

Dyer, Gwynne. *Climate Wars: The Fight for Survival as the World Overheats*. Mississauga, Ont: Random House of Canada, 2009.

Earth Charter. Available at www.earthcharterinaction.org/content/pages/Read-the-Charter.html.

Engel, J Ronald, Laura Westra and Klaus Bosselmann, eds. *Democracy, Ecological Integrity and International Law*. Newcastle Upon Tyne: Cambridge Scholars Publishing, 2010.

Ewing, Brad, Steven Goldfinger, Anna Oursler, Anders Reed, David Moore, and Mathis Wackernagel, *Ecological Footprint Atlas 2009* www.footprintnetwork.org/images/uploads/Ecological_Footprint_Atlas_2009.pdf (accessed March 15, 2010).

Framework Convention on Climate Change COP 13 http://unfccc.int/meetings/cop_13 (accessed February 19, 2010).

—. COP 15 www1.cop15.meta-fusion.com (accessed February 19, 2010).

Global Footprint Network. www.footprintnetwork.org/en/index.php/GFN/.

Gulick, John. "Arrighi's Tapestry of East and West." *Journal of World Systems Research*, XV, no 2 (2009): 244.

Hardin, Garrett. *Living Within Limits: Ecology, Economics and Population Taboos*. New York: Oxford University Press, 1993.

—. "The Tragedy of the Commons." *Science* 162:1243-8.

Kauffman, Stuart. *Reinventing the Sacred*. New York: Basic Books, 2008.

Klare, Michael T. *Rising Powers, Shrinking Planet: The New Geopolitics of Energy*. New York: Henry Holt & Co., 2008.

—. *Blood and Oil: The Dangers and Consequences of America's Growing Dependency on Imported Petroleum*. New York: Henry Holt & Co., 2004.

—. *Resource Wars: The New Landscape of Global Conflict*. New York: Henry Holt & Co., 2001.

Korten, David C. *The Great Turning: From Empire to Earth Community*. San Fransciso: Berrett-Koehler Publishers, 2006.

Kunstler, James Howard. *The Long Emergency: Surviving the End of Oil, Climate Change, and Other Converging Catastrophes of the Twenty-First Century* (New York: Grove Press, 2005)

Levin, Kelly and Rob Bradley, "Comparability of Annex I Emission Reduction Pledges," Working Paper (Washington, DC: World Resources Institute, February 2010), 2 http://pdf.wri.org/working_papers/comparability_of_annex1_emission_ reduction_pledges_2010-02-01.pdf (accessed February 19, 2010).

Miller, Matthew and Luisa Kroll, "Bill Gates No Longer Richest Man," *Forbes*, March 10, 2010 www.forbes.com/2010/03/09/worlds-richest-people-slim-gates-buffett-billionaires-2010-intro.html (accessed March 13, 2010.

Mouawad, Jad. "Deal for South American Oil Fields Extends China's Global Quest for Energy, *New York Times*, March 14, 2010 www.nytimes.com/2010/03/15/business/global/15oil.html (accessed March 14, 2010).

New Economics Foundation. *The Great Transition: A Tale of How it Turned out Right*. Available at: www.neweconomics.org/search/apachesolr_search/The%20Great%20Transition.

Pendleton, Andrew. "After Copenhagen." Open Democracy www.opendemocracy.org (accessed March 14, 2010).

Press Briefing, Plurinational State of Bolivia, December 12, 2010 http://pwccc.files.wordpress.com/2010/12/press-release-history-will-be-the-judge.pdf (accessed December 12, 2010).

Rees, William E. "Globalization and Extended Eco-Footprints: Neo-colonialism and (Un)sustainability," Chapter 24 in *Democracy, Ecological Integrity and International Law*, eds. Laura Westra and Klaus Bosselmann. Newcastle Upon Tyne: Cambridge Scholars Publishing, 2010, 467-489.

Sassen, Saskia. *Territory, Authority and Rights: From Medieval to Global Assemblages*. Princeton: Princeton University Press, 2006.

Sen, Amartya. *Development as Freedom*. New York: Alfred A. Knopf, 1999.

Simms, Andrew. *Ecological Debt: Global Warming and the Wealth of Nations*, 2nd ed. London: Pluto Press, 2009.

Shiva, Vandana. *Soil Not Oil: Climate Change, Peak Oil and Food Insecurity,* Boston: South End Press, 2008.

—. *Earth Democracy: Justice, Sustainability and Peace*. Boston: South End Press, 2005.

Southern People's Ecological Debt Creditors Alliance www.ecologicaldebt.org/What-is-Ecological-Debt/.

United Nations Framework on Climate Change COP 13 http://unfccc.int/meetings/cop_13/items/4049.php.

—. COP 15 http://unfccc.int/meetings/cop_15/items/5257.php/.

Verhagen, Frans. "The Tierra Fee and Dividend System: Using a Carbon-Based International Monetary System to Combat Climate Change," unpublished paper presented at the 2nd UNITAR/Yale Global Conference on Environmental Governance and Democracy, Strengthening Institutions to Address Climate Change and Advance a Green Economy, Yale University, 17-19 September 2010.

Wilkinson, Richard and Kate Pickett, *The Spirit Level: Why Equality is Better for Everyone*. London: Penguin Books, 2009.
World People's Conference on Climate Change and the Rights of Mother Earth. http://pwccc.wordpress.com/.
Yardley, Jim. "For India's Newly Rich Farmers Limos Won't Do," *New York Times*, March 18, 2010, A1.

Notes

[1] Antonio Gramsci in Quintin Hoare and Geoffrey Noell-Smith, *Selections from the Prison Notebooks of Antonio Gramsci* (New York: International Publishers, 1971).

[2] Although Sassen sees the current period as qualitatively distinct from the past 700 years of Westphalian history, her neglect to factor into her analysis either the inherent nature of capitalism or the global environmental crisis leads her to be far more tentative than I am about the shape of the future. Nevertheless, her analysis suggests the enormity of the changes we are seeing and points to some ways in which we might forsee a more positive future.

[3] Pendleton sees a striking similarity between the interstate dynamics in the trade talks and those exhibited at Copenhagen. "The similarity between the WTO and UN climate negotiations, he has said, "can be demonstrated by a simple test: replace 'liberalisation' by 'decarbonisation' and (in relation to cuts) 'tariffs' by 'emissions.'"

[4] Arrighi (2010, 28-29) defines hegemony as "the power of a state to exercise functions of leadership and governance over a system of sovereign states. . . . This power is something more and different from 'dominance' pure and simple. It is the power associated with dominance expanded by the exercise of 'intellectual and moral leadership.'"

[5] Despite the fact that oil demand has been shrinking in developed nations because of the global economic slowdown and efforts to conserve energy, Chinese consumption is surging. The International Energy Agency reported that China's oil demand jumped by "an astonishing" 28 percent in January of 2010 (Mouawad, 2010).

[6] China, for example, has .85 hectares per person, while India has only .37 compared to 4.43 for the US and 17.08 for Canada, near the top of the list. Tables 4, 5 in Ewing, et. al.(2009).

[7] Map 4 Ecological creditor and debtor countries, 1961 and 2006. Ibid., 33.

[8] The mission statement of the ALBA, or Bolivarian Alternative to the Americas, a plan for regional integration whose membership now includes Bolivia, Cuba, Dominica, Honduras, Nicaragua and Venezuela exhibits these values. It calls for reciprocal trade relations between member states, based on cooperation, solidarity and complementarity, with the goal of fighting poverty and inequality.

[9] "Ecological debt," a term coined by developing countries, is a claim against the rich countries for centuries of resource extraction, resource depletion, intellectual

appropriation, disproportionate energy use and atmospheric pollution associated with colonial, neocolonial and neoliberal interstate relations. For a thorough discussion see the website of the Southern People's Ecological Debt Creditors Alliance.

[10] BBC poll finds widespread dissatisfaction with capitalism, November 9, 2009 http://news.bbc.co.uk/2/hi/8347409.stm (accessed December 1, 2010).

[11] Meeting with Bolivian ambassador to the U.N. Pablo Solon attended by the author February 16, 2010.

CHAPTER EIGHT

SUSTAINABILITY IMPERATIVES

HELMUT BURKHARDT

Abstract

The conditions for the sustainability of a civilization are complex. Integrity of the individual, the community, the ecosystem, and of the cosmic environment are all seen as necessary conditions for sustainability.

Introduction

Sustaining human life is the overarching issue for all; it is at the heart of work and play of individuals and at the core of social action, politics, and ethics. Could humans go the way of dinosaurs and the many other species that have disappeared? The human struggle for survival is assisted by greater intelligence, use of tools, and of technology. However, these survival instruments can backfire, and become the cause of the extinction of our own species, and many others. We cannot take the long term survival of Homo sapiens for granted.

Indeed, our technology-based civilization is presently facing existential threats at several levels. A growing number of individuals suffer from shortages of food and water, and ill health. There is a wide spread fear of epidemics. -- Human society is plagued by self-imposed dangers such as weapons of mass destruction and climate change. -- The ecosystem is on the verge of collapse as the climate conditions are close to tipping point. Individual integrity, societal integrity, ecological integrity, and the integrity of the cosmic environment are the conditions sine qua non for sustainability; the imperatives.

Personal Integrity

Health and satisfying Maslow's basic needs are required for the integrity for an individual. Health in this context means a state capable of sustaining life. The basic needs, as established by Maslow, are physical needs such as food, water, and shelter; mental needs of sensory inputs and communication; and spiritual needs which include self-esteem, ethical conduct, and community recognition (Maslow).

The primary values (priorities) of individuals are to sustain their own life, and to propagate their own genes. The primitive egocentrism based on the law of force can be observed widely in nature and is commonly interpreted as Darwin's theory of survival of the fittest. However, the best chance for survival and achieving personal integrity lies in an enlightened egocentrism were the individual grants rights to others and submits to the rules of life in a community. Kroopotkin wrote about mutual aid in 1902 (Alexeivich 1998). New studies show that Charles Darwin's notion of fitness to survive included love, compassion, and cooperation, not just brute strength (Alexeivich 1998; Axelrod 2006).

Threats to an individual's integrity arise not only from within through loss of health, but also from events external to the individual such as societal and ecological corruption and, albeit with a low probability, from cosmic events.

Societal Integrity

Primary social values (priorities) are to protect the human community and its members from harm and to supply the necessary resources by whatever means are available. In practice, most societies are anthropocentric in this primitive way. However, enlightened anthropocentrism offers the best long term chances for sustaining a human community. Enlightened anthropocentrism recognizes the legitimate right of other human communities to exist, and recognizes the vital importance of other species for a healthy ecosystem, of the web of all life.

The principal self-inflicted threat to societal integrity comes from the weapons of mass destruction, particularly nuclear weapons (Willens 1984). **Figure 8-1** illustrates the scale of the threat to life from existing nuclear weapons, which can release thousands of times more energy than all the weapons used on both sides of WW2.

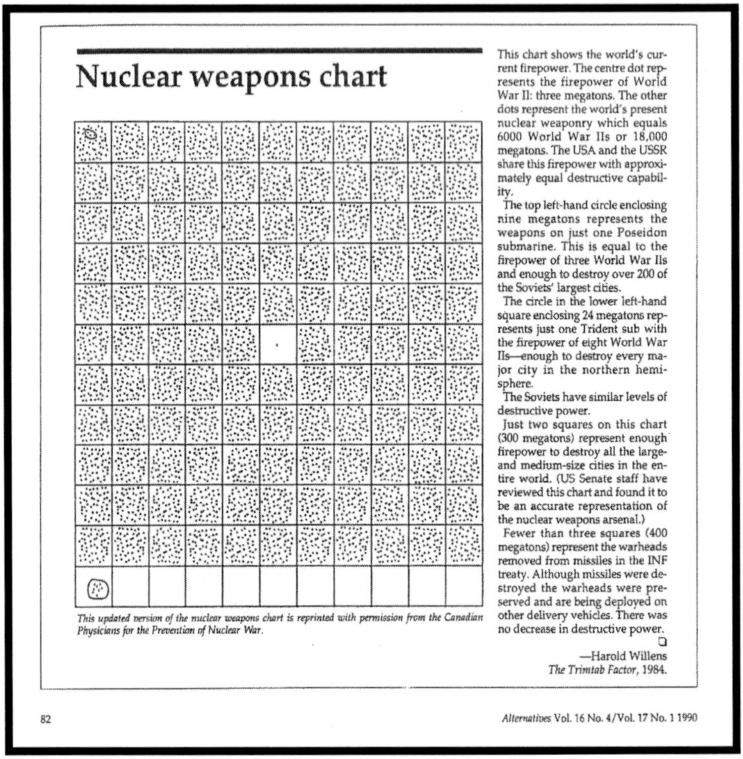

Nuclear weapons chart

This chart shows the world's current firepower. The centre dot represents the firepower of World War II: three megatons. The other dots represent the world's present nuclear weaponry which equals 6000 World War IIs or 18,000 megatons. The USA and the USSR share this firepower with approximately equal destructive capability.

The top left-hand circle enclosing nine megatons represents the weapons on just one Poseidon submarine. This is equal to the firepower of three World War IIs and enough to destroy over 200 of the Soviets' largest cities.

The circle in the lower left-hand square enclosing 24 megatons represents just one Trident sub with the firepower of eight World War IIs—enough to destroy every major city in the northern hemisphere.

The Soviets have similar levels of destructive power.

Just two squares on this chart (300 megatons) represent enough firepower to destroy all the large- and medium-size cities in the entire world. (US Senate staff have reviewed this chart and found it to be an accurate representation of the nuclear weapons arsenal.)

Fewer than three squares (400 megatons) represent the warheads removed from missiles in the INF treaty. Although missiles were destroyed the warheads were preserved and are being deployed on other delivery vehicles. There was no decrease in destructive power.

—Harold Willens
The Trimtab Factor, 1984.

This updated version of the nuclear weapons chart is reprinted with permission from the Canadian Physicians for the Prevention of Nuclear War.

82 *Alternatives* Vol. 16 No. 4/Vol. 17 No. 1 1990

Figure 8-1. The explosive energy of the global arsenal of nuclear weapons. Harold Willens Nuclear Weapons Chart graphically illustrates the senseless overkill in the existing nuclear weapons. Each point represents the energy of all weapons exploded in WW2 on all sides including the nuclear bombs on Hiroshima and Nagasaki. The explosion of nuclear weapons in a few of the squares would suffice to create a nuclear winter that would likely lead to extinction of humans and of many other species.

If only a fraction of the available nuclear explosives were used, the resulting nuclear winter (Wikipedia "Nuclear Winter") would prevent food production for several years, and the present global food reserves, if not destroyed and inaccessible after the explosions, would last at best for some 59 days. Therefore, it is imperative that the traditional reliance of national security on military strength must be discontinued. Security by the law of force is no longer an option; it must be replaced by security by force of law.

History shows that governed entities such as municipalities, provinces/states, and nations have an acceptable level of internal peace. The European Union, if it remains successful, demonstrates that governance at the continental level is possible. Appropriate global governance with legislative, jurisprudence, and executive powers can offer credible levels of security to nations and their citizens. Without a doubt, humanity has the technical means needed for global governance. It is a question of collective intelligence and political will to verify a world without war.

Here are several ideas on how a functioning global governance could work. Similar to democratic national governments it must have legislative, juridical, and executive powers.

a) Replace the UN General Assembly of nations, with a General Assembly of Global Districts as the international law making body. Global Districts of some ten million inhabitants will represent the local interests better than the self-serving sovereign nations who do as they please, commit acts of aggression, and plunder the global commons. The General Assembly of Global Districts will have some 600 members.

b) Change the composition of the Security Council, and abolish the veto. Give permanent seats in the Security Council to Global Blocks, unions of 100 million people. This new Security Council will have some 60 members. All must support unconditionally a world constabulary to enforce international law. This Security Council will make an effective global executive power.

c) Global jurisprudence already exists in the International Criminal Court, and the International Court of Justice. An Environmental Court of Justice ought to be added to rule on matters arising from the Earth Charter. All members of the global community must be pressured to recognize the authority of the International courts.

d) Let the principle of subsidiarity guide the design of nested hierarchy of local to global governance; subsidiarity requires that the lower levels of governance are imbedded and harmonized with the higher levels. This means that the mandate of the global level of governance is to guard the global commons and manage other global affairs. The mandate of the Block governance is to guard the block commons and manage other block affairs, and so on down to the individual. Clearly define each one of the complex set of relations between various levels of governance shown in the Peace Matrix (Burkhardt et. al. 2006), shown in **Figure 8-2.**

e) Be tolerant; let every law abiding individual, family, municipality, district, nation, and global block strive for happiness and survival within their own culture; stop missionary zeal for exporting ideologies such as communism, capitalism, religion, democracy, or lifestyle. – Charles Darwin stated in the Origin of Species: “there is survival potential in diversity”.
f) Encourage women and youths to continue their engagement in the political process using current technology to enhance their influence.

A practical and just distribution of the rights and responsibilities in the 64 relations are of utmost importance for understanding and for managing the affairs of all these social entities and individuals in a peaceful manner. All should be clearly defined in order to avoid confusion and misunderstandings in the highly complex interactions: Peace is possible when all these 64 relations are harmonized.

Illustrative examples: 'nw' represents the actions/responsibilities of a nation vis-a-vis the world union and ‘wn’ represents the actions/responsibilities of the world union vis-a-vis a nation state. Similarly, 'ni', and 'in' name the interactions between individual citizen and the nation state. ‘The Responsibility to Protect’ (R2P) (United Nations, “Responsibility to Protect”) falls into the ‘wi’ category. Pairs of identical letters categorize the relations between entities at the same level of governance. ‘nn’ represents the interactions between nations, and ‘ii’ the interactions between individuals.

Practical examples: if Quebec and Canada do not see eye to eye on what the rights and responsibility of the province towards the nation are ('dn' relation), and what the rights and responsibilities of the nation are towards the province ('nd' relations), then there will be problems. In Kosovo and Yugoslavia violence resulted in the 'dn' and 'nd' categories. In this case, 'wn' and 'wd' interactions were needed to settle the dispute. What are the tasks and authorities of the provincial police force and what are the tasks and authorities of the municipal police force? How can these 'dm' and 'md' relations be harmonized?

War is not the only threat to societal integrity. Other threats come from inside, through loss of integrity of its members, and from outside, through loss of integrity of the life-supporting ecosystem. For example, infertility can arise through cultural corruption, or through environmental pollution. The internal threat to sustainability can be observed in rich industrialized nations; they have an average of 1.4 children per woman, while an average of 2.1 children per woman is required for sustaining a population (Riche 2004).

Ecological Integrity

Ecological sustainability at the global and local levels is determined by the nexus of population, consumption and technology. One well defined measure for sustainability is the deviation of the ecological footprint from the ecological capacity. The Global Footprint Network (Global Footprint Network, "World Footprint") has analysed the footprint and the ecological capacity of many nations, covering a large part of the world population. According to this study, only a few nations are still sustainable, while most are not; altogether, the footprint of humankind is already surpassing the Earth's ecological capacity by more than 30%. This means that humankind lives unsustainably off the global ecological capital. What is the solution to this problem with ecological sustainability?

\	E	W	B	N	D	M	F	I
E	ee	ew	eb	en	ed	em	ef	ei
W	we	ww	wb	wn	wd	wm	wf	wi
B	be	bw	bb	bn	bd	bm	bf	bi
N	ne	nw	nb	nn	nd	nm	nf	ni
D	de	dw	db	dn	dd	dm	df	di
M	me	mw	mb	mn	md	mm	mf	mi
F	fe	fw	fb	fn	fd	fm	ff	fi
I	ie	iw	ib	in	id	im	if	ii

Figure 8-2. Peace Matrix. In a nested hierarchy of eight levels of governance there are 64 categories of interactions that need to be harmonized between the quasi autonomous, governed, self-organizing social entities denoted by capital letters. E: Environment, Ecosystem; W: World Union; B: Global Block, continental unions or appropriate aggregate of nation with 100 million inhabitants. N: Nation; D: Global District, i.e. appropriate neighbourhood aggregates of small nations, provinces, states, or territories with some 10 million inhabitants; M: Municipalities; F: Family; I: Individual. The pairs of small letters indicate the relationship between the entities. The relationships named by the letter pairs can be matter energy trade relations, economic relations, information relations, responsibility relations, or rights relations.

George Montbiot (Monbiot) and David Satterthwaite (Satterthwaite) state that overconsumption by the rich is causing the large footprint of humankind, and population size does not matter much. There is truth in this view for the existing distribution of wealth. Share International (Share

International) gives the present population and consumption statistics in a simplified form, which indicates that at present 80% of the world's poor population consume only 20% of the resources, while 20% of the population consume 80% of the resources. However, in a just and equitable society of the future, where the poor are legitimately consuming more, and the rich are substantially reducing consumption, population numbers will be relevant. The long term outlook requires further analysis.

Ecological sustainability analysis can be quantified by the use of the simplified 80/20 global wealth distribution and the Ehrlich-Holdren equation (Ehrlich and Holdren 1971) for the human environmental impact 'I':

$$I = P\,A\,T = X\,T \quad . \qquad (1)$$

Where 'P' is the number of humans, 'A' is the per capita consumption of goods, and 'T' is the environmental damage caused by the production of one unit of consumer goods with the current technology. X = P A is the total global consumption of goods,

According to the Global Footprint Network, using current technology, today's global resource consumption X needs to be reduced in order to reduce the global ecological footprint to fit the Earth's ecological capacity and to limit the environmental impact to a sustainable level. The sustainable level of total resource consumption 'X_s' is:

$$X_s = 0.77\ X. \qquad (2)$$

Using the simplified 80/20 global population and consumption statistics, and AP = X, given in (1). The global average per capita consumption 'A_{av}', the per capita consumption of today's rich countries 'A_r', and the per capita consumption 'A_p', of today's poor can be determined. The global average per capita consumption is: $A_{av} = X/P$. $A_r\ 0.2\ P = 0.8\ X$, and $A_p\,0.8\ P = 0.2\ X$, yield the following results:

$$A_r = 4\ A_{av}\ ,\ A_p = (1/4)\ A_{av},\ \text{and}\ A_r = 16\ A_p\ . \qquad (3)$$

The sustainable global average per capita consumption 'A_{sav}' is less than our present global average. With (2) it is:

$$A_{sav} = 0.77\ A_{av}\ . \qquad (4)$$

Assuming a globally just and equitable society, where the rich consume less and the poor consume more until the per capita consumption of all humans converges toward a common average, equation (1) represents the connection between population, per capita consumption, and technology.

Keeping present day carbon based technology, the relation between a sustainable world population 'P_s' and the resource consumption per capita A in relation to per capita resource consumption of today's rich is with (2), (3), and (5):

$$P_s\, A = X_s = 0.77\, X = 0.77\, P\, A_{av} = 0.77\, P\, 0.25\, A_r = 0{,}19\, P\, A_r, \text{ or:}$$
$$P_s\, (A / A_r) = 0.19\, P \tag{5}$$

This relation between a sustainable world population P_s, the per capita consumption as a fraction of the capita consumption of today's rich (A / A_r), and the present world population P, is represented in the red series of **Figure 8-3**. The red bars represent the sustainable population under the assumption that today's carbon based energy technology is used to drive the economy.

The Global Footprint Network states that the carbon footprint is half of the total ecological footprint. Therefore, the sustainable population can be doubled by changing energy technology to all renewable energy resources with zero carbon emissions:

$$P_s\, (A / A_r) = 0.38\, P \tag{6}$$

The green bars of **Figure 8-3** represent the sustainable world population under the assumption that the economy is driven totally by renewable energy resources, with zero carbon emissions.

It is interesting to discover that for maintaining ecological integrity in a just and sustainable world using carbon based technology with today's population of 6.5 billion, the rich would have to reduce per capita consumption to 20% of today's level, while the poor could increase consumption by some 300%. The world average per capita consumption would have to be reduced from the present 25% to 20% of the per capita consumption of today's rich. If renewable energy technology was introduced, together with reduced consumption, some increase in world population seems feasible.

However, on closer inspection the sustainability criteria presented by the Global Footprint Network, which are the basis of the above analysis, seem to be too optimistic. The global footprint analysis does not sufficiently include the loss of biodiversity. Like all monocultures, a human monoculture on planet Earth carries a high risk of collapsing.

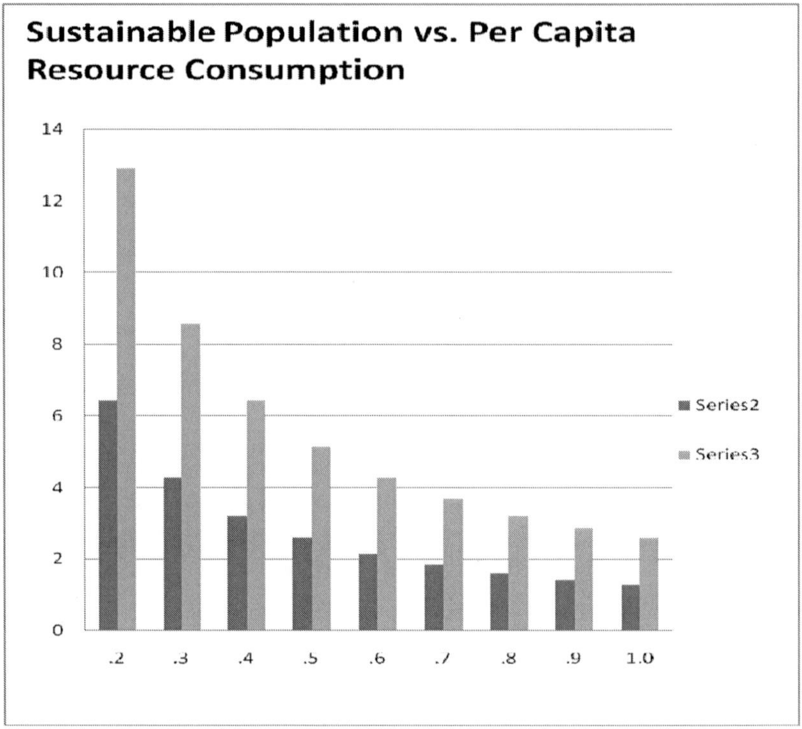

Figure 8-3. The vertical axis gives the sustainable world population in billions as a function of the ratio of per capita consumption of resources to the per capita consumption of today's rich (A/A_r). The red series assumes present day carbon based energy technology, the green series assumes zero carbon, total renewable resource based energy technology.

Cosmic Integrity

Last but not least, the often forgotten cosmic conditions are important for sustaining life on Earth. Even though the cosmic givens are beyond human influence, they ought to be remembered.

Life on Earth depends on the input of the right amount of sunlight with the right spectral composition. It is likely that solar activity does not change radically, even in millions of years, keeping up the production of biomass through photosynthesis, and a fairly steady weather. However, this cosmic equilibrium is threatened by self-inflicted climate change through CO_2 emissions, or by increased UV radiation through atmospheric pollution that damages the protective ozone layer in the atmosphere.

Large meteorites could deal a fatal blow to the web of life on Earth, albeit there is only a small probability of such a cosmic event. A near Earth asteroid was discovered in 2004 and named Apophis (Wikipedia, "The Near Earth Asteroid Apophis").

It caused some concern as the probability of this object, of more than 200m in diameter, hitting the Earth in 2029 was determined to be 2.7%, which put it on level 4 of the Torino Scale. The Torino Scale, with numbers from 1 to 10, is a combination of energy and probability of hitting the Earth; it categorizes the impact hazard of near Earth objects (Wikipedia, "The Torino scale categorizing the impact hazard of near Earth objects"). On the basis of newer observations Apophis has been downgraded to the level 1 on the Torino Scale. Collision avoidance strategies with nuclear weapons and other methods are being developed (Wikipedia, "Avoidance strategies of collision with near Earth objects").

The Earth is blessed with a liberal amount of life-supporting water like no other planet in our solar system. However, humanity is already suffering from a water shortage.

Conclusion

Sustaining human civilization on Earth requires the integrity of individuals, communities, ecosystems, and cosmic conditions. All of these are mutually dependent, and can be partly influenced by technology under the control of humankind. Radical changes in the political global architecture are required to achieve societal integrity, and the nexus of population, consumption, and technology must be addressed for achieving ecological integrity.

References

"Maslow's hierarchy needs" www.deepermind.com/20maslow.htm (Accessed 2009-10-29).

Alexeivich, Kropotkin Peter, *Mutual Aid: A Factor of Evolution* (London: Freedom Press, 1998 paperback ed.).

Axelrod, Robert, *The Evolution of Cooperation* (New York: Basic Books, 2006 Revised Edition).

Burkhardt, Helmut, Julia Morton-Marr, and Rose Dyson, "Handbook on Critical Global Issues and Viable Solutions" (paper presented at the World Peace Forum, UBC, Vancouver BC, June 23-28, 2006). (http://c-g-i.info/handbook.html, accessed 2010-08-28).

Ehrlich, P.R and J. Holdren, "The Impact of Population Growth", *Science,* 171, (1971): 1212.

Global Footprint Network, "World Footprint". www.footprintnetwork.org/en/index.php/GFN/page/world_footprint/ (Accessed 2009-10-27).

Monbiot, George, "The Population Myth", www.countercurrents.org/monbiot300909.htm (Accessed 2009-09-30).

Riche, Martha Farnsworth, "Low Fertility and Sustainability", World Watch Magazine, Sept/Oct (2004) (Available at www.worldwatch.org/node/561 accessed 2010-08-28).

Satterthwaite, David, "The implications of population growth and urbanization for climate change". http://eau.sagepub.com/cgi/reprint/21/2/545 (Accessed 2009-10-27).

Share International, "The 20/80 partition of global wealth". www.share-international.org/archives/environmental/en_amfair.htm (Accessed 2009-10-27).

United Nations, "Responsibility to Protect", www.responsibilitytoprotect.org (accessed 2010-08-28).

Wikipedia, "Avoidance strategies of collision with near Earth objects". http://en.wikipedia.org/wiki/Asteroid_deflection_strategies (Accessed 2009-10-28).

Wikipedia, "Nuclear Winter", http://en.wikipedia.org/wiki/Nuclear_winter#2007_study_on_global_nuclear_war (Accessed 2010-08-28).

Wikipedia, "The Near Earth Asteroid Apophis", http://en.wikipedia.org/wiki/99942_Apophis#Close_approaches (Accessed 2009-10-27).

Wikipedia, "The Torino scale categorizing the impact hazard of near Earth objects", http://en.wikipedia.org/wiki/Torino_Scale (Accessed 2009-10-28).

Willens, Harold, *The trimtab factor: how business executives can help solve the nuclear weapons crisis* (New York: W. Morrow, 1984). The Nuclear Weapons Chart is available at www.cafeweltgeist.org/ihtec/Resources/NuclearWeaponsChart.jpg (Accessed 2010-08-28).

PART III

INTERNATIONAL LAW: ECOLOGICAL INTEGRITY AND HUMAN RIGHTS

INTRODUCTION

KLAUS BOSSELMANN

Like science and ethics, the law is a significant contributor to advancing ecological integrity. A number of countries have developed legislation and policies around the concept of ecological integrity and the body of legal research into ecological integrity is constantly growing. Yet, legal systems are reactive in character, often resistant to innovation and not at the forefront of social change. This is particularly true of international law which lacks many of the features of modern national legal systems.

This part of the book examines several innovative trends in international law. Each of the following five chapters aims not only to achieve a better understanding of the current law, but also its potential for change.

Joseph Dellapenna sets the scene with an account of what international law is and how it works. With many of its legal characteristics and practical effects still in doubt, Dellapenna considers international law a "primitive system" which cannot solve the world's problems by itself. It is, however, an essential element of any solution. For more effectiveness, international law needs a fully developed institutional framework.

The inefficiency of international law is of particular concern it relates to issues of ecological integrity. Transnational environmental harm, for example, is at odds with national boundaries determining any legal responsibilities. In her chapter, Sara Seck shows the detrimental consequences of the mismatch between ecological and legal boundaries, but does so through the Third World Approach to International Law (TWAIL), questioning the global nature of current international law. With its emphasis on home state regulation and local autonomy, TWAIL makes the point that the "global" effectively appears in the "local".

TWAIL is part of the wider search for effective law and governance. In recent years, this search has been extended to include global constitutionalism. This is the subject of Bosselmann's chapter. As a more positive effect of globalisation, national constitutions have increasingly followed similar ideas, principles and values. If there is indeed a trend towards greater global commonality, the possibility of a global constitution –

shaped around democracy, human rights and responsibilities - may arise. To this end, the Earth Charter offers practical inspiration and guidance.

As an aspect of global constitutionalism and global governance, Owen McIntyre describes the emerging concept of Global Administrative Law (GAL). With reference to human rights-based approaches to water, he shows how GAL permeates diverse areas such as international water resources law, international environmental law, and international investment law, as well as national public and constitutional law. Like water, the related legal entitlements and responsibilities come in many forms, both national and transnational. McIntyre makes the case for a globalised concept of water law and, at the same time, for emerging GAL.

The final chapter by Kathryn Kintzele sets forth the previous chapters with its emphasis on a commonality of principles and values. Citing from twenty new or newly-amended constitutions from around the world, Kintzele shows how ethical principles of environmental conservation have found their way into constitutions, law-making and judicial opinions.

While the world's legal systems are still far from adopting a new ethic, visionary lawyers such as J Weeramantry of the International Court of Justice give such a new ethic its voice: "The first principle of modern environmental law is the principle of trusteeship of Earth's resources." And further: "It would not be wrong to say that the love of nature, the desire for its preservation… are among those pristine and universal values which command international recognition."

CHAPTER NINE

THE FORMS OF INTERNATIONAL LAW

JOSEPH W. DELLAPENNA

For those who are not familiar with international law, just what it is or how it operates is often a puzzle. Some will doubt whether there even is such a thing, asking if international law really is law. This question was asked emphatically by John Austin in the nineteenth century (Austin 1955: 122-125). To answer this question, one must consider the forms that international law takes and how it functions. This analysis begins with a consideration of how law works in general.

What Makes Law Law?

Whether international law really is law is a question about legitimacy, of whether international law and international legal processes are accepted as justifying the prescription and enforcement of norms (Franck 1990). Legitimacy is partly a function of perceived fairness in procedures and partly a function of perceived rightness of result. Societies where governance is perceived as legitimate can carry on for decades or centuries without making particularly good decisions; when societies reject the legitimacy of their institutions, its governance structures collapse rather quickly from what often appear to be rather slight shocks (Dellapenna 2010).

Questions about the legitimacy of international law arise when people have in mind a specific model of how law works in describing something as a law and some claim of right or obligation as legal. This model envisions a legislature acting formally to create a highly determinate rule enforced by a policeman: "The ultimate expression of law isn't order, it's a prison" (Jackson 1972: 119). This notion is called "legal positivism" because it focuses attention solely on "positive" law, law that is formally enacted and formally enforced. John Austin, one of the foremost proponent of legal positivism in the nineteenth century, defined law as "the command of a sovereign" enforced by a sanction (Austin 1955: 133, 201).

By this theory, the practice of law pertains to identifying the commands of a sovereign and properly using the commands to achieve a desired result. Austin concluded that international law simply cannot be law, but was merely "positive morality," because there was no identifiable sovereign issuing commands nor any formal enforcement mechanisms for such commands (Austin 1955: 134-142).

Most people who live under highly developed national legal systems are probably comfortable with the positivist description of what law is and how it operates. This model, however, does not explain what we call "law" even in national legal systems. Consider traffic laws. In the United States, nearly everyone drives faster than the speed limit and there are not enough police to compel people to drive at or below that limit. Any attempt to do so would fail because too many are violating the law. The best achievable result is to keep most people driving not very much faster than the speed limit through selective enforcement targeted at those who violate the limit too egregiously. Contrast speed limits with traffic lights. People in the United States seldom simply drive through red lights (although they sometimes cheat a little). Yet if nearly everyone were to disregard them, the laws on driving through those lights could no more be enforced than speed limits. Most people do not drive through red lights from self-interest: Driving through a red light is more dangerous than speeding, and would be suicidal if nearly everyone did so. Yet the response to someone driving through a red light is not simply that the act is dangerous. People see driving through a red light as anti-social behavior that is and ought to be illegal. H.L.A. Hart, the twentieth-century's leading legal positivist, described the decision to obey traffic signals, and the sense of moral outrage against those who do not, as legal and not merely a moral because drivers refer to the law to explain their actions and thoughts (Hart 1961).

Now consider the more subtle situation with the law applicable to a contract. Contracts, voluntarily defined and assumed obligations, are an essential feature of modern life. Without contracts, modern economies would be impossible. Every developed state has a well-developed and usually highly technical law of contracts. Business people, consumers, and others often know nothing about these technicalities, or, worse, "know" something about these technicalities that is wrong. Yet business does not suffer because contracts are enforced not so much by formal law as by informal sanctions based on the sense of the relevant community, often with radically different results than prescribed by formal legal processes. The Uniform Commercial Code, which now provides the law of sales of goods throughout the United States, embraces this reality by relying heavily on the concept of "reasonableness" both to determine whether a contract has been formed and what its provisions mean. Provisions such as this can hardly be characterized as "commands of a sovereign" without seriously distorting the actual functioning of the legal system. Such rules accept that the parties themselves form a community that creates law for itself—a truth that is seldom explicit in the common law but that is the central tenet of the civil law tradition (Kahn-Freund *et al.* 1979: 315-348). In other words, the true basis of contracts and commercial law is the social sense of legitimacy granted to or withheld from particular voluntary conduct, just as it is with speed limits and traffic lights.

The Austinian paradigm of law that so many now seem to think of as "natural" moreover is a relatively recent idea; it essentially strips the notion of law down to organized coercion or the threat of organized coercion. This is a wholly inadequate notion of what law is and how law operates. As A.L. Goodhart observed, "It is because a rule is regarded as obligatory that a measure of coercion may be attached to it; it is not obligatory because there is coercion" (Goodhart 1953: 17; Hart 1961: 20-25). Even modern positivists have conceded as much when they embrace a normative explanation of positive law without an identifiable "sovereign" or the presence of a "command" or a "sanction." Hans Kelsen has developed a widely influential positivist theory where legitimacy derives from a "*grundnorm*" (a "basic norm" or "basic law") that just is, or at least is derived from social notions that are not explicable in legal terms (Kelsen 1992: 58-63). H.L.A. Hart developed a positivism without reliance on a coercion theory of law, positing a "habit of obedience" as the source of law and legitimacy (Hart 1961: 77-96). These theories seem inadequate to capture the sense of legitimacy that underlies law, yet they are closer to the reality of what law is and why it is effective than simple notions of command or sanction that are popularly thought of as constituting law.

Clifford Geertz, a noted anthropologist, reached a similar conclusion when he described "law" as an organic mechanism whereby certain claims of right are elevated to the status of socially established norms and other claims of right are denied that standing; it allows society to make sense of things (Geertz 1983: 175). When normative judgments are truly accepted as law, generally few violate the norms and those who do will pay a higher price than someone who violates a mere social or moral convention. The price might be exposure to official coercion, but it might also entail social enforcement such as public censure or even ostracism.

This leaves us with a question: What is the function of formal law, of law on the books? History provides an answer. Informal law functions successfully when each person in a particular community knows the others and what they are doing, each depends on the others for a wide range of social supports, and each realizes that overreaching too far or too often will cost them the social supports that he or she needs to survive or thrive. As a society becomes larger and social interactions become less personal, the complex web of mutual reciprocities that ensures compliance with customary law breaks down. Formal law—written law with specialized processes to make and enforce it—arises in response to that breakdown (Dellapenna 2000). Formal law provides adequate certainty and predictability to people in such a society. This was as true of Hammurabi's Babylon or the Rome of the *Decemviri* as it was of medieval Islam or modern Europe. A good example is the process, described by David Trubek and his co-authors, whereby during the last 20 years, under the impact of competition from English, Dutch, and American law firms, the French method of dealing with hostile corporate takeovers through informal arrangements among a few leading men broke down to be replaced by a highly formal set of legal rules and institutions that mirror the similar institutions that were created as much as 90 years earlier in the United States and perhaps 10 years earlier in the United Kingdom (Trubek *et al.* 1994).

Opportunities to create certainty, or at least the appearance of determinate outcomes knowable in advance, multiplied enormously with the invention of the printing press. That invention made possible the mass distribution of "law" and married formal law to the centralized state by making centralized control possible, but only if legal actors (lawyers, jurists, and knowledgeable lay people) were required to follow the letter of the law. From this possibility arose the characteristic form of modern law, nationally unified legal systems that claim a monopoly over legal questions (Dellapenna 2000). From these beginnings, intended to enable autocratic rulers to rule *by* law, emerged the modern notion of the rule *of* law (the *Rechtstaat*) (Berhman & Witte 1989). Certainty and predictability are important values, particularly for those of who seek to make firm plans for the future. Formal law also serves the valuable social end of ensuring that the state itself abides by the law created by the state and society. Yet law is more nuanced and sophisticated than just formal law.

International Law as Law

International law operates similarly to national law and is experiencing some of the same pressures to change and develop greater formalism. International law until recently involved only a small and structureless society of states. The United Nations was created at the end of World War II with only 51 members. Its current membership approaches 200. Changes in such international structures have transformed the international legal system from the relatively simple forms of the past to an increasingly diverse and complex community of actors who too often no longer know much about each other. The United Nations and other international organizations also count as full players ("legal persons") in the international legal system (Alvarez 2006). Rapidly proliferating non-governmental and other official and semiofficial participants are also now playing a distinct albeit subordinate role (Depuy & Vierucci 2008). Even natural and artificial persons (people and corporations) are now to some extent as participants in the international legal community (Clapham 2010).

The growing international community introduces into international legal processes a larger variety of sharply differentiated cultural traditions than before. The differences are further accentuated by the division of nations into contending ideological camps (Study Committee 2006). This is precisely the setting in which participants can be expected to welcome more specialized and more formal legal structures (Dellapenna 2000). That has certainly happened at the regional level—e.g., the Association of Southeast Asian States, the European Union, or the North American Free Trade Association—as well as globally for specialized forms of activity—e.g., the International Atomic Energy Agency, the International Civil Aviation Organization, and the World Trade Organization. Still, the international legal system remains institutionally underdeveloped and decentralized (Boyle & Chinkin 2007).

International law largely remains a primitive legal system (Dinstein 1986). The international legal system lacks the superstructure of specialized institutions—executive, legislative, judicial, and administrative—found in modern national legal systems. But to conclude from this lack that international law is not really law is to confuse particular institutional arrangements with what law really is and how it really operates. Modern legal systems function in far more complex ways than a simplistic focus on "positive law" suggests, a way that implicates a link to custom as a primary source of law. Why do you obey a law? If it is because the legislature enacted it, why should we care what a legislature enacts? If that is because the constitution says so, why should we tolerate the rule of the living by the dead? The answer is and must be custom, just as for why speed limits are not enforced effectively. Moreover, the absence of formalized courts, legislatures, and executives no more indicates an absence of law in the international system than the absence of those institutions indicated the lack of law in pre-industrial societies the world over (*e.g.,* Nader 1990). International law's less formal processes similarly are law and must be examined carefully to learn its capabilities and its limitations.

Modern political and legal theory today generally bottoms all law on the consent of the governed (Estlund 2008; Przeworski 2008). While the roots of the centrality of consent to international law are different from the roots of the centrality of consent for national law, international law too rests on the consent of the participants—classically nation states. Some today argue for rooting international other than in the consent of states, including religion revelation (Weeramanty 1988), natural law (Janis 2008: 62-69), or the consent of the global population rather than of states (Alkoby 2008; Baker 2010; Villalpando 2010). None of these alternatives has gained any traction among governments (Xue 2007), international organizations, or international tribunals and thus will not be considered here. Instead, the following sections will follow the short, definitive statement of the forms of international law found in Article 38 of the *Statute of the International Court of Justice:*

> 1. The Court, whose function is to decide in accordance with international law such disputes as are submitted to it, shall apply:
>
> a. international conventions, whether general or particular, establishing rules expressly recognized by the contesting states;
>
> b. international custom, as evidence of a general practice accepted as law;
>
> c. the general principles of law recognized by civilized nations;
>
> d. subject to the provisions of Article 59, judicial decisions and the teachings of the most highly qualified publicists of the various nations, as subsidiary means for the determination of rules of law.
>
> 2. This provision shall not prejudice the power of the Court to decide a case *ex aequo et bono*, if the parties agree thereto.

The remainder of this chapter will consider these classic sources of law in the order listed in the statute of the court.

International Conventions

The term "international conventions" includes any form of voluntarily international agreement. Even a unilateral oral declaration can create an obligation under international law if the government on the behalf of which the declaration was made intends to adopt the obligation. There is thus no particular form required and, at least in international tribunals rather than national courts, compliance with national legal requirements is unnecessary so long as the government's representative meant the declaration or exchange to create an international obligation (Lipson 1991).

The relation of international conventions to the consent of participating states is clear. International conventions, like private contracts, provide the law for the parties thereto. So clearly is this understood that nations usually comply with their international conventions without requiring any enforcement of the convention—unless the state has a proper basis for questioning its validity or meaning or an internationally recognized legal excuse for non-compliance. These possibilities are enshrined in two well-known Latin maxims: *pacta sunt servanda* (agreements are to be observed) and *rebus sic stantibus* (so long as things remain the same, in other words, unless there is a sufficiently substantial change in circumstances (Lukashuk 1989; Vagts 2004). Bringing these two apparently contradictory principles together is an implied obligation to interpret and perform in good faith. There can be complex questions about whether the states had actually reached an agreement, what the agreement means, and whether there is a legally valid excuse for non-compliance. Many international conventions also resolve the problem of enforceability by providing for compulsory arbitration or other means.

Today, the rules for international conventions for the most part are found in the *Vienna Convention on the Law of Treaties.* Left unanswered by the *Vienna Convention* is why treaties are legally binding or how they create law. Their binding nature must come from some underlying body of law, which ultimately cannot be just another treaty. The *Vienna Convention* itself does not claim to provide the legal basis for enforcing other conventions. Just as with national law, the underlying law that gives legal effect to international conventions is customary international law, (Jia 2010). In fact, the *Vienna Convention* itself has now been widely recognized as binding, at least for most of its provisions, as customary law on nations that have not ratified it (Brownlie 2003: 579-580). Unlike national legal systems, however, customary international law is not merely a deep background to conventions. It is a direct source of legal obligation and as the underlying matrix in which international conventions are imbedded.

Customary International Law

Customary international law is more complex and uncertain than formal international conventions. Customary international law consists of the practices of states undertaken out of a sense that the practice is required by law (*opinio juris sive necessitatus,* often simply stated as *opinio juris*) (Wolfke 1993). If the two elements combine, law results regardless of how long, or how briefly, the practice continued. This is similar to the standards for finding a binding custom in national law. Customary law is binding because the participating states have expressly or implicitly consented to the rule (Wolfke 1993: 160-167). References to law connect a customary practice to a sense of legitimacy, and thus constitute the practice as law in a highly decentralized and institutionally undeveloped system like international law (or among subsistence farmers or nomadic tribesmen).

A modernized analogy first suggested some 90 years ago by Pitt Cobbett makes clear the process by which customary international law develops (Cobbett 1922: 1-5). Suppose, as a result of global climate change, people settle a newly thawed island in two villages. A meadow emerges between the two villages, but there is no road. People initially wander at will from one village to the other. Gradually, most people come to follow a particular path. Perhaps it is the shortest route, or perhaps it is the easiest route, or perhaps it is the route most convenient to the heaviest walkers—walkers whose tread wears a path more decisively into the land. As a definite path emerges, it gradually becomes a road. Eventually, everyone agrees that this road is the right way to travel from village to village. When, at some point, people begin to object that those who follow other paths are trespassers, we have a legal and not merely a factual claim. If that claim is accepted on the island, we have a customary rule of law, even if no one can say precisely when this became the law.

Customary international law emerges in an analogous process, developing through a process of claim and counterclaim between states. When a state undertakes an action affecting other states, other states either acquiesce in the action or take steps to oppose it, usually at first through rhetorical strategies; if the matter is important enough, an objecting state will eventually escalate its opposition by imposing sanctions up to the possibility of military operations. Often the states will reach a consensus, expressed through an exchange of diplomatic notes or otherwise, about what each state is entitled to do in the circumstances. Even without a consensus, a pattern of practice will emerge that allows prediction of how states will behave. If nothing more were involved, one might question whether this were anything that could properly be called law, yet beginning with the simplest rhetorical strategies and continuing through to outright war, states involved in a controversy refer to international law as a primary justification of their claims and their practices (Byers 20006). Diplomats know very well the difference between appeals to law, appeals to morality, and appeals to expedience; they often express these different propositions at appropriate points in their statements and assertions.

Customary international law works satisfactorily if there are only a few participants in a particular international process (a regional or special custom) or when general customary international law operates without major controversy, either because there is a broad consensus on what is proper or because states are unwilling to challenge the few states with a strong interest in the matter (Koh 1997). Yet to determine whether customary international law exists and what its content is requires examination of a wide variety of sources on state practice; finding evidence regarding the reasons for the practice often is even more challenging. A widespread pattern of treaties or other international agreements may demonstrate that a practice is so widely followed as law that it has become a customary rule binding even on states not a party to the treaty (Wolfke 1993: 68-72). Under some circumstances, even an unratified treaty is indicative of customary law. Resolutions of the General Assembly and of other international organizations can be evidence that states consider a rule to be a legal obligation, leaving the question of whether state practice is consistent with it, although diplomats, lawyers, scholars, and even international tribunals often overlook that question (Wolfke 1993: 79-84, 100-104). Even unilateral acts of states can demonstrate that the particular state embraces a customary rule of law (Rubin 1977).

The process of determining customary international law, even when successful, is "inelegant" (Janis 2008: 56-57). It often leaves gaps and ambiguities in the law. Treaties and other international agreements only sometimes fill these gaps or clarify the ambiguities. Despite the obvious difficulties in determining the precise content of customary international law, the system has been remarkably successful. No form of international life could exist without a shared set of largely self-effectuating norms (Rittberger & Mayer 1993). Only by focusing exclusively on the relatively few, albeit highly dramatic, situations where international law has failed can one gain the impression that the system is ineffective.

Successful areas of customary law often are codified under United Nations or other auspices, a process made possible precisely because the rules are so seldom questioned and are so generally followed. Even when a body of customary law has been codified, however, parts (even a great deal) of it often survive as customary law. An example is the virtual outlawry of chemical weapons despite the inability of the international community to ratify a treaty dealing with more than a small part of that concern (Koplow 1990).

Customary international law empowers international actors by legitimating their claims, but it also constrains them by limiting the claims they can make (Norman & Trachtman 2005). Customary international law is, in some respects, ill-fitted to perform its functions because frequently it is ill-defined and uncertain. These are characteristics of all customary law, and not just customary international law. Identifying when a practice has crystallized as customary law and the precise content of a customary rule is difficult, requiring research into what often are obscure sources. Turning as it does on a question of motive, examination of the primary evidence for a customary rule is often inconclusive. That, and the lack of a neutral enforcement mechanism, makes exclusive reliance on customary international law both undesirable and impossible.

General Principles of Law

International agreements cover a growing area of international legal concerns, while customary international law covers much of the rest. Neither body of law, however, is comprehensive, leaving gaps that must be filled through recourse to "general principles of law recognized by civilized nations" (Schachter 1991: 50-55). At one time the phrase "civilized nations" was meant to limit reference to nations within the European cultural sphere. Today, the source of general principles of law could be more aptly described as those found in "representative legal systems," which in turn are drawn from all corners of the globe. Oscar Schachter has developed the theory of general principles further, suggesting that there are five types of general principles applied by international decision makers: (1) principles of domestic law "recognized by civilized nations"; (2) general principles of law "derived from the specific nature of the international community"; (3) principles "intrinsic to the idea of law and basic to all legal systems"; (4) principles "valid through all kinds of societies in relationships of hierarchy and co-ordination"; and (5) principles of justice founded on "the very nature of man as a rational and social being."

Recourse to general principles of law reflects a conclusion that if all (or nearly all) national legal systems embrace a point of law, nations can hardly complain if that point of law is applied in their relations with other nations. General principles could almost be seen as special sort of custom, with the incorporation of the legal principle into national law as state practice and its use as implicit consent to the validity of the legal principle. Yet not all of Schachter's schematic of "general principles" can be easily linked back to some sort of implicit consent and therefore some commentators describe "general principles" as a form of "nonconsensual" international law (Janis 2008: 58-62)—something of an oxymoron for classical international law. Some commentators have gone so far as to assimilate the concept of "general principles" to "customary international law" (Tunkin 1974: 200), which would render the concept of general principles of law superfluous.

Demonstrating that a general principle of law exists requires a comparative inquiry into the legal practices of representative legal systems (no one could claim to have examined all legal systems), and no inquiry at all into whether a state has actually intended or consented to having that principle applied in their interstate relations. Such consent as one might conclude applies is strictly implicit or even assumed. While general principles of law can be useful for filling gaps in international law, their utility is limited. The wider the range of legal systems examined, the less specific agreement among them on legal principles is likely to be (Janis 2008: 60-61). Thus, an arbitral tribunal held that limitations on the timeliness of the presentation of a claim applied between states even without proof of an agreement or custom because the temporal limitations were a general principle of law. Some scholars have attempted to identify "global administrative law principles" by recourse to general principles of administrative law in national legal systems (Wuertenberg & Karacz 2009). When state practice internationally is not consistent with even clearly identified general principles applied in national law, presumably state practice (customary law, perhaps) prevails over the general principles.

The Work of the Most Highly Qualified Publicists

International legal processes, particularly regarding determining whether there is a binding custom or general principle of law, requires extensive surveys of state practice, national laws, and the reasons underlying the practices and national laws found in sources are in scattered and obscure sources and difficult to interpret. Legal advocates and tribunals seldom have the time or resources to conduct thorough surveys, and for advocates any such survey is likely to be suspect as self-serving. The international legal system turns to the work of leading scholars of international law (the "most highly qualified publicists" as the *Statute* of the court phrases it) for evidence of what the law is, as opposed to what they think the law should be. The distinction between what the law and what it should be can be a fine one to say the least.

This possibility opens up two interesting aspects for finding secondary sources of international law: recourse to international or national tribunals; and private organizations for the study of international legal issues. International law does not have a formal rule of precedent. Nonetheless, judges or arbitrators generally are experts on international law so that their opinions are the "opinions of highly qualified publicists." Other courts, diplomats, and scholars frequently refer to such opinions of evidence of what international law is (Romano 2009: 763-764). These decisions do not make law or bind other courts, or even the same court. Given the prestige of the particular members of a court, the opinion might, however, carry great weight in determining whether there is in fact a customary rule of law or an applicable general principle.

Another particularly influential form of expert opinion is a report or "codification" of one or another of the international associations of legal experts that have flourished since the nineteenth century. Leading examples are *l'Institut de droit international* and the International Law Association. While these groups have no official standing as lawgivers, the importance of the opinions of the "most highly qualified publicists" gives them an importance that would be remarkable for a similar group in a national legal system. Their opinions represent the work of members of considerable stature who worked on the projects and approval of the end result carries the imprimatur of a large and diverse body of expert opinion.

Proceeding *ex Aequo et Bono*

The second paragraph of article 38 of the *Statute* of the court provides that, if the parties to a dispute agree, the court can decide a case *ex aequo et bono* ("according to what is fair and good") (Trakman 2008). This is an invitation to decide according to the decision maker's personal notion of justice, that is, on non-legal grounds of fairness. Such an invitation is quite different from recourse to "equity" or "equitable" as the standard under an international rule of law. "Equity" or "equitable" under a rule of law is a legal standard for which, over time, specific criteria develop and must be applied much as any other rule of law might be applied (Janis 2008: 71-82; Schachter 1991: 55-61). Because of the freedom accorded to decision makers, deciding *ex aequo et bono* always requires express consent by the state parties to the dispute. Parties have rarely conferred such a power on an international tribunal, and never on the International Court of Justice.

Summary

Even when a norm of customary international law has been determined with some certainty, the customary form of enforcement—claim and counterclaim among states—does not provide a neutral enforcement mechanism. There is no mandatory recourse to a tribunal. Without a neutral enforcement mechanism, there is always the suspicion that national interest overrides a real commitment to law. Without a neutral enforcement mechanism, international law ultimately has nothing better to offer for violations than the law of the vendetta, which international lawyers call *rertorsion.* The institutional limitations of international law have always been most clear during periods of major crisis. This is true as well for treaties if the treaty does provide an adequate enforcement mechanism. A related problem for treaties as they proliferate is the creation of overlapping and not entirely consistent legal regimes that cause conflicts and disputes about which treaty regime is to take precedence (Study Committee 2006). International law is not an illusion, but it is a primitive system with definite limits on its effectiveness (Dinstein 1986). As a result, while international law by itself cannot solve the world's problems, international law is an essential element of any solution (Franck 1990). What is necessary to make international law more effective is a fully developed institutional framework, although such a framework is not likely unless there is a severe crisis. To get beyond the limitations of current international law, states must combine the sophisticated insights of international lawyers with the practical structures of political actors through institutions for managing or resolving conflicts before they escalates to injurious levels.

References

Alkoby, Asher 2008. "Theories of Compliance with International Law and the Challenge of Cultural Difference," *Journal of International Law & International Relations* 4:151-198.

Alvarez, José E. 2005. *International Organizations as Law-Makers,* Oxford University Press: Oxford, UK.

Austin, John 1954 (H.L.A. Hart ed., orig. date 1861).*The Province of Jurisprudence Determined*, Noonday Press, New York, NY.

Baker, Roozbeh (Rudy) B. 2010. "Customary International Law in the 21st Century: Old Challenges and New Debates," *European Journal of International Law* 21:173-204.

Berman, Harold J. & John Witte, jr. 1989. "The Transformation of Western Legal Philosophy in Lutheran Germany," *Southern California Law Review* 62:1573-1660.

Boyle, Alan & Christine Chinkin 2007. *The Making of International Law,* Oxford University Press: Oxford, UK.

Brownlie, Ian 2003. *Principles of Public International Law* (6th ed.), Oxford University Press: Oxford, UK.

Byers, Michael (ed.) 2000. *The Role of Law in International Politics: Essay in International Relations and International Law,* Oxford University Press: Oxford, UK.

Clapham, Andrew 2010. "The Role of the Individual in International Law," *European Journal of International Law* 21:25-30.

Cobbett, Pitt 1922. *Leading Cases on International Law,* Sweet & Maxwell, Ltd.: London, UK.

Dellapenna, Joseph W. 2000. "Law in a Shrinking World: The Interaction of Science and Technology with International Law," *Kentucky Law*

Journal 88:809-883.

—. 2010. "Behind the Red Curtain: Environmental Concern in the Fall of Communism," in J. Ron Engel, Laura Westra, and Klaus Bosselmann (eds.) 2010, *Democracy, Ecological Integrity, and International Law* ch. 4, Cambridge Scholars Publishing: Newcastle-on-Tyne, UK.

Dinstein, Yoram 1986. "International Law as a Primitive Legal System," *New York University Journal of International Law & Politics* 19:1-32.

Dupuy, Pierre-Marie & Luisa Vierucci 2008. *NGOs in International Law,* E. Elgar: Cheltenham, UK.

Estlund, David M. 2008. *Democratic Authority: A Philosophical Framework,* Princeton University Press: Princeton, NJ.

Franck, Thomas M. 1990. *The Power of Legitimacy among Nations,* Oxford University Press: New York, NY.

Geertz, Clifford 1983. *Local Knowledge: Further Essays in Interpretive Anthropology,* Basic Books: New York, NY.

Goodhart, Arthur L. 1953. *Law and the Moral Law,* Stevens: London, UK.

Hart, H.L.A. 1961. *The Concept of Law,* Clarendon Press: Oxford, UK.

Jackson, George 1972. *Blood in My Eye,* Random House: New York, NY.

Janis, Mark Weston 2008. *International Law* (5th ed.), Aspen Publishers: New York, NY.

Jia, Bing Bing 2010. "The Relations between Treaties and Custom," *Chinese Journal of International Law* 9:81-109.

Kahn-Freund, Otto, Claudine Lévy, & Bernard Rudden 1979. *A Source-Book on French Law* (2nd ed.), Clarendon Press: Oxford, UK.

Kelsen, Hans (Bonnie Litschewski Paulson & Stanley Paulson trans.) 1992. *An Introduction to the Problems of Legal Theory*, Clarendon Press: Oxford, UK.

Koh, Harold Hongju 1997. "Why Do Nations Obey International Law?," *Yale Law Journal* 106:2599-2659.

Koplow, David A. 1990. "Long Arms and Chemical Arms: Extraterritoriality and the Draft Chemical Weapons Convention," *Yale Journal of International Law* 15:1-76.

Lipson, Charles 1991. "Why Are some International Agreements Informal?," *International Organizations* 45:495-537.

Lukashuk, I.I. 1989. "The Principle Pact Sunt Servanda and the Nature of Obligation under International Law," *American Journal of International Law* 83:513-518.

Nader, Laura 1990. *Harmony Ideology: Justice and Control in a Zapotec Mountain Village,* Stanford University Press: Stanford, CA.

Norman, George & Joel P. Trachtman 2005. "The Customary International Law Game," *American Journal of International Law* 99:541-580.

Przeworski, Adam 2010. *Democracy and the Limits of Self-Government,* Cambridge University Press: Cambridge, UK.

Rittberger, Volker & Peter Mayer (eds.) 1993. *Regime Theory and International Relations,* Clarendon Press: Oxford, UK.

Romano, Cesare P.R. 2009. 'Deciphering the Grammar of the International Jurisprudential Dialogue," *New York university Journal of International Law & Politics* 41:755-787.

Rubin, Alfred P. 1977. "The International Legal Effects of Unilateral Declarations," *American Journal of International Law* 71:1-30.

Schachter, Oscar 1991. *International Law in Theory and Practice,* M. Nijhoff: Dordrecht, Neth.

Study Comm. of Int'l L. Comm'n 2006. "Fragmentation of International Law: Difficulties Arising from the Diversification and Expansion of International Law," UN Doc. A/CN.4/L.862 (2006).

Trakman, Leon 2008. "*Ex Aequo et Bono:* Demystifying an Ancient Concept," *Chicago Journal of International Law* 8:621-642.

Trubek, David M., Yves Dezalay, Ruth Buchanon, & John R. Davis 1994. "Global Restructuring and the Law: Studies of the Internationaliztion of Legal Fields and the Creation of Transnational Arenas," *Case Western Reserve Law Review* 44:407-498.

Tunkin, G.I. (William Butler trans.) 1974. *Theory of International Law,* Harvard University Press: Cambrige, MA.

Vagts, Detlev T. 2004. "*Rebus* Revisited: Changed Circumstances in Treaty Law," *Columbia Journal of Transnational Law* 43:459-476.

Villalpando, Santiago 2010. "The Legal Dimension of the International Community: How Community Interests Are Protected in International Law," *European Journal of International Law* 21:387-419.

Weeramantry, C.G. 1988. *Islamic Jurisprudence: An International Perspective.* St. Martin's Press: New York, NY.

Wolfke, Karol 1993. *Custom in Present International Law* (2nd rev. ed.), M. Nijhoff: Dordrecht, Neth.

Wuertenberger, Thomas D. & Maximilian C. Karacz 2009. "Using an Evaluative Comparative Law Analysis to Develop Global Administrative Law Principles," *Michigan State Journal of International Law* 17:567-599.

Xue, Hanqin 2007. "Chinese Observations on International Law," *Chinese Journal of International Law* 6:83-93.

CHAPTER TEN

ECOLOGICAL INTEGRITY AND THIRD WORLD APPROACHES TO INTERNATIONAL LAW

SARA L. SECK

Introduction

International legal scholars who see within international law the potential to address global ecological integrity issues often draw upon the natural law tradition as an analytical tool. This chapter will explore whether another analytical tool – Third World Approaches to International Law, or TWAIL – could enrich a natural law approach to international legal analysis of eco-integrity problems. In particular, this chapter will demonstrate how TWAIL can serve to draw attention to the importance of prioritizing the perspectives of subaltern communities who tread lightly upon the earth, and at the same time explicitly highlight the problematic nature of over-consumption by the rich.

International law divides global ecological issues into different categories of harm, depending upon the spatial dimensions of the problem. This paper will focus on ecological integrity issues that arise in the transnational harm context, that is, in cases where despite the "activity and physical damage" all occurring within a single host state, there is a clear "transnational involvement" due to the export of capital from a state of origin or home state.[1] A focus on transnational harm is useful in part because proposals put forward in home states to regulate transnational corporate conduct for compliance with international human rights and environmental norms are often met with claims this would be a neo-colonialist or imperialist violation of host state sovereignty. These claims can only be met by exploring what Third World states – and Third World legal scholars – would have to say in response.

This chapter will first examine the different ways in which international law categorizes global ecological harms, and then provide further detail on the problem of transnational harm. Secondly, TWAIL will be situated within a selection of theoretical or methodological approaches invoked in the analysis of global ecological problems under international law. Finally, the chapter will elaborate the contributions that TWAIL might make when seeking solutions to global ecological integrity problems in the transnational harm context.

Ecological Problems within International Law

States, usually thought of as the primary participants in the international legal system, are spatially defined, with state sovereignty equated with exclusive control over defined territory.[2] As a consequence, a spatial perspective is important for an international legal analysis of global ecological problems. This is despite the fact that neither ecosystems, nor the global economy, respect state borders. International environmental law provides spatial categories that assist in understanding how states may exercise jurisdiction in order to address ecological problems.

The simplest category of ecological harm is local or domestic harm, that is, harm that is limited to within state territorial boundaries. It has been described by Klaus Bosselmann as "intraterritorial" pollution that arises where an impacted ecosystem is entirely within the boundaries of the state.[3] Unless a state has agreed to accept environmental obligations for its territory under a specific treaty, international environmental law "knows no obligation not to pollute the 'national' environment, nor is there an obligation to protect it."[4] Indeed, the first clause of Principle 21 of the *Stockholm Declaration*,[5] restated as Principle 2 of the *Rio Declaration*,[6] clearly affirms this freedom: "States have, in accordance with the Charter of the United Nations and the principles of international law, the sovereign right to exploit their own resources pursuant to their own environmental and developmental policies …".

A second category of ecological harm is transboundary harm, defined by Xue Hanqin as "border-crossing damage via land, water or air in dyadic State relations"[7] and by Bosselmann as "pollution that originates (wholly or in part) within the area under the jurisdiction of one country and which has effects in the areas under the national jurisdiction of another country".[8] Common examples of transboundary or transfrontier harms are where pollutants enter a river or lake shared by more than one state, or air pollution crosses state borders. As the second clause of *Rio Declaration* Principle 2 indicates, obligations do arise under international law environmental law in relation to transboundary harm, as well as common areas harm as discussed below: "… [States have] the responsibility to ensure that activities within their jurisdiction or control do not cause damage to the environment of other States or of areas beyond the limits of national jurisdiction."[9]

A third category of ecological harm recognized under international environmental law is harm to common areas, described by Xue as global commons harm and understood to impact areas "located beyond national jurisdiction and control", that is, beyond any state's territory.[10] Bosselmann prefers to leave "global" for another category (described below), and includes within common areas harm those spaces that international law recognizes as humankind's common heritage (seabed), common concern (Antarctica) or shared resources (high seas, airspace).[11]

"Global harm" is a fourth type of ecological harm proposed by Bosselmann, but he detaches this category from state-centric understandings of international law. According to Bosselmann, global harm is neither confined to national jurisdiction nor to areas beyond national jurisdictions, but includes both and "embraces the earth as a whole."[12] For Bosselmann, the inclusive civil society initiated *Earth Charter* presents an "ethical framework for sustainable development" that sees "the global environment and global civil society – not states – as referential points for rights and duties."[13]

A fifth spatial category of ecological harm, which will be the focus of this chapter, is acknowledged by Xue but not by Bosselmann. Xue, citing Ballarino, defines transnational harm as arising where the "activity and physical damage all occur within one country," nevertheless there is clearly "a transnational involvement" as:

> "capital (including technological know-how) has been exported from another country in order to make possible the activity which has caused environmental damage and, presumably, any profits realized from such exported capital will be returned in one way or another to its country of origin."[14]

International law distinguishes between transboundary and transnational harm, such that the obligations flowing from Principle 21 are not seen as relevant to cases of transnational harm. However, according to Xue, "[a]t a time when transnational corporations are more and more inclined to move their business to developing countries", in part to take advantage of lax environmental regulations, this distinction "is not reflective of reality."[15]

It is important to recognize that other categories of ecological harms could arise simultaneously with transnational harm, and also that transnational harms may be multi-jurisdictional. For example, a mining development in a host state (A) may be financed by equity capital raised on a stock exchange in a second state (B) and the mine may release toxins into an river that flows through state (C). This example would raise both transnational and transboundary ecological harm issues. A multi-jurisdictional example of transnational harm would be where the mining development is operated by a joint venture of companies incorporated in states A, B, and D.

Transnational Harm and the Extraterritoriality Problem

A topical Canadian example of a transnational ecological harm problem is that of Canadian mining companies operating internationally.[16] Canada is the largest source of equity financing for global mining in the world, and more mining companies are based in Canada than anywhere else.[17] While mining companies may at times be implicated in large-scale environmental disasters after operations have begun,[18] a more frequent problem is fierce opposition to a proposed project by local, often indigenous communities.[19] In many cases, those who oppose the mine do so out of fear that if the project is allowed to go ahead, the local environment will suffer irreparable harm. As they are reliant upon the land for their health and livelihoods, threats to local ecological integrity are of paramount importance to these communities. Despite the fact that companies have often obtained the necessary permits from various levels of government, these communities protest the arrival of the company, leading to violence and in some cases even to the death of community activists.[20]

It is often assumed by those who view global mining as a route to development for poor countries that these types of protests are designed to ensure that local communities receive a share (or larger share) of the profits and benefits that mining will bring. Accordingly, the solution most often proposed is that of consultations between communities and companies, leading ideally to the signing of impact and benefit agreements. However, sometimes the issue is not one of shared profits but of irreparable ecological harm, which, in the eyes of community members, has no price. As a result, the only process that could address community concerns would be one that sought their consent. Yet, even the existence of potential ecological harm is contested by companies and governments that point to scientific environmental impact studies which conclude there are no serious threats of significant environmental harm, while at the same time disregarding local and traditional ecological knowledge as a valid source of information.[21] The question then becomes whose perception of ecological reality counts, and, from this, who gets to decide on the appropriate level of precaution?

Various local, international, and transnational forums have been called upon to resolve these types of disputes, often with limited success.[22] One possibility that is increasingly raised is whether home states like Canada should play a role in regulating and adjudicating claims against transnational corporate actors like mining companies. Importantly, home states provide many potential points of control, including corporate laws, stock exchanges, export credit agencies, as well as prescriptive environmental laws which could be structured to give local communities the opportunity to voice their concerns over proposed projects, seeking to prevent or remedy ecological harm. Yet transnational home state mechanisms are often said to require the exercise of extraterritorial jurisdiction, which, in the human rights context, is frequently equated with a violation of host state sovereignty unless exercised reasonably.[23] Even those who acknowledge the permissibility of home state regulation and adjudication of transnational corporate conduct do not often accept that this could be an obligation under international law.

The key question, then, is whether according to an analysis of international law, home state regulation is a neo-colonialist infringement of host state sovereignty, or, alternately, whether it is a permissible if not mandatory exercise of jurisdiction that protects ecological integrity of concern to vulnerable local communities? The premise of the chapter is that there is an urgent need to respect the wishes of those who wish to continue to tread lightly upon the earth. This alone would not be enough to protect global ecological integrity, which will require the reversal of global over-consumption patterns of the rich; but when vulnerable communities who are sensitive to ecological limits wish to preserve their local ecology, it is critical that all possible layers of governance are sensitive to their call, including potential points of control within home state jurisdiction.

International Law Theory and Methodology

It is clear that a positivist analysis of the rules of international law applicable to cases of transnational harm suggest that home state mechanisms are inherently problematic, absent host state consent, at least unless the exercise of home state jurisdiction is "reasonable". This is due to the effect of the doctrine of sovereign equality coupled with the doctrine of non-interference in territorial affairs of other states, and reflected in the recognition of the sovereign rights of states to exploit natural resources in accordance with host state environmental and developmental policies as discussed above. The precise scope of what would be considered "reasonable" is the subject of much discussion at present, particularly in the business and human rights context.[24] Yet there is also increasing international legal analysis that suggests that, at least as far as social, economic and cultural rights are concerned, these obligations extend extraterritorially and include an obligation to regulate non-state actors like corporations.[25] For the purpose of this chapter, the key question is whether invoking a different international legal theory, methodology, or approach leads to a different, or at least more nuanced conclusion on the extraterritoriality question.[26]

Scholars such as Laura Westra have explicitly drawn upon natural law to analyze international law and ecological integrity. For example, in Westra's work on environmental justice and the rights of indigenous peoples,[27] she first proposes a biological/ecological integrity model of First Nations' (indigenous) rights as the best possible antidote against "eco-footprint crime."[28] Westra then argues from the natural law tradition in favour of *erga omnes* obligations for the defence of indigenous peoples' rights. This argument includes an explicit recognition of, in Westra's own words, the "wrong turn" taken by Emmerich de Vattel in 1872 in applying natural law to states, resulting in a "total misunderstanding of both natural law and of the concepts of individuals and communities within it."[29] Westra supports her argument by drawing upon Justice Weeramantry's well known separate opinion in the *Gabcikovo-Nagymorous* case,[30] and Immanuel Kant's cosmopolitanism.[31] Westra's reliance upon the natural law tradition to argue in favour of recognition of *erga omnes* obligations is thus supportive of home state obligations to regulate and adjudicate to protect ecological integrity concerns of indigenous peoples.

Reliance upon natural law theory is also implicit in Klaus Bosselmann's environmental governance work proposing a new approach to territorial sovereignty.[32] He notes with regard to intraterritorial, transboundary and common concern environmental spaces that: "environmental limitations (to state sovereignty under international law) do not protect the environment per se, but territorial interests in the environment"; furthermore, "the degree of environmental limitations depends on the space affected by pollution"; and finally, "territorial sovereignty is not even restricted for the gravest forms of pollution."[33] Accordingly, he then proposes a concept of environmental governance placed within state sovereignty, and defined in two steps.[34] First: "just as private property cannot be defined without its social dimensions, states' sovereignty cannot be defined without its international dimensions. Both are neither absolute, nor independent from the system they are operating in."[35] Second, Bosselmann proposes that we "add the ecological dimension to concepts such as property and sovereignty, respectively. Just as property is determined by ecological realities, so is sovereignty."[36]

Bosselmann's proposal can also be applied to the problem of home states and transnational harm. He suggests that the new "sovereign right of states to sustainably use their resources pursuant to environmental governance" applies to extraction of mineral resources, a legitimate exercise of territorial sovereignty, provided that the associated environmental effects do not violate ecological sustainability.[37] He explicitly recognizes that if the trusteeship role he is proposing is derived from the community of states (a positivist approach), then it would be limited to environmental components of "global importance or global commons". However, deriving the trusteeship role from the environment per se (implicitly a natural law approach) would include the entire environment, not just "globally important 'components'."[38] His analysis would thus suggest that when host states extract minerals they must do so in a way that does not violate ecological sustainability, and by extension that other states, and home state actors like transnational corporations must not assist host states in the violation of their duties.

While arguments drawn from natural law theory seem persuasive, scholars such as Upendra Baxi have critiqued natural law-based arguments for human rights, a point that Westra notes in her work.[39] This concern, mirrored in the work of other TWAIL scholars, is largely due to the way in which natural law was used by colonial powers to suppress the uncivilized "other". When the troubling history of natural law is combined in the transnational harm context with opposition to extraterritorial regulation that is rooted in claims of its imperialist or neo-colonialist nature, it is clear that at least for transnational harm problems, a natural law analysis must be supplemented by another analytic tool.

Before turning to TWAIL, however, another method of international legal analysis must be addressed in order to overcome the pessimistic possibility that international law in fact offers no emancipatory potential for host state local communities who do not wish to engage in externally driven mining development. This would be conclusion of at least some Marxist scholars. For example, China Miéville, perhaps the most pessimistic international law scholar today, carefully distinguishes three forms of "denial" about international law: those who do not believe that international law is really law; those who do not believe that international law is ultimately determinative of state policies (it is not effective); and those (like himself) who are skeptical about whether international law can be used to systematically improve the world (it is effective, and cannot further a just world order).[40]

According to Miéville, the social content of international law is found in the struggle among capitalist states for domination over the rest of the world in order to provide resources for capital. Imperialism "outlasts the transition to universalized juridical sovereignty … not because postcolonial sovereignty is incomplete," but because "the power dynamics of political imperialism are embedded within the very juridical equality of sovereignty." He concludes that he sees "no prospect of any systematic progressive political project or emancipatory dynamic coming out of international law," as "the very social problems which liberal cosmopolitan writers want to end are the result of the international system, which is the international legal system." Indeed, any attempts "to reform law can only ever tinker with the surface of institutions" for a world that is "structured around international law cannot but be one of imperialist violence."[41]

Third World Approaches to International Law

Many TWAIL scholars engage with Marxist or commodity form or class analysis of international law.[42] However, rather than rejecting the emancipatory potential of international law, TWAIL scholars seek to make the people of the Third World the ultimate decision makers when identifying and interpreting international legal rules.[43]

TWAIL has been described as "a broad dialectic (or large umbrella) of opposition to the generally unequal, unfair, and unjust character of an international legal regime that all-too often (but not always) helps subject the Third World to domination, subordination, and serious disadvantage."[44] According to Obiora Okafor, TWAIL offers both theories of, and methodologies for, analyzing international law and institutions, and is most usefully thought of as a broad approach.[45] TWAIL offers various models or frameworks for describing the behaviour of a related set of social phenomena, including Antony Anghie's work on colonial origins of international law, and Karin Mickelson's scholarship on international environmental law and the third world.[46] However, it is less clear whether TWAIL articulates a unified or general framework for analyzing, critiquing and reconstructing international law and institutions. A theoretical school should be self-consistent, however, Okafor argues that it need not be *entirely* self-consistent, in the sense that all members of the school agree on everything.[47] TWAIL theories are connected by an overarching central set of ideas, a broadly shared approach. Thus although diverse, TWAIL scholars are:

> "solidly united by a shared ethical commitment to the intellectual and practical struggle to expose, reform, or even retrench those features of the international legal system that help create or maintain the generally unequal, unfair, or unjust global order...a commitment to centre the *rest* rather than merely the *west*, thereby taking the lives and experiences of those who have self-identified as Third World much more seriously than has generally been the case."[48]

According to Okafor, TWAIL also offers a "body of methods" used in the "activity of international legal analysis". These are TWAIL's methodological insistence:

> "on global (as opposed to merely West-centric) historicisation; on identifying continuities amidst discontinuities that we behold; on centering the Third World (i.e. "the Rest and not merely the West"); on being wary of glib universality narratives; and on focusing on the under-studied resistance of third world peoples."[49]

Together, these methods allow TWAIL scholars to "more effectively write the Third World into international legal history and analysis". Okafor concludes that TWAIL is best defined more broadly, as an 'approach' or 'school of thought', both of which incorporate theory and methodology in their definitions, although with more emphasis on methodology.[50]

De-Territorializing the Third World

Historical Third World scholarship has highlighted the colonial origins of international law, revealing how despite international law's universal claims, it was used to justify, manage and legitimize the subjugation and oppression of Third World peoples. Colonialism was central to the formation of international law, and neo-colonialism continues to be central to the structure of international law today through contemporary initiatives such as the discourse of development that presents Third World peoples as deficient and in need of international intervention.[51] Yet who are the Third World with which TWAIL scholars are concerned?

While some scholars have claimed that the expression "Third World" is no longer a useful analytical category in a post-cold war era, TWAIL scholars reject this.[52] For example, Okafor suggests the significance of the expression is tied to the group of states and populations which self-identify as Third World. They coalesce "around a historical and continuing experience of subordination at the global level that they feel they share – not the existence and validity of an unproblematic monolithic third-world category."[53] The expression therefore needs to be thought of with a more "flexible geographic sensibility."[54] There is then "a sense in which states or societies or even scholars must *choose* whether or not to self-identify as Third World."[55] Karin Mickelson notes that the validity of the term "Third World" has been increasingly called into question given the growing diversity among the various countries that have been identified by the label.[56] Mickelson sees the definition of Third World as not only descriptive but also normative, in the sense that the disadvantage experienced by Third World countries is seen as an intolerable situation that demands a response. This definition, she argues, draws attention to justice posited by social movements.[57] For example, in her work on climate change, Mickelson highlights the existence of a North within the South and a South within the North, citing the indigenous peoples in the North as one of the most vulnerable populations to climate change.[58] For Rajagopal, decentering the Third World from its physical geography will enable a focus on the various levels at which power operates to subjugate, followed by engagement in oppositional practices that challenge those power structures.[59] Dominant discourse would then be compelled to rethink the relationship between the local and the global, for if the Third World is not geopolitically defined, it becomes possible to "think of transnational linkages among the oppressed."[60]

TWAIL and International Environmental Justice

According to Karin Mickelson, the "colonial background of international law is one that international environmental law shares."[61] Mickelson sees the failure of international environmental lawyers to confront the differing perspectives of the South and North as "a central, if not *the* central, debate regarding the conceptual foundation of their discipline."[62] As a result, international law as a discipline has failed to respond to Third World concerns in a meaningful way, relegating these concerns to the margins instead of incorporating them into the core of the discipline.[63] Two reasons for the failure are the tendency to provide an a-historical account of the evolution of international law,[64] and the South's portrayal as a grudging participant, rather than an active partner, in environmental regimes.[65]

Specifically, international environmental law has accommodated Third World concerns at the margins through acceptance of principles such as common but differentiated responsibilities, and the inclusion of technology transfer and financial assistance provisions as inescapable features of international diplomacy. Yet the South has not been recognized as an active partner "in an ongoing effort to identify the fundamental nature of environmental problems and appropriate responses thereto."[66] Instead, there is a tendency to see the developmental aspects of IEL as "concessions" to the Third World, or a political compromise, based upon the assumption that the Third World would "take a stand against the environment," rather than understanding that at the time of such significant events as the Stockholm conference, the Third World was seeking fundamental change and recognition that "environmentalism" itself was/is open to interpretation.[67] Mickelson suggests that while an environmentalism of the rich might have the luxury of valuing the environment for its own sake, quite apart from its value to humans, the environmentalism of the poor "originates as a clash over productive resources," with the result that "issues of ecology are often interlinked with questions of human rights, ethnicity and distributive justice."[68]

Mickelson uses the phrase "South-North" to refer to an alternate way of conceptualizing the relationship between developed and developing nations.[69] She cites the role of the South in the context of hazardous waste regime building as an example of developing countries taking a lead role in attempting to develop effective international regimes, only to meet with the unsatisfactory (and somewhat paradoxical) arguments from developed countries of the importance of freedom of movement to the sovereign rights of receiving states.[70] She explores how the fundamental principle of common but differentiated responsibilities can "reflect totally different ways of thinking about the respective ways of South and North in addressing environmental degradation."[71] The North's perspective reflects the different financial and technological capacities, and the imbalance of the consumption of resources, thus a focus on *ability to pay*. The South's perspective, on the other hand, reflects an understanding of *responsibility to pay*:

> "an acknowledgment of the historic, moral, and legal responsibility of the North to shoulder the burdens of environmental protection, just as it has enjoyed the benefits of economic and industrial development largely unconstrained by environmental concerns. Implicit in the latter view is a sense that the North has received a disproportionate share of the benefits of centuries of environmentally unsustainable development, and the underprivileged in the South have borne many of its costs."[72]

In examining climate change as a "problem of global injustice"[73] Mickelson highlights the resistance to viewing climate change as being (at least in part) about "sustainable development and international equity."[74] She proposes two theories for reconceptualising the relationship between South and North: 'ecological debt' and 'environmental space.'[75] The theory of ecological debt views the North as "owing an enormous amount to the peoples of the South, who have borne many of the costs of environmentally unsustainable development but have reaped few of its benefits."[76] Proponents of this idea suggest that the solution is to bring about a "more equitable and sustainable allocation of global resources" by means of a process called "contraction and convergence."[77] The North would make changes to reduce its environmental impacts. The South would increase development to meet the aspirations of its peoples.[78] In contrast, the key concepts in the theory of 'environmental space' are ecological limits and equity.[79] According to this theory, there are limits to the amount of environmental pressure that the Earth's ecosystems can handle without irreversible damage to the systems, within the context of equal rights to resource consumption and concern for the quality of life for all peoples internationally.[80] The difference between this idea and ecological debt is that environmental space does not take into account historical responsibility for resource depletion.[81]

Both ecological debt and environmental space critique the existing distribution of resources as unfair and demonstrate this unfairness by highlighting the discrepancy in consumption patterns between the North and South.[82] Both theories allocate the burden of resolving international environmental problems to the North. Developed countries must radically shift their current consumption levels, while developing countries must attempt to live within their own environmental space, which will require reassessing current understandings of development.[83]

TWAIL, Ecological Integrity and Transnational Harm

If the theory of ecological debt is applied to the problem of transnational harm, then developing countries should be entitled to engage in development to meet the aspirations of their people, while developed countries should bear the burden of reducing consumption patterns. This would suggest that developed countries have no business regulating to prevent or remedy environmental harm in developing countries. Yet, if the theory of ecological space is applied to the transnational harm problem, then regulation and adjudication of transnational environmental claims could be seen as contributing to the protection of the ecological space of vulnerable communities whose space is at risk of being infringed by the colonialist consumption patterns of the rich. Recognition of ecological debt could then offer support for the creation of forums, including in home states, for the adjudication of past transnational infringements of ecological spaces, and prevention of future infringements.

The unilateral exercise of home state jurisdiction in the human rights and environment realm creates a curious problem from a TWAIL perspective. On the one hand, as I have examined elsewhere,[84] if in reality home states only ever exercise jurisdiction to promote internal economic goals, then unilateral home state regulation, even while claiming to be designed to protect global ecological integrity, is innately problematic as an imperialistic infringement of host state sovereignty. Moreover, if home state regulation designed to prevent and remedy harms were to become routine state practice that contributed to the development of customary international law norms, it could unintentionally serve to reinforce the neo-colonialist tendencies of international law (even if it also supported global ecological integrity at the same time).[85] On the other hand, to the extent that neo-colonial tendencies are already embedded within the structure of international law, the public international law rules of jurisdiction which suggest that extraterritoriality is illicit and a violation of international law could themselves be neo-colonialist. The language of extraterritoriality thus shields home states from pressure to take action to ensure home state TNCs respect the rights of citizens in Third World host states. It also shields the home state from the fear that another home state might take action to protect the human or ecological rights of its own Third World peoples, including indigenous peoples.

Notably, TWAIL scholars who focus on human rights issues often complain that home state courts have been reluctant to exercise "justice jurisdiction" over TNC conduct that has violated the rights of communities within developing countries, while at the same time according protection to developed state investors.[86] Moreover, according to Rajagopal, despite the problematic reliance of human rights discourse upon the state as the primary duty-holder, human rights should not be dismissed. The problem with human rights theory is that it is linked with the colonial origins of the doctrine of sovereignty, for the state is given a predominant role as the source and implementer of the normative framework.[87] Consequently, the "radical democratic potential in human rights" must be sought out, "by paying attention to the pluriverse of human rights, enacted in many counter-hegemonic frames."[88]

Interestingly, as the definition of Third World is de-coupled from territory, a curious parallel emerges with Bosselmann's call for recognition of an international law for the global environment that is defined not by states, but by civil society. Clearly "civil society" is not co-extensive with the "subaltern social movements" with which Rajagopal suggests we should be concerned. Yet Rajagopal's call accords with TWAIL scholar Obijiofor Aginam who draws attention to the need to pay closer attention to local customary ecological norms and practices.[89] Aginam specifically highlights indigenous conservation practices in Africa and South America as examples of the creativity of postcolonial societies in the South, and argues that "global environmentalism must move beyond the narrow confines of state-centric interests to holistically incorporate ecologically sound practices of indigenous societies."[90] Aginam claims that the pressures of globalization, fuelled by the operations of transnational corporations, including oil operations in Nigeria and Ecuador, has led to "an erosion of multicultural ecological values."[91] He proposes a "South-North" ecological dialogue that is "sensitive to the customary enforcement of ecological norms and practices … handed down from past generations in the South" as a means of counteracting the "hegemonic globalism" and "complicity of international law in impoverishing conceptions of sustainable development and in suffocating Third World contributions."[92] Like Westra, Aginam favours the evolution of an "effective framework for *erga omnes* obligation on environmental issues."[93]

TWAIL and natural law are drawn together in Aginam's analysis. This should come as no surprise if recognition is given to the "ecology of knowledges" and holistic versions of natural law rooted in diverse cultures around the world.[94] What might this mean in practice? An analysis that draws upon TWAIL suggests support for home state regulation by overcoming the neo-colonialism embedded within the jurisdictional rules of public international law. In particular, TWAIL supports home state regulation that gives voice to all local communities who wish to protect ecological integrity through resistance to global mining. This would include giving voice to resistance by the South within the North – including indigenous peoples. However, TWAIL appears to necessitate a distinction between regulation that enables host state individuals and local communities to seek redress from ecological harm (and to seek to prevent harm in the first place), and regulation that imposes home state environmental values on communities in host states without their participation, consultation or consent. Notably, when local communities support mining development instead of resisting, it may be that home states should resist the imposition of ecological integrity values against their wish. Even if regulation is ostensibly designed in the interest of global ecological integrity, giving voice to protest by Third World peoples may be essential for legitimacy. Direct application of substantive home state environmental laws thus appears problematic.

But is this necessarily so? And what does it mean to give voice to protest? In litigation that is brought after the fact of environmental harm, the claim is frequently made that environmental standards adopted by a company with the explicit agreement of the host state are inadequate precisely because the standards permit greater environmental harm than would be allowed in the home state. This raises the question of why the default standard should be one that defers to inadequate host state standards, in which disempowered local communities may have had no say, rather than home state standards in which these communities equally had no say, but may be more in keeping with global ecological integrity. Deference to the higher standard is supported by Bosselmann's proposal that sovereignty over natural resources must include within it a commitment not to violate ecological sustainability. Equally, if the standards at issue are informed by substantive international norms, perhaps in the area of environmental health, then it is not clear why the presumption should be that local communities did not wish to be governed by them. Further analysis of international legal theories and methods may be necessary to assist in better understanding the participation of Third World voices in the creation of international norms.

There is a danger that interpreting TWAIL to provide support for substantive extraterritorial regulation might reflect wishful thinking on the part of an international environmental lawyer. Mickelson cautions that environmental lawyers need to "acknowledge that their vision of international environmental law reflects one version of environmentalism."[95] This requires an acknowledgement that you are advocating for a particular interest – the environment. Secondly, international environmental lawyers must "inhabit the same historical reality" as the Third World – thus Maurice Strong's desire for an *Earth Charter* to hang on every child's bedroom wall unintentionally reflects a lack of appreciation for the reality that children in many parts of the world do not have bedrooms.[96] In Mickelson's words: even "the most wonderfully inspiring document in the world will not mean anything as long as there are these terrible disparities between those who have and those who have not."[97]

It would be both hypocritical and imperialist for home states that engage in ongoing destructive consumption patterns to be able to limit the potential consumption of poor communities in developing countries by denying them the possibility of development without their consent. A TWAIL analysis also suggests that the real question we should be asking is whether state-created institutional structures of the global economic order must regulate the TNC conduct that they facilitate so as to give voice to communities protesting the infringement of their environmental space. Only if these voices can be heard loud and clear in the hallways of power might we hope to shame the rich into reducing excessive consumption.

Acknowledgements

I would like to thank Jennifer Buttkus, Aleesha Khan and David Vaughan for valuable research assistance, and the Social Sciences and Humanities Research Council of Canada for financial support in the form of a Standard Research Grant.

Notes

[1] Xue Hanqin, *Transboundary Damage in International Law* (Cambridge, Cambridge University Press, 2003) at 9-10, citing T. Ballarino, "Private International Law Questions and Catastrophic Damage," *Recueil des Cours*, vol. 220 (1990-1), p. 293.
[2] See generally Colin Warbrick, "States and Recognition in International Law" in Malcolm D. Evans, ed., *International Law* 2nd Ed. (Oxford University Press, 2006) 217; Thilo Maraugn, "The Changing Role of the State" in Daniel Bodansky, Jutta Brunnée, & Ellen Hey, eds., *The Oxford Handbook of International Environmental Law* (Oxford University Press, 2007) 727 at 730.
[3] Klaus Bosselmann, "Environmental Governance: A New Approach to Territorial Sovereignty" in Robert J. Goldstein, ed., *Environmental Ethics and Law* (Ashgate Publishing Ltd., 2004) 293 at 299-300.
[4] *Ibid.* at 299. See also Dan Tarlock, "Ecosystems" in Daniel Bodansky, Jutta Brunnée, & Ellen Hey, eds., *The Oxford Handbook of International Environmental Law* (Oxford University Press, 2007) 574 at 582.
[5] U.N. Conference on the Human Environment, 16 June 1972, *Stockholm Declaration*, U.N. Doc. A/CONF.48/14.
[6] U.N. Conference on Environment and Development, 3-14 June 1992, *Rio Declaration,* U.N. Doc. A/CONF.151/26.
[7] Xue, *supra* note 1 at 3.
[8] Bosselmann, *supra* note 3 at 300.
[9] *Rio Declaration, supra* note 6.
[10] Xue, *supra* note 1 at 15.
[11] Bosselmann, *supra* note 3 at 302-303.
[12] *Ibid.* at 303.
[13] *Ibid.* at 304. See further *ibid.* at footnote 91 and accompanying text, noting that the Earth Charter was not designed as a draft covenant, but as a stand-alone 'peoples' charter, to avoid the 'pact with the devil' potential of a treaty solution. For a history of the drafting of the Earth Charter, see Brendan Mackey, "The Earth Charter, Ethics and Global Governance" in Laura Westra, Klaus Bosselmann & Richard Westra, eds., *Reconciling Human Existence with Ecological Integrity* (Earthscan, 2008) at 61.
[14] Xue, *supra* note 1 at 9, citing T. Ballarino, "Private International Law Questions and Catastrophic Damage" *Recueil des Cours*, vol. 220 (1990-1), p. 293.
[15] Xue, *ibid.* at 10.

[16] See generally Sara L. Seck, "Home State Responsibility and Local Communities: The Case of Global Mining" (2008) 11 Yale Hum. Rts. & Dev. L. J. 177; and most recently Bill C-300, *An Act Respecting Corporate Accountability for Mining, Oil and Gas Corporations in Developing Countries*, online: www2.parl.gc.ca/HousePublications/Publication.aspx?Docid=3658424&file=4.

[17] Government of Canada, *Building the Canadian Advantage: A Corporate Social Responsibility (CSR) Strategy for the Canadian International Extractive Sector* (March, 2009) at 3, online: www.international.gc.ca/trade-agreements-accords-commerciaux/assets/pdfs/CSR-March2009.pdf.

[18] See for example the *Cambior* case as discussed in Sara L. Seck, "Environmental Harm in Developing Countries Caused by Subsidiaries of Canadian Mining Companies: the Interface of Public and Private International Law" (1999) 37 Can. Y.B. Int'l L. 139 at 154-168.

[19] See for example Amanda M. Fulmer, Angelina Snodgrass Godoy & Philip Neff, "Indigenous Rights, Resistance, and the Law: Lessons from a Guatemalan Mine" (2008) 50:4 Latin American Politics and Society 91.

[20] *Ibid.* See also Brant McGee, "The Community Referendum: Participatory Democracy and the Right to Free, Prior and Informed Consent to Development" (2009) 27 Berkeley J. Int'l L. 570; Lisa J.Laplante & Suzanne A. Spears, "Out of the Conflict Zone: The Case for Community Consent Processes in the Extractive Sector" (2008) 11 Yale Hum. Rts. & Dev. L. J. 69.

[21] Deborah McGregor, "Linking Traditional Knowledge and Environmental Practice in Ontario" (2009) 43 J. Can. Stud. 3; Sheila D. Collins, "Interrogating and Reconceptualizing Natural Law to Protect the Integrity of the Earth" in J. Ronald Engel, Laura Westra & Klaus Bosselmann, eds., *Democracy, Ecological Integrity and International Law* (Cambridge Scholars Publishing, 2010) 445.

[22] See for example Fulmer *et al., supra* note 19; Adam McBeth, "Crushed by an Anvil: A Case Study on Responsibility for Human Rights in the Extractive Sector" (2008) 11 Yale Hum. Rts. & Dev. L. J. 127; Natalie L. Bridgeman & David B. Hunter, "Narrowing the Accountability Gap: Toward a New Foreign Investor Accountability Mechanism" (2008) 20 Geo. Int'l Envtl. L. Rev. 187.

[23] See discussion in Sara L. Seck, "Conceptualizing the Home State Duty to Protect Human Rights" in Karin Bhuman, Mette Morsing & Lynn Roseberry, eds., *Corporate Social and Human Rights Responsibilities: Global Legal and Management Perspectives* (Palgrave Macmillan, forthcoming 2010) 25.

[24] John G. Ruggie, United Nations Human Rights Council, 8th Session, *Promotion and Protection of All Human Rights, Civil, Political, Economic, Social and Cultural Rights, Including the Right to Development: Protect, Respect and Remedy: A Framework for Business and Human Rights*, UN. Doc. A/HRC/8/5, (April 7, 2008): para. 19; John G. Ruggie, United Nations Human Rights Council, 14th Session, *Business and Human Rights: Further Steps Toward the Operationalizing of the "Protect, Respect and Remedy" Framework*, UN. Doc. A/HRC/14/27 (April 9, 2010): paras. 46-49.

[25] F. Coomans and M.T. Kamminga, ed., *Extraterritorial Application of Human Rights Treaties* (Antwerp: Intersentia, 2004); John Knox, "Diagonal Environmental

Rights," in Mark Gibney & Sigrun Skogly, ed., in *Universal Human Rights and Extraterritorial Obligations* (Philadelphia: University of Pennsylvania Press, 2010) 86; Mark Gibney & Sigrun Skogly, eds., *Universal Human Rights and Extraterritorial Obligations* (Philadelphia: University of Pennsylvania Press, 2010); Robert McCorquodale and Penelope Simons, "Responsibility Beyond Borders: State Responsibility for Extraterritorial Violations by Corporations of International Human Rights Law," (2007) 70 Mod. L. Rev. 598.

[26] For the purpose of this chapter, a methodology for the analysis of international law will be defined as: "the application of a conceptual apparatus or framework – a theory of international law – to the concrete problems faced in the international community". A methodology may apply more than one theory of international law. See Anne-Marie Slaughter & Steven R. Ratner, "Appraising the Methods of International Law: A Prospectus for Readers" (2004) 36 Stud. Transnat'l Legal Pol'y 1 at 3-4.

[27] Laura Westra, *Environmental Justice and the Rights of Indigenous Peoples: International & Domestic Legal Perspectives* (Earthscan, 2008).

[28] *Ibid.* at 19.

[29] *Ibid.* at 57.

[30] *Ibid.* at 61-62. See ICJ *Gabcikovo-Nagymaros Dam (Hungary v. Slovakia)*, Judgment September 25, 1997, ICJ Reports 7 (1997), Judge C. Weeramantry Separate Opinion.

[31] Westra, *ibid.* at 63-64.

[32] Bosselmann, *supra* note 3.

[33] *Ibid.* at 305.

[34] *Ibid.* at 305-313.

[35] *Ibid.* at 310.

[36] *Ibid.* at 310.

[37] *Ibid.* at 311.

[38] *Ibid.* at 312. Note that Bosselmann refers here to the "laws of physics".

[39] Westra, *supra* note 27 at 5, citing Upendra Baxi, "Voices of suffering and the future of human rights" (1998) Transnational Law & Contemporary Problems; and at 60, citing Upendra Baxi, "Geographies of injustice: Human rights at the alter of convenience" in Craig Scott, ed., *Torture as Tort* (Oxford: Hart Publishing, 2001).

[40] China Miéville, *Between Equal Rights: A Marxist Theory of International Law* (Leiden: Brill, 2005) at 17, 26-27.

[41] See citations in Sara L. Seck, "Unilateral Home State Regulation: Imperialism or Tool for Subaltern Resistance?" (2008) 46 Osgoode Hall L. J. 565 at 590-591, and subsequent critique of Mieville's analysis. Mieville concludes: "The chaotic and bloody world around us *is the rule of law*."

[42] See e.g. B.S. Chimni, "An Outline of a Marxist Course on Public International Law" (2004) 17 Leiden J. Int'l L. 1; Obiora Chinedu Okafor, "Marxian embraces (and de-couplings) in Updendra Baxi's human rights scholarship: a case study" in Susan Marks, ed., *International Law on the Left: Re-examining Marxist Legacies* (Cambridge University Press, 2008) 252.

[43] Anne-Marie Slaughter & Steven R. Ratner, "The Method is the Message" (2004) 36 Stud. Transnat'l Legal Pol'y 239 at 248-249. As international law provides Third World peoples with no real voice, TWAIL scholars themselves "must imagine or somehow approximate the actual impact of specific rules or practices on their daily lives and define or interpret those rules accordingly."

[44] Obiora Chinedu Okafor, "Newness, Imperialism, and International Legal Reform in Our Time: A TWAIL Perspective" (2005) 43 Osgoode Hall L.J. 171 at 176. See also Makau Mutua, "What is TWAIL?" (2000) 94 Am. Soc'y Int'l L. Proc. 31; Karin Mickelson, "Taking Stock of TWAIL Histories" (2008) 10 Int'l Community L. Rev. 353.

[45] Obiora Okafor, "Critical Third World Approaches to International Law (TWAIL): Theory, Methodology, or Both?" (2008) 10 Int'l Community L. Rev. 371-378.

[46] *Ibid.* at 374.

[47] *Ibid.* at 375. Otherwise, certain theories, such as Marxist or feminist theory, which are comprised of differing strains of thought, could not exist.

[48] *Ibid.* at 376.

[49] *Ibid.* at 377.

[50] *Ibid.* at 377-378.

[51] See generally Antony Anghie, *Sovereignty, Imperialism and the Making of International Law* (Cambridge: Cambridge University Press, 2005).

[52] Okafor, "Newness", *supra* note 44 at 174.

[53] *Ibid.*

[54] *Ibid.* at 175.

[55] *Ibid.*

[56] Karen Mickelson, "Rhetoric and Rage: Third World Voices in International Legal Discourse" (1997-1998) 16 Wis. Int'l L.J. 353-419 at 357.

[57] *Ibid.* at 360.

[58] Karin Mickelson, "Beyond a Politics of the Possible? South-North Relations and Climate Justice" (2009) 10 Melbourne J. Int'l L. 411 at 419.

[59] Balakrishnan Rajagopal, "Counter-hegemonic International Law: rethinking human rights and development as a Third World Strategy", (2006) 27 Third World Q. 767 at 768; Balakrishnon Rajagopal, "Locating the Third World in Cultural Geography" (1998-1999) Third World Legal Studies 1 at 19. See further Rajagopal's comprehensive definition of Third World as: an ideological model; a geopolitical model (the national allegory of the dominant discourse of development); the historical deterministic model ; and the popular representational model (as a set of images). *Ibid.* at 1-19.

[60] *Ibid.* at 20.

[61] Karin Mickelson, "South, North, International Environmental Law, and International Environmental Lawyers" (2000) 11 Y.B. Int'l Envtl. Law 52 at 57-58. See also Karin Mickelson, "Critical Approaches" in Daniel Bodansky, Jutta Brunnée & Ellen Hey, eds., *The Oxford Handbook of International Environmental Law* (New York: Oxford University Press, 2007) 262 at 274; Karin Mickelson, "Beyond a Politics", *supra* note 58.

---

[62] Mickelson, "South, North", *ibid.* at 53.
[63] *Ibid.* at 54.
[64] *Ibid.*
[65] *Ibid.* at 54, 60.
[66] *Ibid.* at 54.
[67] *Ibid.* at 62-66.
[68] *Ibid.* at 65, citing Guha and Martinez-Alier, *Varieties of Environmentalism: Essays North and South* (1997).
[69] Karin Mickelson, "Co-opting Common Heritage: Reflections on the Need for South-North Scholarship" in Obiora Chinedu Okafor & Obijiofor Aginam, eds., *Humanizing Our Global Order: Essays in Honour of Ivan Head* (University of Toronto Press, 2003) 112 at 113. She credits Ivan Head with first using this phrase.
[70] *Ibid.* at 66-67.
[71] *Ibid.* at 70; "Critical Approaches", *supra* note 61 at 274. See also Co-opting, *ibid.* at 115-119 for Mickelson's critique of the "co-option" of the "common heritage of mankind" (humankind) principle by international environmental lawyers.
[72] Mickelson, "South, North", *supra* note 61 at 70.
[73] Mickelson, "Beyond a Politics", *supra* note 61 at 413.
[74] *Ibid.* at 415.
[75] Karin Mickelson, "Leading Towards a Level Playing Field, Repaying Ecological Debt, or Making Environmental Space: Three Stories about International Environmental Cooperation" (2005) 43 Osgoode Hall L.J. 137 at 137.
[76] *Ibid.* at 153.
[77] *Ibid.* at 157.
[78] *Ibid.*
[79] *Ibid.* at 158.
[80] *Ibid.* at 158-159.
[81] *Ibid.* at 160.
[82] *Ibid.* at 162.
[83] *Ibid.* at 166.
[84] Seck, "Unilateral Home", *supra* note 41.
[85] *Ibid.*
[86] Chimni, "An Outline", *supra* note 42 at 20. See generally Baxi, *supra* note 39; Upendra Baxi, "Mass Torts, Multinational Enterprise Liability, and Private International Law," in Vol. 276, *Recueil des Cours* (Hague Academy of International Law, 1999); Muthucumaraswamy Sornarajah, "Linking State Responsibility for Certain Harms Caused by Corporate Nationals Abroad to Civil Recourse in the Legal Systems of Home States" in Craig Scott, ed. *Torture as Tort* (Hart Publishing, 2001) 491.
[87] Balakrishnan Rajagopal, *International Law from Below: Development, Social Movements and Third World Resistance* (Cambridge: Cambridge University Press, 2003) at 186-187. Thus, despite its "nominal anti-sovereignty posture", human rights remains a "state-centred" discourse, and protest or resistance movements inside societies are ignored. *Ibid.* at 187.

[88] Rajagopal, "Counterhegemonic", *supra* note 59 at 768.
[89] Obijiofor Aginam, "Saving the Tortoise, the Turtle and the Terrapin: The Hegemony of Global Environmentalism and the Marginalization of Third World Approaches to Sustainable Development" in Obiora Chinedu Okafor & Obijiofor Aginam, eds., *Humanizing Our Global Order: Essays in Honour of Ivan Head* (University of Toronto Press, 2003) 12.
[90] Aginam, *ibid.* at 14.
[91] *Ibid.* at 25.
[92] *Ibid.*
[93] *Ibid.*
[94] Collins, *supra* note 20. Collins attributes the phrase "ecology of knowledges" to Boaventura de Sousa Santos, "Beyond Abyssal Thinking: From Global Lines to Ecologies of Knowledges," *Eurozine* (18 February, 2008) www.eurozine.com. See also the separate opinion of Judge Weeramantry in the Gabcykovo-Nagymorous case, *supra* note 30.
[95] Mickelson, "South, North", *supra* note 61 at 80.
[96] *Ibid.* at 80.
[97] *Ibid.* at 80.

CHAPTER ELEVEN

GLOBAL CONSTITUTIONALISM AND THE PROSPECT OF A GLOBAL CONSTITUTION

KLAUS BOSSELMANN

The idea of a global constitution, as utopian as it seems, is not new. Immanuel Kant's "Perpetual Peace"[1], for example, outlines principles and articles towards a universally binding code. In his essay, Kant makes a strong case for cultural interdependence, a peaceful world order and the rule of law. In "The Law of World Citizenship", he specifies that goods, ideas and knowledge should be shared across all cultures, but never with the use of force. Essentially, Kant advocates a democratization of states and a concept of world citizenship based on common human values and concerns. When could this idea ever have been more topical than today?

Yet, we live in world of division and diversity. The world is socially, culturally, politically and legally diverse. This makes the idea of a unifying global constitution questionable. However, ecologically the world is one. This makes the idea of unity within diversity plausible. That all people and cultures share one Earth and possibly one destiny is more than a thought. Rather, it describes an ever more daunting reality. And with the reality of a common destiny the need to think about a global legal order becomes more urgent.

Interestingly, over the last few years there has been quite an outburst of literature on global constitutionalism. Written mostly by international lawyers (and only a few environmental lawyers), this literature suggests a degree of harmonious development among the world's national constitutions that would allow the new notions of international constitutional law and global constitutionalism.[2] Broadly speaking, the thesis is that the increasing interdependence of nation states has already created new forms of transnational institutions[3] and necessitates a transnational way to think about institutions that so far have been strictly national. Examples include the institutions of 'democracy' and 'constitution'.

To be sure, the concept of global constitutionalism is one thing and the idea about a global constitution something quite different. While global constitutionalism can be described as a new form international jurisprudence, a global constitution is a pure idea, an idea however worth pursuing given our totally globalized world. With states largely failing in their effort to control global affairs, the idea well stimulate serious legal research.

This chapter summarizes some of the relevant literature to collect a few building-blocks for a future global constitution.[4] These building blocks include the idea of a constitution (I.), some proponents (and critics) of global constitutionalism (II.), the question of an "international environmental constitution" (III.), the "greening" of international law (IV.), the legitimacy of a global constitution (V.), and a quasi-constitutional coherence of various international environmental agreements culminating in the Earth Charter (VI.).

I. What Constitutes a 'Constitution'?

There is some uncertainty in the discourse on global constitutionalism as to the precise meaning of the word 'constitution', in addition to the related terms 'constitutionalism' and 'constitutionalization'. The Oxford dictionary defines a constitution as 'the body of fundamental principles or established precedents according to which a nation or other organisation is acknowledged to be governed'.[5] Daniel Bodansky explains a constitution as: "At the most general level, a higher level of law, typically of an enduring nature, setting forth the fundamental rules of a political community. These rules can be of two types – procedural (establish the basic political institutions of a community and define how other norms are created, interpreted, changed and enforced; what HLA Hart called 'secondary rules' and Bardo Fassbender refers to as 'metarules') and substantive rules (such as human rights protections defining the core values)".[6] The literature largely follows such a distinction between procedural and substantive rules.

Some procedural features of a constitution include:

- Is in written form (not always, but more often than not)[7]
- Supercedes ordinary law[8]
- Is created pouvoir constituant (revolutionary; usually start a new political order)[9]
- Is entrenched – more difficult to change than ordinary law[10]

- Aims primarily to regulate governmental rather than private actors; represents public rather than private law[11]
- Takes precedence in conflicts with other norms[12]
- Is typically of an enduring nature[13]
- Is framed in a general way to allow a considerable amount of flexibility in order to avoid becoming outdated.[14]

Not all of these features are required for a typical constitution. Bodansky encapsulates the extent of possibilities captured by the term: "Constitutions can vary along many dimensions. They can be enacted (as in the case of the U.S. Constitution) or arise through a customary process (as in the case of the British Constitution). They can have an extra-legal status or the status of ordinary or fundamental law. They can be enforced through political or judicial processes. They can be federal, unitary or confederate. They can be defined narrowly to include just the formal rules in a particular document or more broadly to include all the fundamental rules of a political community, formal and informal, written and unwritten." [15] With all these possible elements "the constitutional character of governance is not a matter of yes or no, but of more or less".[16]

The substantive elements and functions are even more ambiguous. Anne Peters offers three answers to the question which functions and contents must be present to call a given body of law a 'constitution'. First, she says, the broadest notion of a constitution is that which refers to the bulk of laws organising and institutionalising a polity. Second, in a more narrow sense, the answer could be those rules and principles that fulfil typical constitutional functions; the traditional constitutional functions being to constitute a political entity as a legal entity, to organise it, to limit political power, to offer political and moral guidelines, to justify governance, and to contribute to integration. Last, the narrowest answer is the notion that a constitution is a value-laden concept, encapsulating the idea of democracy, the separation of powers, the protection of human rights, and a minimum of social security guarantees.[17]

Jeffrey L. Dunoff and Joel P. Trachtman also divide the possible functions of a constitution into three. First they say constitutional norms can be termed 'enabling'. By this they mean that some constitutional norms enable the production of ordinary international law, that is to say they 'enable constitutionalisation'. Next, some international constitutional norms constrain the production of ordinary international law, or they 'constrain constitutionalisation'. Last, some international constitutional norms respond to domestic constitutional deficiencies, particularly where the deficiency arises from or is exacerbated by increased globalisation and increasing density in international law. This function is referred to as 'supplemental constitutionalisation'.[18]

Bodansky agrees that the functions of a constitution are both to constitute and constrain a power. Constitutive rules define the basic values, institutions and decision-making processes of a political community. By including in these rules limitations on governmental authority, constitutions serve an important legitimating function.[19] In the essence of limiting power, guaranteed human rights protection is often a substantive element of a constitution.[20] Importantly, Cottier and Hertig observe that the constitution's legitimising of political authority is ultimately derived from the citizens.[21]

A further effect of constitutionalising governance is the strengthening of law as a creator of "government by laws and not by men".[22] Thus the law is guaranteed and certain, following the rule of law. Dunoff and Trachtman add that "one of the most important functions constitutionalisation can play is to provide mechanisms addressing how different constitutions relate to one another, and how they can be coordinated."[23] In this way, the notion of constitutionalism is both descriptive and prescriptive.[24]

II. From International Law to Global Constitutionalism

Not surprisingly, international lawyers disagree as to whether an existing core set of international principles and institutions amounts to ingredients of a global constitution. Some may say that an international constitutional order already exists, pointing to the United Nations Charter, for example, while others may consider any consensus-building among states as an never-ending process. Somewhere in between, we may think of a 'global constitution' as a possibility that arises when international consensus-building has reached a certain degree of maturity.

Assuming its constitutional character, the UN Charter could be seen as articulating the fundamental rules of the international system, such as the principles of sovereign equality and self-determination, the prohibition of the use of force, and the obligation to promote and respect human rights. These fundamental rules would then represent core values of an emerging international political community, which are outside politics and no longer dependent on states as such. Also the rules of the Charter are intended to provide a stable framework of governance of indefinite duration. Accordingly they are entrenched, their amendment requiring ratification by two-thirds of the member states, including all the permanent members of the Security Council. To this end, Article 103 makes the Charter hierarchically superior to other treaties. Thus a number of constitutional features are evident in the Charter.[25]

On the other hand, many operating rules of international law are incorporated in other documents and treaties such as the Vienna Convention of the Law of Treaties or the International Law Commission Articles on State Responsibility. Further, the globalised world of 2010 looks very different from the post-war period when the United Nations were established. Crucially, a consensus between Western liberal constitutions (featuring individual human rights, democratic decision-making, the separation of powers and judicial review etc.) and Non-Western constitutions cannot be assumed.

This last point raises the important issue that any global constitution must be global by nature and not informed solely by Western values and conceptions. Onuma Yasuaki, for example, stresses a transcivilisational perspective towards constitutionalism and global legal order.[26] A transcivilisational perspective is a 'theoretical device by which we can recognize and appreciate various ways of thinking of diverse peoples and seek to identify values and virtues that are perceived as legitimate by as many people as possible'.[27] Such a viewpoint is necessary, Yasuaki argues, since less than twenty percent of the world share the Westernised point of view.[28] Cottier suggests a multilevel governance system that interweaves various forms of governance may be a solution to coping with pluralism in cultures and values around the world.[29]

The terms 'thick' and 'thin' may be useful when assessing the constitutional qualities of international law. Thin would refer to certain constitutional aspects of an international document or agreement. Anne Peters calls such a viewpoint a 'micro-constitutionalist analysis'.[30] Basic constitutive factors could be creating institutions, specifying the rules that guide and constrain those institutions, and entrenching those rules through amendment procedures. Thick would refer to more substantial patterns of constitutionalism. For example when documents and agreements form the law of an emerging international community with shared responsibility and solidarity; or when the law is of a high status not dependent on state consent; or when documents and agreements represent a unified, coherent system, rather than a series of discrete regimes, and provide for a process of judicial review.

Whether constitutionalism is found in international law then can be viewed from these two perspectives. Some scholars assert that elements constituting a thin form of constitutionalism are evidence of an international constitution; others say the thick sense of a constitution must be present. Alec Stone Sweet and Cottier are two of the former,[31] while Bodansky,

Andreas Follesdal, Petra Dobner and others look for evidence of substantial coherence to detect genuine constitutionalism.

III. Is There an International Environmental 'Constitution'?

Thus far, the discourse on international constitutionalism has been largely between constitutional law and international law specialists. Environmental lawyers are only just entering the debate. Among the few environmental law scholars are Daniel Bodansky and Alexandre Kiss.

Tracing 36 years of development of international environmental law, Alexandre Kiss concludes that a constitution in the "usual sense of the term" cannot be deduced.[32] He explains there is neither a global instrument nor an independent global institution overarching the whole area and "even environmental protection is integrated into a vast complex: sustainable development".[33] Still, he argues that there are international functions and principles that satisfy such needs, at least, potentially.

Kiss points out that a major issue in constitutionalising environmental law is that environmental law in particular relies heavily on scientific information that is constantly updated.[34] Thus the international law regime protecting the environment needs to be flexible enough to be amended regularly. This is a potential conflict with the nature of constitutionalisation having an enduring quality.

Bodansky agrees that an international environmental constitution does not exist. Global environmental agreements appear, at best, as thin or weak forms of constitutions as they establish ongoing systems of governance to address specific issues that involve the creation of institutions, rules and procedures. However, they are not constitutions in the thick sense because they fail to establish institutions that are independent from states. Nor would those institutions have independent legal personality.

Further, international environmental agreements do not in themselves control the institutions they establish. For example, the International Convention for the Regulation of Whaling sets out how the International Commission on Whaling operates. It specifies that that the IWC should operate on a scientific basis. Yet this requirement did not prevent the IWC from adopting a moratorium on commercial whaling. What tends to happen is that the treaty text itself will contain the governance arrangement, describing the basic institutions and decision making procedures of the regime, but then the majority of norms and regulations created by that governing body will be captured in protocols, annexes or schedules, attached to the treaty.[35] In this way international environmental agreements can generate bureaucracy of own decision-making power, but will not be able to replace the consensus of states upon which the entire regime rests. In other words, states will always remain in charge.

Bodansky does not consider multilateral environmental agreements as having a truly global nature, describing them as 'still very much state-driven'[36] (such as those addressing climate change, biological diversity, or hazardous chemicals). Further, states generally retain the right of exit, meaning they can withdraw from a treaty.[37] The role of non-State actors in standard setting and the compliance process, although significant, is generally extra-constitutional because it takes place through informal processes, outside of the formal arrangements provided for by the regime's foundational instrument.[38]

Nor can international environmental law as a whole be perceived as a form of a global constitution. Noting a number of characteristic features which distinguish this area from classical international law, such as widespread use of the framework convention and protocol approach, rapid amendment procedures, a distinctive system of treaty bodies and non-compliance procedures that are more political than judicial in nature, Bodansky asserts that these distinctive features of international environmental law 'do not amount to a constitution in any meaningful sense of the term'.[39] They do not establish secondary rules about how international environmental law is developed and enforced. To the contrary, the concept of a constitution is undermined by some prominent features of international environmental law, such as the use of politically-oriented non-compliance procedures. Instead of creating a cohesive system of unified law, 'the distinctive mechanisms of international environmental law represent a toolbox that states can use when addressing a variety of new problems'.[40]

A more promising candidate to fill the role of an international environmental constitution may be the collection of general principles of international environmental law, such as the duty to prevent transboundary harm, the polluter pays principle, the precautionary principle, the principle of common but differentiated responsibility or a fundamental principle of sustainability. But these general principles are still too vague and ambiguous to represent a core value system for the international community. Furthermore, a general system of governance such as a world environmental agency, for example, is still lacking. Such an agency or organisation would need to provide a common framework for the multiplicity of existing international environmental regimes, consolidate common rules that organise its different regimes, demonstrate a separation of powers and provide a platform for judicial review.[41]

IV. The 'Greening' of International Law

Current international environmental law includes a plethora of global treaties, covenants and documents. From the creation of the United Nations in 1945 under the United Nations Charter to the UN Declaration on the Rights of Indigenous Peoples in 2007 or the 2009 Copenhagen Accord, numerous documents have been developed in response to what today is perceived as the global ecological crisis. The following section sketches the key principles of some of the more well-known arrangements.

Dominating the plethora of documents are the general themes of human rights and the environment. Even the 'UN Global Compact' with its focus on business, is about aligning business with principles of human rights, labour rights, the environment and anti-corruption. There is a large cross-over in the subject matter of the documents. Almost all start with lengthy preambles reiterating commonalities such as basic human rights, freedom and self-determination. Recurring goals include the conservation and preservation of the world's ecosystems and biodiversity (Stockholm Declaration, World Conservation Strategy, UN World Charter for Nature, Caring for the Earth, United Nations Framework Convention of Climate Change (UNFCCC), Rio Declaration, Agenda 21, Earth Charter, amongst others), the reduction of poverty (UN Millennium Declaration, Earth Charter, Johannesburg Declaration), the need for peace and security (UN World Charter for Nature, UN Millennium Declaration, Earth Charter).

In achieving these goals the concepts of co-operation, equity and interdependence of states are important. For example, proportionate responsibility (or common but differentiated responsibilities) is employed as a method to encourage equity (featuring in the UNFCCC and Kyoto Protocol, and the Rio Declaration). Often, soft law agreements recognise existing inequalities and specifically address them, for example the UN Millennium Declaration includes a goal entitled 'Meeting the Special Needs of Africa'.[42]

Sustainability has been a common thread throughout the years. From the World Conservation Strategy in 1980, to the Rio Declaration of 1992, to the Earth Charter of 2000, to the draft 'Universal Declaration of the Rights of Mother Earth' of 2010[43], the concern for environmental fundamentals has been ever present. A trend that may be emerging is a change from (anthropocentric) sustainable development to (non-anthropocentric) sustainability. State-negotiated documents are usually anthropocentric; for example the Rio Declaration has its first principle as 'Humans are at the centre of concerns for sustainable development. They are entitled to a healthy and productive life in harmony with nature'.

Principle 2 recognises the rights of sovereign states to 'exploit' their own resources. Somewhat more ambitiously, the UN World Charter for Nature states: 'Mankind is a part of nature and life depends on the uninterrupted functioning of natural systems which ensure the supply of energy and nutrients.'[44] Legal concepts, however, tend to protect the environment so far valuable to humans and not as an inherent necessity of ecological interdependencies.

Tension therefore arises as the documents often attempt to balance the three pillars of sustainability: economic, social and environmental.[45] In numerous documents we see elevation of the social and economical pillars – even in documents supposedly based on the environment, for example the Rio Declaration.[46] While some documents, such as the UN World Charter for Nature, place the environment first (asserting development must occur within ecological limitations), other agreements are more indifferent and often place economic concerns in the centre. Taken together, they give a bewildering account of either vague or contradictory international objectives.

A paradigmatic change came with the Earth Charter in 2000. For the first time at international level, human destiny is linked with the destiny of the planet. The Earth Charter's first principle acknowledges that all beings are interdependent and every form of life has value regardless of its worth to humans. This holistic, non-anthropocentric view is also reflected, for example, in Ecuador's new constitution involving the affirmation of the rights of nature.[47] The rights of nature and the appreciation of *Sumak kawsay* (the 'good life') are rooted in respect of *Pachamama* ('Mother Earth') as the source of all life. Similarly, the Bolivian constitution of 2009 reflects an ethno-ecological vision of people and the planet.[48] Along with the 2010 'Universal Declaration of the Rights of Mother Earth'[49] all these recent documents suggest a transition in environmental law thinking towards the strong form of sustainability.[50]

It is worth noting who drafted and influenced the major documents. A majority were created under the auspices of the United Nations, and many others by the International Union of Conservation and Nature (IUCN). While these are organisations reflective of global interests, it merits questioning to what extent they actually do. The emphasis of many of the documents is based on rights rather than duties, which may reflect Western influence. It should be noted that even the Universal Declaration of Human Rights has been criticised by Muslim communities who prefer to elevate their own Universal Islamic Declaration of Human Rights or the Cairo Declaration of Human Rights in Islam. This clash illustrates a potential lack of consensus in values as reflected in the UDHR, one of the foundational documents of contemporary international law and governance.

Ethical influences on the drafters and therefore the concepts within the documents affect procedural elements. For example, the Western judicial concept is prevalent in many of the global agreements.

A further problem is that treaties are voluntary for each state to sign and ratify. Consider, for example, that the world's largest economy, the United States, has never ratified key treaties such as the Rome Statute (treaty) on the International Criminal Court, the UN Convention on the Law of the Seas, the Convention on Biological Diversity or the Kyoto Protocol. The United States' reluctance alone illustrates the vulnerability of legally binding agreements. Soft law agreements, on the other hand, do not require ratification, in its legal sense, and can potentially be more inclusive and effective than treaties, conventions and other legally binding documents.[51]

What has been the success of soft law documents? Many have been reaffirmed in later documents; for example the Rio Declaration affirms the Stockholm Declaration, and the Johannesburg Declaration reaffirms both the Stockholm and Rio Declarations. There is also a reaffirmation of the UDHR – the Vienna Declaration and Program of Action. These reaffirmation documents suggest that the aspirations of the earlier documents were not being realised. And yet, some agreements have certainly inspired worldwide progress. The Stockholm Declaration is said to have paved the way for the environmental policies of the European Community, which eventually led to the Kyoto Protocol.[52] Many governments subsequently created Ministries for the Environment or national agencies for environmental monitoring after Stockholm. The World Conservation Strategy formed the basis for preparation of National Conservation Strategies in over fifty countries.[53] Agenda 21 has been used all around the world for promoting and implementing more sustainable policies and practices (see e.g. the Local Agenda 21 movement). And the Earth Charter – not even a genuine soft law document[54] - has been endorsed by many thousand civil society groups, several international organisations, including UNESCO and IUCN, and a number of national governments, most recently the Portuguese parliament.[55]

V. Democratization of International Law

Considering the feasibility and practicalities of a global constitution reveals numerous issues to be resolved. In particular, problems of legitimacy are apparent. While the drafting of a global constitution as such is still idealistic, the discourse on global constitutionalism is growing and ideas around a constitution-type global document have emerged.

First, some legitimacy problems should be addressed.

Occasionally, concern is expressed that new global constitutional law would undermine existing domestic constitutions.[56] Fears include that supranational law may circumvent the domestic constitutional lawmaking structure, for example in the way international customary law is argued to have a direct effect in American courts; or that it may change the dynamics of the domestic lawmaking process.[57]

Ernest Young is therefore wary that domestic law has checks and balances in it that ensure its legitimacy, and those features are somewhat different to the checks and balances that legitimise international law. Aspects of international law can conflict with domestic law, however, concerns of inconsistencies as expressed by Young and other scholars, such as Anne Peters, Thomas Cottier and Mattias Kumm[58], suggest a new ground of legitimacy for international constitutional law.

The legitimacy of traditional international law is based on the will and consent of sovereign states. That is because since the Treaty of Westphalia in 1648, international law has been largely framed around the rights and duties of individual sovereign states. Because it is a mostly horizontal system, of inter-state agreements, state consensus is essential for entering and validating an agreement. Such a method cannot work in a vertical system. In a vertical system, there is a hierarchy of community interests and values that sit above the individual interests of states. The purpose of a discourse on a global constitution is to protect those higher community interests and values. To that end, state consensus cannot, or not alone, be a legitimating factor of a global constitution. Brun-Otto Bryde explains this point with reference to human rights, saying protection of human rights in international law has meaning only if it is recognized as a system of obligations of states owed to an authority different from states themselves.[59] Similarly, with regard to the environment, protection requires the interests of States to be subordinated to the interest of mankind as a whole.[60]

The constitutional mechanisms of *jus cogens* and *erga omnes* have already come into play to replace state consensus. *Jus cogens* signifies a norm is of such importance that states cannot disregard it – whether they are in a treaty regarding the norm or not. For example, many human rights are *jus cogens*, therefore any violation of a human right is a crime by a state. *Obligations erga omnes* compliment *jus cogens* by creating obligations on all states to protect against and punish crimes which violate *jus cogens*. Some say that since these two devices are generally recognised in international law, the constitutionalisation of international law is already developing. These mechanisms signify a value system that places state interests subordinate to them. The two legitimise international law, providing specific rules for all states to abide by.

It is generally agreed that democracy forms the fairest basis for political decision-making and thus ensuring a democratically based international law system legitimises the system. Virtually all theories of democracy require that members of a political community decide for themselves the contents of the rules that govern their collective life.[61] Current international law, on the other hand, suffers from a much-debated democracy deficit.[62] Bryde[63], Follesdal[64], Delbrück[65] and others[66] all suggest that all peoples must be involved in the international law making process in order to construct a system that centres on the common interests of humanity. These common interests are not or not adequately represented when parochial state governments monopolise international decision making. Thus for democracy to be applied on the international plane, transnational common interests must be voiced additionally to, and disconnected from, state voices.

An international constitution may have enhanced legitimacy by having a governing body. Bryde suggests the European Union as a suitable model[67], while others emphasize multilayered governance beyond current EU structures.[68] To this end, Anne Peters distinguishes a desirable global system of governance from any moves towards global government.[69]

With respect to the possibility of a global constitution, the role of the judiciary can be enhanced to act as a check on international law, thus safeguarding legitimacy. However, it must be remembered that the adversarial judicial system is a western concept, and may not appeal to all cultures. On this note, some scholars warn, a global constitution must be 'transcivilisational'.[70] Douglas Johnston explains, "the political reality is that such a project, to be useful, must be shared across all regions, so that allegations of cultural bias in the field of international law can be confronted."[71] The goals of legal uniformity and universality may have to be reconciled with the value of cultural diversity. It is imperative that rules of an international system are not established by the wealthy elite; that the rights of poorer countries are ensured.[72] Law requires underlying values of a given society or community upon which its legitimacy relies.[73] Therefore it is crucial to a global constitution to determine whether common ethical foundations and shared beliefs exist worldwide. Arguably they do as not only the Earth Charter testifies.[74] Human rights, after all, reflect a universally shared value, and even economic international law illustrates bottom line shared values.[75] People may disagree on specifics, but the quest for shared generalities is a possibility. Laws around the world show similarities in basic values, for example, in their shared belief in the peaceful process of negotiation, discourse and decision-making and settlement of disputes.

There are numerous ways in which legitimacy may be given to a global constitution.[76] If a set of standards is agreed upon for international law, as a global constitution aspires to do, then domestic constitutions will know very clearly the parameters of the international constitution and the interplay it has on domestic constitutional law.

Notwithstanding the many considerations that may speak in favour or against a global constitution from a legal point of view, there have always been strong voices within civil society. In fact, the basic values of any constitutionalism have been shaped and promoted by civil society. A prime example of these efforts is the history of the 1948 Universal Declaration of Human Rights (UDHR).

The morals and values of human rights can be traced through the history of many religious beliefs and cultures. In Europe, the philosophers of the Enlightenment developed theories of natural law that paved the way for the modern idea of fundamental human rights. The English Bill of Rights (1689), the American Declaration of Independence (1776) and the French Declaration of the Rights of Man and of the Citizens (1789) reflected this idea and inspired constitutionalism of the 19th and 20th century.

Following the breakdown of human rights protection in Nazi Germany, the United Nations Charter "reaffirmed faith in fundamental human rights, and dignity and worth of the human person" and committed all member states to promote "universal respect for, and observance of, human rights and fundamental freedoms for all without distinction as to race, sex, language or religion".[77] The very fact, however, that these freedoms had just been violated by a presumed “civilized” European nation, required more than mere reaffirmation. A more specific declaration of specific rights was needed and this was seen primarily as a task for the international community, not individual states. Under the leadership of Eleanor Roosevelt this concern of civil society was given strong expression in the UDHR, a soft law document that became inspirational not only for human rights conventions and domestic constitutions, but also for the idea of a global constitution.

During 1948/49 a group of humanists, social scientists, philosophers and lawyers around University of Chicago professors drafted a *Preliminary Draft of a World Constitution*[78] which was translated into 40 languages. Significantly, the first chapter of this *Draft World Constitution* is structured as a "declaration of duties and rights" – not just rights – and contains the following sub-chapter c:

> "The four elements of life - earth, water, air, energy - are common property of the human race. The management and use of such portions thereof as are vested in or assigned to particular ownership, private or corporate or national or regional, of definite or indefinite tenure, of individualist or collectivist economy, shall be subordinated in each and all cases to the inherent interest of the common good."

The concern behind this draft was that international social justice cannot be achieved on the basis of existing concepts of property including the right to use and misuse it, as Elisabeth Mann Borgese points out.[79] She asserts that the *Draft World Constitution* established a legal order to protect the Earth and its resources as a common heritage and property of humankind. The *Draft World Constitution*, also known as the "Chicago World Constitution", has influenced many pioneers of international environmental law.[80]

For the development towards a possible global constitution no subject area could be more inspiring than the global environment. People and nations all depend on it and cannot sustain themselves without some collective understanding that, and how, the global environment must be protected. The mentioned Chicago World Constitution and all other models and ideas for a global constitution have proven to be most powerful when they articulate this common concern, and fate, of humankind.

Arguably, no other document has articulated this concern more strongly and more inclusively than the Earth Charter. Considering its origins, scope and legal importance[81], the Earth Charter might be considered as a benchmark and model for a future global constitution.

VI. The Earth Charter - A Model Global Constitution?

In essence, the Earth Charter seeks to inspire in all people a new sense of global interdependence and shared responsibility for the well-being of the whole human family, the greater community of life, and future generations. The Earth Charter is centrally concerned with the transition to sustainable ways of living and sustainable human development. Ecological integrity is one major theme. Additionally, the Earth Charter recognizes that the goals of ecological protection, the eradication of poverty, equitable economic development, respect for human rights, democracy, and peace are interdependent and indivisible. It provides, therefore, a new, inclusive, integrated ethical framework to guide the transition to a sustainable future. The Charter comprises 16 principles under 4 headings: Respect and Care for the Community of Life; Ecological Integrity; Social and Economic Justice; Democracy, Nonviolence and Peace.

Endorsement of the Earth Charter by individuals or organizations signifies a commitment to the spirit and aims of the document. The Earth Charter has currently been endorsed by over 5000 organisations including businesses, NGOs and universities as well as by a number of states and international organisations.[82]

The wide consensus that was reached in determining its contents is likely a key factor in the acceptance and success of the Charter. The reputation and credibility of the Earth Charter rest largely on its transnational, cross-cultural and inter-denominational approach. The Charter provides an action-oriented environmental ethic, and a global vision that recognises the interconnection of humanity's environmental, economic, social, cultural, ethical and spiritual ambitions. The Charter has guided soft law agreements on sustainable development and also influenced implementation processes of treaties such as the Convention on Biodiversity and the Framework Convention on Climate Change.[83] However, while widely accepted, the Earth Charter itself cannot be seen as soft law and a global constitution that builds on the Earth Charter would need to go a long way to eventually become legally binding.[84]

Summarizing some key features, we can highlight the foundational and hence constitutional qualities of the Earth Charter.

The ambition of the Earth Charter is *"to bring forth a sustainable global society founded on respect for nature, universal human rights, economic justice, and a culture of peace"* (Preamble, first paragraph). The Charter itself organizes its 77 principles around four main themes: *"Respect and Care for the Community of Life"* (Principles 1 – 4), *"Ecological Integrity"* (5 – 8), *"Social and Economic Justice"* (9 – 12) and *"Democracy, Nonviolence, and Peace*" (13 – 16).

The first two themes contain the environmental principles, the other two themes contain the social and economic principles of a sustainable global society. We can say, therefore, that the Earth Charter reflects the concept of sustainable development with its three pillars of environmental, social and economic equity. This three-pillar model is commonly accepted by states, expressed in many soft law documents[85] and promoted within business.

The Earth Charter, however, takes an important further step. It not only defines the three pillars, it also organizes them in a particular way. Environmental concerns are perceived differently from social and economic concerns. The environment is not merely the resource base for human consumption, not just one of three factors to be considered. It is the basis of all life, i.e. the community of all beings including human beings. This shift from a narrow human-centred to a broader life-centred perspective reflects ecological wisdom and marks the difference between "weak" and "strong" sustainability.[86]

The moral response following from strong sustainability is clearly spelled out in the Preamble: "It is imperative that we, the peoples of Earth, declare our responsibility to one another, to the greater community of life, and to future generations". The new element, that the Earth Charter promotes unlike any other international document, is our responsibility to the greater community of life. In jurisprudential terms, such three-fold moral responsibility translates to a three-dimensional concept of (distributional) justice: justice towards another, the ecological community and future generations.[87] Such broadened sense of justice is at the heart of a global "covenant for democracy".[88]

The notion of a covenant for democracy reflects both, normative and institutional aspects of a constitution. The normative aspects around the first two themes describe humans' commitment to the Earth, the normative aspects around the other two themes describe humans' commitment to each other, in particular, human rights and responsibilities.

The institutional aspects are contained in part three and four. They are all organised around strengthening of "democratic institutions at all levels" (Principle 13). The key ingredients of democratic governance are described as "transparency", "accountability", "participation in decision-making" and "access to justice" coupled with a strengthening of local communities (Principle 13.e.). The importance of education to enable informed and active citizenship is expressed in Principle 14. Further institutional requirements for a tolerant, non-violent and peaceful society are formulated in Principles 15 and 16. Throughout the Earth Charter the legitimacy of democratic institutions, including states, is assumed, but not without qualifications. Citizens – empowered in their communities and as part of civil society – must be the sovereign of decisions affecting all. This is currently not the case. A constant process of disempowerment, aggravated through global neo-liberalism, is the main reason for global environmental, social and economic decline.

The Earth Charter aims for turning this around. As states have largely failed to respond to the ecological crisis, the strengthening of citizenship and civil society is our only hope. In a situation where the entire human species is struggling to overcome its own ecological blindness, a new covenant is needed. Only an act of self-empowerment and renewed confidence can achieve what "democracy" holds as a promise. And only then can "*demos*" and "*kratos*" be reconciled.

The Earth Charter ends with "The Way Forward" containing this central statement:

> "Life often involves important values. This can mean difficult choices. However, we must find ways to harmonize diversity with unity, the exercise of freedom with the common good, short-term objectives with long-term goals. Every individual, family, organization, and community has a vital role to play. The arts, sciences, religions, educational institutions, media, businesses, nongovernmental organizations, and governments are all called to offer creative leadership. The partnership of government, civil society, and business is essential for effective governance."

This statement sums up what a global constitution must encapsulate. We are all together in the struggle and need to form a global partnership to care for the Earth and one another. To this end, the Earth Charter can serve as a model, if not for a global constitution, surely for the substance of what global governance needs to aspire to.

Notes

[1] Immanuel Kant, *Zum Ewigen Frieden. Eine philosophische Skizze* (1795). (Perpetual Peace: A Philosophical Essay, translated with Introduction and Notes by M. Campbell Smith, with a Preface by L. Latta, London: George Allen and Unwin, 1917; as a free ebook available at: http://oll.libertyfund.org/index.php?Option =com_staticxt&staticfile=show.php%3Ftitle=357&Itemid=27.

[2] Anne Peters "Global Constitutionalism in a Nutshell" in *Weltinnenrecht: Liber Amicorum Jost Delbrück,* ed. by Klaus Dicke, Stephan Hobe et al. (Duncker & Humblot, Berlin, 2005) 535-50; Christian Walter "Constitutionalising (Inter)national Governance – Possibilities for and Limits to the Development of an International Constitutional Law" (2001) 44 German Y.B Int'l L.192; Philip Allott 'The Emerging Universal Legal System' in *New Perspectives on the Divide Between National and International Law* Janne Nijman and Andre Nollkaemper (eds) (Oxford University Press Inc., New York, 2007) 63-83; Bardo Fassbender "The United Nations Charter as Constitution of the International Community" (2008) 36 Colum. J. Transnat'l L. 530; Anne-Marie Slaughter & William Burke-White "An International Constitutional Moment" (2002) 43 Harvard International Law Journal 1; Ernst-Ultrich Petersmann "How to Constitutionalise International Law and Foreign Policy for the Benefit of Civil Society?" (1998) 20 Mich. J. Inte. L. 1; Gunther Teubner 'Societal Constitutionalism: Alternatives to State-Centred Constitutional Theory?' in *Transnational Governance and Constitutionalism* Christian Joerges, Inger-Johanne Sand and Gunther Teubner (eds) (Hart Publishing,

Oxford and Portland Oregon, 2004) 3-28; Jan Klabbers "Constitutionalism Lite" (2004) 1 Int'l Org. L. Rev. 31; Inger-Johanne Sand, 'Polycontextuality as an Alternative to Constitutionalism' in *Transnational Governance and Constitutionalism* Christian Joerges, Inger-Johanne Sand and Günther Teubner (eds) (Hart Publishing, Oxford and Portland Oregon, 2004) 43-65; and Christian Joerges 'Constitutionalism and Transnational Governance: Exploring a Magical Triangle' in *Transnational Governance and Constitutionalism* Christian Joerges, Inger-Johanne Sand and Gunther Teubner (eds) (Hart Publishing, Oxford and Portland Oregon, 2004) 339-375. See also David Kennedy "The Mystery of Global Governance" in *Ruling the World?* Jeffery L Dunoff and Joel P Trachtman (eds) (Cambridge University Press, New York, 2009) 37-68; Andreas L Paulus 'The International Legal System as a Constitution' in *Ruling the World?* Jeffery L Dunoff and Joel P Trachtman (eds) (Cambridge University Press, New York, 2009) 69- 109; Martti Koskenniemi "Constitutionalism as Mindset: Reflections on Kantian Themes about International Law and Globalization" (2007) 8 Theoretical Inquiries in Law 9; Daniel Halberstam "Constitutional Heterarchy: The Centrality of Conflict in the European Union and the United States"; Miguel Poiares Maduro "Courts and Pluralism: Essay on a Theory of Judicial Adjudication in the Context of Legal and Consitutitional Pluralism" both in *Ruling the World?* Jeffery L Dunoff and Joel P Trachtman (eds) (Cambridge University Press, New York, 2009) 326-379; Karolina Milewicz "Emerging Patterns of Global Constitutionalization: Toward a Conceptual Framework" (2009) 16 Ind. J. Global Legal Studies 413-436; Petra Dobner and Martin Loughlin (eds), *The Twilight of Constitutionalism?* (Oxford University Press, New York, 2010).

[3] Such as the United Nations and the European Union.

[4] An earlier version of this chapter appeared as "Outlook: The Earth Charter – A Model Global Constitution?" in Klaus Bosselmann and J. Ronald Engel (eds.), *The Earth Charter: A framework for global governance* (KIT Publishing, Amsterdam, 2010).

[5] Amongst other meanings in *The New Zealand Oxford Dictionary* Tony Deverson and Graeme Kennedy (eds) (Oxford University Press, Victoria, Australia, 2005) 233.

[6] Daniel Bodansky "Is there an International Environmental Constitution?" (2009) 16 Ind. J. Global Legal Stud. 565 at 569-570. For additional analysis of the terms constitution and constitutionalism see Samantha Besson "Whose Constitution(s)? International Law, Constitutionalism, and Democracy" in *Ruling the World?*, above n 1, at 376-380.

[7] Anne Peters "Compensatory Constitutionalism: The Function and Potential of Fundamental International Norms and Structures" (2006) 19 Leiden J. Int'l L 579 at 595, at 584.

[8] Ibid.

[9] Ibid 585.

[10] Bodansky, above n 6, at 571.

[11] Ibid.

[12] Ibid.

[13] Ibid at 569.
[14] Ibid at 573.
[15] Ibid at 570.
[16] Ibid at 571.
[17] Peters "Compensatory Constitutionalism", above n 7, at 585.
[18] Jeffrey L. Dunoff and Joel P. Trachtman "A Functional Approach to Global Consitutionalism" in *Ruling the World?,* above n 1, at 8-11.
[19] Bodansky, above n 6, at 572.
[20] Ibid.
[21] Thomas Cottier & Maya Hertig "The Prospects of 21st Century Constitutionalism" in (2003) 7 Max Planck Y.U.N L. 261 at 280.
[22] Ibid. Bodansky cites John Adams (1774) for saying 'a government of laws, not of men' in the Massachusettes Constitution, see Bodansky, above n 6, at 573.
[23] Jeffrey L. Dunoff and Joel P. Trachtman, above n 1, at 28.
[24] Anne Peters & Klaus Armingeon "Global Constitutionalism from an Interdisciplinary Perspective" (2009) 16 Ind. J. Global Legal Stud. 385 at 389.
[25] For a discussion on the constitutionality of the UN Charter see Bardo Fassbender "The United Nations Charter as Constitution of the International Community" (2008) 36 Colum. J. Transnat'l L. 530; Bardo Fassbender "Rediscovering a Forgotten Constitution: Notes on the Place of the UN Charter in the International Legal Order" in *Ruling the World?,* above n 1, at 133-147; and Michael W Doyle "The UN Charter – A Global Constitution?" in *Ruling the World?,* above n 1, at 113-132.
[26] Onuma Yasuaki "A Transcivilizational Perspective on Global Legal Order in the Twenty-first Century: A Way to Overcome West-centric and Judiciary-centric Deficits in International Legal Thoughts" in *Towards World Constitutionalism* Ronald St John Macdonald and Douglas M Johnston (eds) (Martinus Nijhoff Publishers, Leiden, 2005) 151-189.
[27] Ibid at 153.
[28] Ibid.
[29] See generally Thomas Cottier "Multilayered Governance, Pluralism, and Moral Conflict" (2009) 16 Ind. J. Global Legal Stud. 647.
[30] Peters, above n 7, at 594.
[31] Alec Stone Sweet "Constitutionalism, Legal Pluralism, and International Regimes" (2009) 16 Ind. J. Global Legal Stud. 621.
[32] Alexandre Kiss 'The Legal Ordering of Environmental Protection' in *Towards World Constitutionalism,* above n 26, at 574.
[33] Ibid.
[34] Ibid.
[35] Bodansky, above n 7, at 577-578.
[36] Ibid at 577.
[37] The current crises of the IWC system and the Kyoto regime are examples in case.
[38] Ibid 577.
[39] Ibid 579.
[40] Bodansky, above n 7, at 579.

[41] Notably, the World Trade Organisation displays a number of these features which tells us that the international community of states is quite capapble of establishing legal superstructures if they suit them.
[42] *2000 UN Millennium Declaration* A/Res/55/2 (2000). See section 7, para 27.
[43] Adopted by the World People's Conference on Climate Change and the Rights of Mother Earth on April 22, 2010, in Bolivia; http://motherearthrights.org/.
[44] *1982 UN World Charter for Nature* 48th plen mtg, A/Res/37/7 (1982). See the Preamble, a).
[45] Increasingly a fourth pillar is added: cultural.
[46] Report of the United Nations Conference on Environment and Development A/CONF.151/26 (Vol. I) (1992). See Principle 12.
[47] The official Spanish version of the 2008 Ecuador Constitution can be found at www.asambleanacional.gov.ec/documentos/constitucion_de_bolsillo.pdf. See articles 71-74 and 397.
[48] Vernon Tava, *Resisting Enclosure: The Emergence of Ethno-Ecological Governance in a Comparative Study of Venezuela, Ecuador and Bolivia*, LLM Thesis (The University of Auckland, 2010).
[49] Above n 43.
[50] Klaus Bosselmann, *The Principle of Sustainability: Transforming Law and Gov*ernance (Ashgate, Aldershot, 2008).
[51] Dinah Shelton, Soft Law, *Handbook of International Law* (Routledge Press, 2008); GWU Legal Studies Research Paper No. 322; GWU Law School Public Law Research Paper No. 322. Available at SSRN: http://ssrn.com/abstract=1003387.
[52] John Baylis & Steve Smith *The Globalization of World Politics* (3rd ed, Oxford University Press, Oxford, 2005) 454-455.
[53] Caring for the Earth (IUCN/UNEP/WWF, 1991) 1.
[54] Imke Sagemüller, "The International Recognition and Legal Status of the Earth Charter", in Bosselmann and Engel, above n 4, 41-59; see also Bosselmann and Engel, Introduction, ibid., at 21/22.
[55] www.earthcharterinaction.org/content/articles/530/1/National-Parliament-of-Portugal-embraces-the-EC/Page1.html.
[56] See for example Ernest A. Young with reference to America in "The Trouble with Global Constitutionalism" (2003) 38 Tex. Int'l L.J. 527-545.
[57] Ibid 529. For further discussions on the effect of constitutionalism on State relationships with international law see Andre Nollkaemper "Constitutionalization and the Unity of the Law of International Responsibility" (2009) 16 Ind. J. Global Legal Stud. 535 -563, and Janne Nijman and Andre Nollkaemper, "Beyond the Divide" in *New Perspectives on the Divide Between National and International Law*, Janne Nijman and Andre Nollkaemper (eds) (Oxford University Press Inc., New York, 2007) 341-360. And for an international relations point of view see Jens Steffek "Sources of Legitimacy Beyond the State: A View from International Relations" in *Transnational Governance and Constitutionalism* Christian Joerges, Inger-Johanne Sand and Gunther Teubner (eds) (Hart Publishing, Oxford and Portland Oregon, 2004) 81- 101.

[58] Mattias Kumm "The Cosmopolitan Turn in Constitutionalism: On the Relationship Between Constitutionalism In and Beyond the State" in *Ruling the World?,* above n 1, at 258-324. See further Daniele Archibugi and David Held (eds.), *Cosmopolitan Democracy* (Polity Press, Cambridge, 1995).
[59] Brun-Otto Bryde "International Democratic Constitutionalism" in *Towards World Constitutionalism,* above n 26, at 107.
[60] Ibid.
[61] Dunoff & Trachtman "A Functional Approach to Global Constitutionalism" in *Ruling the World?,* above n 1, at 25.
[62] Andreas L Paulus "The International Legal System as a Constitution" in *Ruling the World?,* above n 1, at 94.
[63] Above 59.
[64] Andreas Follesdal "When Common Interests are Not Common: Why the Global Basic Structure Should Be Democratic" (2009) 16 Ind. J. Global Legal Stud. 585.
[65] Jost Delbrück "Exercising Public Authority Beyond the State: Transnational Democracy and/or Alternative Legitimation Strategies" (2003) 10 Ind. J. Global Legal Stud. 29 at 31.
[66] Jon Elster *Introduction to Constitutionalism and Democracy* (Jon Elster & Rune Slagstad eds., 1988); Carlos S Nino *The Constitution of Deliberative Democracy* (1996); Jürgen Habermas "Constitutional Democracy: A Paradoxical Union of Contradictory Principles?" (2001) 29 Pol. Theory 766; Samantha Besson "Whose Constitution(s)?" in *Ruling the World?,* above n 1, at 389-406; Richard Bellamy *Political Constitutionalism: A Republican Defence of the Constitutionality of Democracy* (Cambridge University Press, New York, 2007) 120-141, 209-259; Jiunn-Rong Yeh & Wen-Chen Chang "The Emergence of Transnational Constitutionalism: Its Features, Challenges and Solutions" (2008) 27 Penn State Int'l L. R. 89.
[67] Bryde, above n 67, at 111.
[68] Cottier, above n 29, at 656; Klaus Bosselmann, "Towards Multilevel Governance for Sustainability: Trends and Challenges", *Contested Futures: Sustainability, Governance and International Human Rights*, (Europe – New Zealand Research Series, University of Auckland, 2010), 1-72.
[69] Peters "Compensatory Constitutionalism", above n 7, at 610.
[70] Yasuaki, above n 26.
[71] Douglas M. Johnston in "World Constitutionalism in the Theory of International Law" in *Towards World Constitutionalism,* above n 26, at 27.
[72]Ibid at 8-10.
[73] Cottier, above n 29, at 659.
[74] Ibid at 661.
[75] Ibid at 667.
[76] See also Daniel Bodansky "The Legitimacy of International Governance: A Coming Challenge for International Environmental Law?" (1999) 93 American J. Int'l L. 596 at 597.
[77] United Nations Charter, preamble and article 56.
[78] www.worldbeyondborders.org/chicagodraft.htm.

---

[79] Mann Borgese, *The Beginnings - The IOI Story* (International Ocean Institute, Malta 1999), 9.
[80] Ibid., 9-13.
[81] Bosselmann and Engel, Introduction, *The Earth Charter*, above n 4, 15-26.
[82] See Earth Charter 'Endorse' at www.earthcharterinaction.org/content/pages/Endorse.html# .
[83] Klaus Bosselmann "Earth Charter (2000)" Max Planck Encyclopaedia of Public International Law, (Max Planck Institute for Comparative Public Law and International Law (Heidelberg and Oxford University Press, 2009) 4.
[84] See, however, Klaus Bosselmann and Prue Taylor, "The Significance of the Earth Charter in International Law", in: Peter Blaze Corcoran (ed.), *Toward a Sustainable World: The Earth Charter in Action* (KIT Publishing, Amsterdam, 2005), 171-173, for the argument that the legal status of documents is less important for global consensus-building than their factual significance.
[85] E.g. *Rio Declaration on Environment and Development,* Report of the *United Nations Conference on Environment and Development*, UN Doc. A/CONF.151/6/Rev. 1 (1992), 31 I.L.M. 874 (1992) [*Rio Declaration*], United Nations Conference on Environment and Development *Agenda 21* UN Doc A/CONF151/26/Rev1 (1992) and *Johannesburg Declaration on Sustainable Development*, Report of the *World Summit on Sustainable Development*, Johannesburg, South Africa, 16 Aug.- 4 Sept.2002, A/CONF.199/20.
[86] Klaus Bosselmann, "Strong and Weak Sustainable Development: Making the Difference in the Design of Law", (13) *South African Journal of Environmental Law and Policy*, 2008, 14-23; Bosselmann, above n 50, 9-41.
[87] Bosselmann, above n 50, 79-109.
[88] J. Ronald Engel, "The Earth Charter as a New Covenant for Democracy", in Bosselmann and Engel, above n 4, 29-40.

CHAPTER TWELVE

THE HUMAN RIGHT TO WATER AND GLOBAL ADMINISTRATIVE LAW: MUTUALLY SUPPORTING CONCEPTS?

OWEN MCINTYRE

Introduction

This chapter explores whether we can better understand the significance of the current, urgent discourse in international law on human rights-based approaches to water and sanitation entitlements as an expression of universally accepted standards of global governance, rather than as an enforceable 'human right' to water in any strict sense of that term. Such an analysis might obviate many of the difficulties which arise in relation to the precise normative status of the human right to water under international law as well as doubts concerning its enforceability by means of the traditional enforcement mechanisms existing under human rights law. Further, an analysis of the human right to water (HRW) in terms of 'global administrative law' (GAL) assists in explaining its potential application in such disparate doctrinal areas of law as international water resources law, international environmental law, and international investment law, as well as in national public and constitutional law. Crucially, the GAL concept helps to address problems which arise with the extension of human rights requirements to the actions of private corporate actors and with the problem of extra-territoriality.

Conversely, a survey of the commonly accredited sources of GAL rules helps to make sense of the wide diversity of mechanisms, in addition to the formal conventional obligations and practice of States, for the generation of rules and standards which inform the procedural and substantive normative content of the HRW concept. Such mechanisms include, for example, the International Standards Organisation, various voluntary codes of corporate conduct, international investment arbitration tribunals provided for under bilateral investment treaties, national systems of administrative law, and of course the institutional machinery which elaborates upon human rights values.

Such an analysis in turn permits concrete conclusions to be drawn on, *inter alia*, the key procedural and due process elements of the HRW concept, the key rule of law values stemming from the movement towards free trade and economic liberalism, the key good governance values, relating particularly to transparency, participation and accountability, and the key human rights values impacting upon the concept. This mode of analysis goes some way towards detailing the rights and obligations created by the HRW concept for a variety of actors, including individuals, vulnerable communities, transnational corporations, investors in water and sanitation services, and of course State agencies.

Global Administrative Law

The emerging concept of Global Administrative Law (GAL) addresses the rapidly changing realities of transnational regulation, which increasingly involves, *inter alia*, various forms of industry self-regulation, hybrid forms of private-private and public-private regulation, network governance by State officials, and governance by inter-governmental organisations with direct or indirect regulatory powers, and 'begins from the twin ideas that much global governance can be understood as administration, and that such administration is often organized and shaped by principles of an administrative law character'.[1] It is proposed that these disparate regulatory regimes, some voluntary and some mandatory, and operating at various levels (sector-specific, national, regional and global),

> 'together form a variegated "global administrative space" that includes international institutions and transnational networks involving both governmental and non-governmental actors, as well as domestic administrative bodies that operate within international regimes or cause transboundary regulatory effects'.[2]

Kingsbury deliberates further on the idea of a 'global administrative space' and explains that it 'marks a departure from those orthodox understandings of international law in which the international is largely inter-governmental, and there is a reasonably sharp separation of the domestic and the international', and that it reflects the practice of global governance, whereby 'transnational networks of rule-generators, interpreters and appliers cause such strict barriers to break down'.[3] To the student of international water resources law, this observation is reminiscent of the

International Law Association's (ILA) 2004 Berlin Rules on Water Resources Law which, though primarily concerned with the rules facilitating inter-State cooperation over shared transboundary water resources, contain a dedicated Article 17 asserting that 'every individual has a right of access to sufficient, safe, acceptable, physically accessible, and affordable water to meet that individual's vital human needs'.[4] Indeed, in remarking on the 'highly decentralized and not very systematic' nature of much of the administration of global governance, Kingsbury observes that '[S]ome entities are given roles in global regulatory governance which they may not wish for or be particularly designed or prepared for',[5] bringing to mind the recent decisions of International Centre for the Settlement of Investment Dispute (ICSID) tribunals which would appear to tacitly support the centrality of human rights concerns to contracts in respect of the provision of water and sanitation services.[6]

Crucially, in respect of the normative content of GAL, and reflective of the key procedural aspects of the HRW concept as articulated by the U.N. Committee on Economic, Social and Cultural Rights in General Comment No. 15,[7] the leading proponents of GAL observe that

> 'These evolving regulatory structures are each confronted with demands for transparency, consultation, participation, reasoned decisions, and review mechanisms to promote accountability. These demands, and responses to them, are increasingly framed in terms that have an administrative law character. The growing commonality of these administrative law-type principles and practices is building a unity between otherwise disparate areas of governance.'[8]

Of course, the function of administrative law generally is to protect individuals by checking the unauthorised, excessive, arbitrary or unfair exercise of public power and, by so doing, to give direction to the practices of administrative bodies, particularly in terms of their responsiveness to broader public interests. Proponents of GAL argue that it can perform a similar function for global administrative structures and point out that many of the types of regulatory measures cited above have resulted from the efforts of global administrative bodies, often stimulated by external criticism, to improve internal accountability and bolster external legitimacy.[9] One needs only to consider the establishment of accountability mechanisms by all major multilateral development banks or the widespread inclusion of mechanisms for NGO participation and representation in the decision-making structures of regulatory bodies. In an attempt to provide a definition of the concept of GAL, the same leading proponents explain that it

> 'encompasses the legal mechanisms, principles and practices, along with supporting social understandings, that promote or otherwise affect the *accountability* of global administrative bodies, in particular by ensuring these bodies meet adequate standards of transparency, consultation, participation, rationality, and legality, and by providing effective review of the rules and decisions these bodies make'.[10]

In addition, they accompany this definition with a broad understanding of the 'global administrative bodies' which generate GAL norms and to which such norms might apply, to include

> 'intergovernmental institutions, informal inter-governmental networks, national governmental agencies acting pursuant to global norms, hybrid public-private bodies engaged in transnational administration, and purely private bodies performing public roles in transnational administration'.[11]

Thus, much of the normative content of the HRW concept and, in particular the procedural rights of individuals and communities contained therein, along with the policies, procedures and decisions of the disparate entities which seek to give effect to the values contained therein, can be viewed through the prism of GAL

As regards the sources of GAL, Benedict Kingsbury, one of the leading scholars in the field, while emphasising that 'there is no single unifying rule of recognition covering all of GAL', includes the conventional sources of public international law, *i.e.* treaties, fundamental customary international law rules, and general principles of law, but also certain principles associated with 'publicness' in law.[12] He suggests that '[p]rinciples relevant to publicness include the [public] entity's adherence to legality, rationality, proportionality, rule of law, and some human rights', which are manifested in 'practices of judicial-type review of the acts of global governance entities, in requirements of reason-giving, and in practices concerning publicity and transparency.'[13] In an account of GAL, which is slightly more sceptical about the difficulty of identifying a universal set of administrative law principles, Harlow systematically identifies and describes four potential sources as a foundation for a global administrative law system:

> 'first, the largely procedural principles that have emerged in national administrative law systems, notably the principle of legality and due process principles; second, the set of rule of law values, promoted by proponents of free trade and economic liberalism; third, the good governance values, and more particularly transparency, participation and

> accountability, promoted by the World Bank and International Monetary Fund; and finally, human rights values.'[14]

Harlow concludes from her examination of all these sources that 'there is considerable overlap between principles found in these different sources'.[15] In addition, Kingsbury includes among the sources of GAL the rules, standards and safeguards developed as a result of processes of so-called 'private ordering', such as the three sets of guidelines adopted in 2007 by Technical Committee 224 of the International Standards Organisation (ISO),[16] though he cautions that such '"[P]rivate ordering" comes within this concept of law only through engagement with public institutions'.[17]

As regards the specific normative content of GAL, Kingsbury identifies certain '[g]eneral principles of public law [which] combine formal qualities with normative commitments in the enterprise of channelling, managing, shaping and constraining political power'.[18] In addition to certain 'more detailed elements, or requirements … particularly review, reason-giving, and publicity/transparency', his indicative list of such general principles of public law includes:

(i) *The Principle of Legality* – requiring that actors within a power system are constrained to act in accordance with the rules of the system;
(ii) *The Principle of Rationality* – requiring the justification of decisions, including that decision-makers give reasons and produce a factual record for decisions;
(iii) *The Principle of Proportionality* – requiring a relationship of proportionality between means and ends;
(iv) *Rule of Law* – requiring particular deliberative and decisional procedures; and
(v) *Human Rights* – requiring protection of human rights values which are intrinsic (or natural) to a modern public law system.[19]

Kingsbury further identifies three broad categories of public global administrative activity to which the rules and principles of GAL might apply, and which in turn generate practices which can give rise to such rules and principles. These include:

(i) The institutional design, and legal constitution, of the global administrative body

(ii) The norms and decisions produced by that entity, including norms and decisions that have as their addressees, or otherwise materially affect:

a. other such public entities
b. states and agencies of a particular state
c. individuals and other private actors

(iii) Procedural norms for the conduct of those public entities in relation to their rules and decisions, including arrangements for review, transparency, reason-giving, participation requirements, legal accountability and liability.[20]

While it is perfectly clear that rules and principles of GAL are relevant to the institutional design, and thus to the legitimate functioning, of the myriad of entities involved in elaborating upon the HRW concept, it is the second and third categories of administrative activity listed above which play a significant role in the development of its normative status and content. Such entities might, for example, include the U.N. Committee on Economic, Social or Cultural Rights, or other global or regional bodies concerned with the interpretation or monitoring of human rights instruments, international standard-setting bodies, such as the International Standards Organisation, and judicial and quasi-judicial organs, such as international investment arbitration tribunals established under the International Centre for the Settlement of Investment Disputes (ICSID) or the accountability mechanisms of multi-lateral development banks (MDBs). As will be illustrated below, the interpretative statements, such as General Comment No. 15,[21] the best practice guidelines, such as the 2007 ISO Guidelines,[22] and the arbitral and quasi-judicial decisions, such as those of recent ICSID tribunals,[23] adopted by such entities lend much-needed support to and substantially inform the HRW concept, while also illustrating the practical utility of the GAL concept as a means of understanding common normative approaches which converge from complex, chaotic and pluralistic origins.

While Harlow includes human rights values as a source of GAL norms, she does so ‘only to the extent that these are procedural in character’.[24] In other words, she highlights the fact that ‘many international human rights texts contain due process rights of a type traditionally developed in and protected by classical administrative law systems’.[25] However, Kingsbury appears to suggest that the substantive normative content of human rights regimes might in some instances be relevant by suggesting that ‘some human rights (perhaps of *bodily integrity*, privacy, personality) are likely to be protected by public law as an intrinsic matter (without textual authority), yet without being subsumed into "rule of law"'.[26] Like the human right to water, the human right to bodily integrity is often closely linked to, and under many human rights texts derived from, the right to health and, indeed, further connected to mutually related standards of protection of the human environment.[27] Therefore, Kingsbury's express reference to bodily integrity implies that substantive human rights values must be relevant to the identification of GAL norms, and *vice versa*. Indeed, though General Comment No. 15 is very largely concerned with informational, participative and other procedural elements of the human right to water,[28] it seems difficult to imagine that substantive human rights values would not be relevant to, and captured by, the general public law principles of proportionality and rationality.[29]

Of course, there are those who have serious misgivings about the GAL phenomenon and highlight its hazards for democracy and traditional political processes, for developing economies, and for the coherence and predictability of applicable legal standards.[30] The key concern is that GAL tends to subvert the traditional democratic processes vital to the legitimacy of law, such as by circumventing the requirement of State consent under international law, by means of which States have traditionally exercised sovereignty. The role of judicial, arbitral and quasi-judicial bodies in particular raise concerns over the juridification of the political process and of 'government by judges' by virtue of a general empowerment of a transnational 'juristocracy'.[31] The undermining of sovereign democratic processes and the emergence of common and universal administrative standards presents a particular risk for developing economies, which may not have had a significant role in generating the practice upon which these standards are based. Harlow suggests that administrative law is largely a 'Western construct',[32] which is protective of Western values and interests and may impact unfavourably on developing economies, leading to a 'double colonization' involving 'a complex process of "cross-fertilization" or legal transplant, whereby principles from one administrative law system pass into another'.[33] She suggests that often '[g]ood governance in this all-embracing sense is, however, simply not obtainable … and, at least for the foreseeable future, it may be necessary and even preferable for them to settle for less costly, "good enough governance"'.[34] Further, due to the non-systematic nature of the processes shaping GAL, the rules and standards invoked as inherent to the GAL concept may often lack clarity and certainty. As Kingsbury points out, the difficulty in identifying universal rules and principles stems from the fact that

> '"[g]lobal administrative law" is not an established field of normativity and obligation in the same way as "international law". It has no great charters,

> no celebrated courts, no textual provisions in national constitutions giving it status in national law, no significant long-appreciated history'.[35]

Similarly, Harlow notes that there is 'no shortage of candidates for a set of universal values' and alludes to the ideological battle raging in this regard between '[h]ard-line economic liberals', '[s]ofter economic theorists' and 'the movement for cosmopolitan law and social democracy'.[36] Indeed, she highlights the considerable disparity of principle that exists '[e]ven within the systems in which modern administrative law [has] developed' and points out that '[a]t least four administrative law families have been identified within the EU alone'.[37] However, as argued below, the inclusive nature of the various institutional structures and processes which have given rise to the HRW concept, as well as the detailed normative guidance adopted thereunder, do much to address such concerns about sovereign legitimacy, normative clarity or Western bias, thus marking out the HRW concept as an exemplar of the GAL phenomenon.

Therefore, rather than attempting to provide a comprehensive and coherent unifying theory of global governance arrangements, the GAL concept is merely an observed phenomenon which seeks to explain the growing commonality apparent among the administrative principles and practices which apply across otherwise disparate areas of governance.[38] This phenomenon has been apparent to observers for some time and one commentator has noted of Lorenz von Stein, an early pioneer, that

> 'the concept of international administrative law (*internationals Verwaltungsrecht*) as originally conceived by Lorenz von Stein in 1866 described an ensemble of legal rules based partially on international sources and partially on domestic sources dealing with administrative activity in the international field as a whole. Von Stein's interest, here as elsewhere, was to capture and describe the reality of public administration rather than its underlying legal basis.'[39]

Human Right to Water

Uncertainty regarding legal bases and status

Despite recent high profile support for the legal status of the human right to water concept from key U.N. bodies, including a U.N. General Assembly Resolution in June 2010[40] and a Resolution adopted by the Human Rights Council in September 2010,[41] considerable uncertainty persists with regard to the true normative status of the human right to water under international law. Although General Comment No. 15

identifies Articles 11 and 12 of the 1966 International Covenant on Economic, Social and Cultural Rights (ICESCR),[42] on the right to an adequate standard of living and the right to the highest attainable standard of health respectively, as the primary legal bases for the HRW concept,[43] it should be remembered that CESCR general comments do not formally impose legal obligations on ICESCR States Parties, let alone other States, and that General Comment No. 15 merely constitutes a non-binding but 'highly authoritative interpretation of the Covenant' and of the legal implications which flow from key relevant Covenant provisions.[44] McCaffrey characterises General Comment No. 15 as being 'more in the nature of a statement *de lege ferenda* rather than *lex lata*' and cautions that the interpretation of Articles 11 and 12 contained therein 'must be accepted by the States parties to the Covenant in order to be binding upon them'.[45] Also, the fact the HRW derives from a number of expressly articulated primary rights may lead to confusion. As Williams puts it, 'various connected rights may implicate different state obligations' and she illustrates this point by explaining that a right to water derived from the right to life, which requires the provision of drinking water, would impose lesser State obligations than a right to water derived from the right to health, which requires the provision of water for both drinking and sanitation.[46] Of course, uncertainty remains as to the true normative status and content of a number of the economic, social and cultural rights listed under the ICESCR, from which the rights to water may be derived, inevitably leading to further confusion as to the implications of the right to water.[47] For example, though some commentators describe the right to food as well established,[48] it might be argued that it raises many questions with regard to the force and extent of such welfare rights under the ICESCR. At any rate, though proponents of an independent right to water argue that it would result in greater interpretive consistency, State compliance, enforcement and remedies for violations,[49] such an independent right could only arise in international law by means of a dedicated treaty instrument or customary international law.[50] It is clear that there does not currently exist a general treaty instrument, nor any proposal for such an instrument, by which States might bind themselves in this regard. Also, it would appear that, despite the sustained declaratory support of international conferences and U.N. agencies, as well as some limited legal and constitutional State recognition,[51] there is as yet insufficient generalised State practice to establish a right to water under customary international law that would bind those States that have not actively and formally recognised the right.[52] Therefore, Williams concludes that '[A]t best, this seems to give the independent right the current status of a normative ideal'.[53] Indeed, overshadowing any discussion of whether the human right to water might exist as an ancillary or independent right is the fact that the ICESCR, and any rights derived therefrom, suffer from a clear lack of immediate enforceability, with Article 2(1) merely requiring each State party 'to take steps … to the maximum of its available resources, with a view to achieving progressively the full realization of the rights recognized in the present Covenant'. As one commentator has noted in relation to such "second generation" rights, '[t]he principal challenge is therefore linking the expectations of individuals as rights-holders with the duties owed by others.'[54] At the level of the practical enforceability of the obligations set out under the ICESCR, McCaffrey points out that the language of Article 2(1) would provide a lawyer acting for a State accused of breaching its obligations with 'ample bases for a defense'.[55]

Provisions of other international human rights instruments are also cited, even less convincingly, as providing support for the normative status of the HRW concept in international law. Article 25 of the 1948 Universal Declaration of Human Rights (UDHR), which proclaims 'the right to a standard of living adequate for the health and well-being of himself and his family, including food …', is often cited,[56] even though, as a U.N. General Assembly Resolution,[57] the UDHR is not binding *per se*. While it is generally accepted that many of the basic human rights contained therein have become part of customary international law, or at least constitute authoritative interpretations of the U.N. Charter's provisions on human rights,[58] and thus bind States generally, such customary status is normally only accorded to the so-called 'liberty rights' contained in the Declaration,[59] rather than the 'welfare rights', of which the right to an adequate standard of living is one.[60] Paragraph 3 of General Comment No. 15 also links the HRW to the right to life set out under Article 6 of the 1966 International Covenant on Civil and Political Rights (ICCPR),[61] even though it remains unclear whether Article 6 'merely protects against arbitrary deprivation of life by the State, or also guarantees against death from such causes as lack of water or food, exposure to the elements, or lack of medical attention'.[62] McCaffrey points out that the view that 'rights such as one to an appropriate means of subsistence belong within the category of economic, social and cultural rights' enjoys the support of respected commentators[63] and is more in accordance with the reality of water services provision and with the true intentions of States parties to the Covenants.[64] Particular provisions of a number of binding international conventions which apply in varying specific contexts are also commonly cited in support of the HRW. These include Article 14(2) of the 1979

Convention on the Elimination of All Forms of Discrimination Against Women (CEDAW),[65] Article 24(2)(c) of the 1989 Convention on the Rights of the Child (CRC),[66] the 1949 Geneva Convention (III) on the Treatment of Prisoners of War,[67] and the 1949 Geneva Convention (IV) on the Treatment of Civilian Persons in Time of War.[68] Though Gleick heralds Article 24 of the CRC as the first ever example in a binding international treaty instrument having general application of 'explicit recognition of the connection between resources, the health of the environment, and human health',[69] McCaffrey is rather more circumspect[70] and cautions that, while the clear need of vulnerable groups for access to sufficient and safe water supplies has been recognised in such instruments, 'this does not necessarily mean that States have recognized a human right to water, with all its implications, either generally or in those specific instruments'.[71] Similarly, in relation to number of regional human rights instruments, such as Article 14 of the 1990 African Charter on the Rights and Welfare of the Child[72] or Article 11 of the 1988 Additional Protocol to the American Convention of Human Rights in the area of economic, social and cultural rights[73], McCaffrey once again cautions that, even where 'safe drinking water' or 'basic public services' are expressly mentioned, '[a] right to water was not recognized per se … however. Rather, the failure to meet basic water needs was found to constitute, or at least contribute to, violations of other rights'.[74] Therefore, though a human rights-based approach may be conceptualised 'in terms of society's obligations to respond to the inalienable rights of individuals',[75] fundamental questions persist about the normative origins and precise legal status of such an approach to water entitlements, adding to the appeal of a GAL analysis of the concept. As the HRW concept has had to be derived from selected provisions of key human rights instruments by bodies charged with their authoritative, if at times progressive, interpretation, it can be regarded to some extent as a creature of GAL. Indeed, it is in keeping with a GAL analysis of HRW that regional bodies with responsibility for monitoring State compliance with human rights obligations have also inferred the existence of a right to water from the core obligations of States under more general regional human rights instruments.[76]

Of course, implicit support for the HRW concept can be found in a diverse range of legal instruments operating at both the international and national levels and covering a variety of areas of activity. In relation to international water resources law, for example, though the U.N. Watercourses Convention addresses the obligations of international watercourse States rather than the rights of individuals, it would appear to support the existence of a State obligation to cater for the basic needs of citizens for water by expressly providing for watercourse States to have 'special regard … to the requirements of vital human needs' over and above all other classes of uses of shared water resources.[77] The 'requirements of vital human needs' would appear to correspond broadly with the obligations of States and the entitlements of individuals under the human right to water.[78] Similarly, though the 1992 UNECE Convention on the Protection and Use of Transboundary Watercourses and International Lakes[79] is a regional instrument concerned primarily with the rights and obligations of States, its 1999 Protocol on Water and Health[80] expressly requires the parties to take 'all appropriate measures for the purpose of ensuring … adequate supplies of wholesome drinking water'[81] and further provides that the parties 'shall pursue the aims of … access to drinking water for everyone …'.[82] At the level of a river basin agreement, the 2002 Water Charter of the Senegal River[83] provides 'a rare example of a treaty referring expressly to a human right to water'.[84] The ILA's 2004 Berlin Rules, which have revised and updated the ILA's seminal 1966 Helsinki Rules, give clear and formal priority to vital human needs[85] and also include a dedicated article on 'The Right of Access to Water'.[86] Thus, a learned body as influential as the ILA expressly links the human right to water to the position, widely acknowledged in international codifications and accepted in State and arbitral practice, that uses required for the satisfaction of vital human needs take priority over other, less urgent uses. Indeed, General Comment No. 15 suggests on numerous occasions the significance of the emergence of the HRW concept for inter-State practice in respect of shared water resources.[87]

Though the recognition given to the HRW concept in national constitutional texts[88], national legislation[89] and the pronouncements of national courts,[90] has often tended to be anything but unequivocal,[91] it has received solid support in numerous seminal declaratory instruments, including the preamble of the 1977 Mar del Plata Action Plan of the United Nations Water Conference,[92] paras. 6.12 and 18.47 of Agenda 21,[93] and Principle No. 4 of the Dublin Statement on Water and Sustainable Development.[94] In recent decades, numerous declaratory instruments have committed governments to improving levels of access to water supply and sanitation.[95] In addition, the UN Commission on Sustainable Development (CSD), has concluded that priority must be accorded to 'the social dimension of freshwater management' and has invited governments to allocate sufficient public financial resources to ensure universal access to water supply and sanitation.[96] Indeed, it would appear that many international financial institutions (IFIs), including most multilateral development banks (MDBs), have for some time included *de facto* recognition of key elements of the right to water in their relevant institutional safeguard policies and procedures, adopted to protect individuals who might otherwise be adversely affected by MDB-funded projects.[97] Thus, the HRW concept conforms to the GAL characteristic of a norm arising from a plurality of sources of practice, but not necessarily satisfying the traditional sources of norms of international law.

Procedural elements of HRW

Significantly as regards a GAL analysis of HRW, it is clear that any elaboration of the concept involves the inclusion of detailed procedural elements regarded as inherent to the concept. In addition to those provisions of global and regional human rights instruments which might be argued explicitly or implicitly to include the human right to water, it is clear that all such instruments would now be interpreted so as to require that States facilitate a participative approach in respect of activities that might impact on human rights, by ensuring the adoption of procedures by which interested individuals or communities likely to be affected can receive and access relevant information, meaningfully participate in decision-making and, if necessary, have access to some appropriate means of legal recourse.[98] Such a participatory approach to guaranteeing human rights would equally apply to activities which might impact on the availability of water resources, particularly where this might arise by virtue of environmental risk, and procedural and participative rights are a very significant element of the normative content of the human right to water as put forward in General Comment No. 15.[99] Indeed, the requirement for States parties to the ICESCR to ensure a participatory and transparent process for the adoption and implementation of a national water strategy and plan of action is included among the non-derogable 'core obligations' of States under General Comment No. 15.[100] Clearly, such procedural requirements, which correlate closely with the procedural and informational requirements of the human right to water as set out under General Comment No. 15, would equally apply under existing regional human rights instruments to any major project or policy initiative, such as the privatisation of a water utility, which threatened the quality or availability of water supply or sanitation services.[101]

These procedural requirements appear all the more widely accepted and applied when one considers that broad informational and participatory rights are generally also included under regional and global environmental instruments. The concept of participation in international environmental law is exemplified by the 1998 UNECE Aarhus Convention[102] and such requirements are also central to the carrying out of an adequate environmental impact assessment (EIA) consistent with the standards established under international law.[103] More generally, in the field of sustainable development, all seminal instruments purport to establish participatory standards which apply not only to States but also to international organisations, including multilateral development banks (MDBs). Participatory rights are absolutely central to Chapter 18 on freshwater resources of Agenda 21[104] Therefore, the accumulated practice of regional human rights enforcement bodies strongly suggests that the CESCR's General Comment No. 15 largely involves a codification of existing State obligations under general international human rights law and general international environmental and sustainable development law, rather than an attempt at the progressive development of participatory principles applying to matters of access to water. The same might be said of the origins and normative basis of the principle of non-discrimination, which forms another essential substantive element of the human right to water as set out under General Comment No. 15 and is also included among the non-derogable 'core obligations' of States.[105] Likewise, the inclusion of special protections for indigenous peoples under General Comment No. 15[106] might be traced to and justified under ILO Conventions 107 and 169.[107] This focus on procedural obligations arising from a diversity of legal sources certainly lends itself to a GAL analysis.

Extension of HRW to private corporate actors

The GAL concept helps to address the difficult issue of extension of human rights norms and values to private corporate actors. The involvement of the private sector in the provision of water and sanitation services gives global administrative law a role in clarifying the altered regulatory obligations imposed on State authorities and private actors by the emergence of the human right to water.[108] Morgan justifies this mode of analysis by explaining that such arrangements comprise 'hybrid blends of public and private actors linked in routines of both formal and informal participation at multiple levels of governance'.[109] In her analysis, she focuses specifically on the issue of participation in decision-making processes that affect vital individual interests to explore '[w]hat are the forms and processes (both formal and informal) that facilitate participation in, or the capacity to participate in, transnational urban water services governance?'[110]

Despite recent interest in the idea of extending the application of key international human rights norms so as to apply directly to corporations,[111] which could make the requirements of the human right to water central to arrangements for the privatisation of water and sanitation services, it is clear that such norms could only as yet dictate corporate behaviour where individual corporations have voluntarily agreed to abide by codes of conduct which explicitly or implicitly require compliance with international human rights norms.[112] High-profile examples of such voluntary initiatives include the U.N. Global Compact, providing a set of 10 core principles which, while not referring explicitly to water rights or entitlements, does provide that companies should comply with international human rights norms, which might include a right to water and sanitation.[113] Similarly, the OECD Guidelines for Multinational Enterprises represent another legally non-binding initiative which can operate to support application of key elements of the human right to water to private companies.[114] In addition, the declarative practice of States tends towards extending responsibility for respecting human rights to private companies involved in the provision of private services. For example, the Draft Declaration on the Right to Access to Essential Services,[115] proposed by France in the context of the 2002 World Summit on Sustainable Development, relates to essential services indispensible for a dignified life, expressly including drinking water and sanitation,[116] and would apply equally to both public and private sector providers. In respect of the development of an appropriate regulatory framework for private sector water service providers, which is informed by international human rights values and widely accepted by States and leading private operators, the adoption in 2007 of three sets of guidelines by Technical Committee 224 of the International Standards Organisation (ISO) represents a significant step.[117] The guidelines set standards for service activities relating to the provision of drinking water supply and sewerage services, which apply equally to both public and private actors, and even attempt to deal with the role of 'cost' or 'price' within the standard of service. With 35 Participating Countries and 17 Observer Countries involved in their development and adoption, in liaison with a range of interested international organisations, including the WHO, World Bank and International Water Association, and leading industry interests, the ISO guidelines are likely to prove influential in determining an acceptable level of service provision where a dispute arises with a private sector provider.

Of potentially far greater significance in this regard is the ongoing U.N. initiative on the issue of human rights and transnational corporations and other business entities which is developing practical recommendations for operationalising a framework for ensuring that private corporations respect human rights. In 2008, the Special Representative of the Secretary-

General (SRSG) proposed a new approach for understanding the issue of human rights and transnational corporations, based on the "protect, respect and remedy" policy framework. The framework is described as resting on three complementary pillars:

> 'the State duty to protect against human rights abuses by third parties, including business, through appropriate policies, regulation, and adjudication; the corporate responsibility to respect human rights, which in essence means to act with due diligence to avoid infringing on the rights of others; and greater access by victims to effective remedy, judicial and non-judicial.'

Though much of the detail of this policy framework continues to be elaborated through the ongoing work of the SRSG, this approach appears already to enjoy considerable support from the Human Rights Council[118] and among States, leading business entities and civil society.[119] In respect of the State's duty to protect, the SRSG's 2009 Report states that this requires each State to ensure the protection of rights 'against other social actors, including business, who impede or negate those rights' and that it 'applies to all recognized rights that private parties are capable of impairing, and to all types of business enterprises'.[120]

Further, recent developments in international investment law, the body of rules that function to provide protection to private sector investors operating in foreign jurisdictions against arbitrary interference with their property or business interests by the sovereign actions of host States, suggest that this area may offer some clarity in respect of the requirements imposed by the human right to water on host States and private actors in cases of water services privatisation. The concept would appear to have received solid, if implicit, support in a number of recent statements of ICSID international arbitration tribunals engaged in the settlement of investor-State disputes over water and sanitation service contracts.[121] In relation to a series of primarily procedural issues, ICSID tribunals have recognised that the provision of water and sanitation services inevitably involves fundamental questions of human rights.[122] Though the precise mechanism by means of which such human rights values came to inform the reasoning of international investment arbitration tribunals was nowhere set out by the tribunals in question, it can be easily explained in terms of the application of GAL standards of good governance.

Conclusion

Therefore, the diversity of legal sources for the HRW concept and the uncertainties surrounding its legal status, along with the procedural character of many of its inherent requirements and the challenges presented by the extension of human rights values to non-State actors, make a GAL analysis a very useful approach to better understanding the HRW discourse. Further, the HRW concept might be presented as an exemplar of the GAL phenomenon. Indeed, the HRW concept would appear to answer most, if not all, of the concerns raised earlier about the GAL concept.[123] First, in response to the allegation that GAL subverts traditional democratic political processes and the principle of State sovereignty, a study published in June 2010 by the French NGO *Coalition Eau* claims that 190 States have thus far declared support for the HRW concept at ministerial level.[124] Indeed, as regards the concern that GAL is a 'Western construct', the same study points out that such ministerial declaratory support can be attributed to 139 developing States and 51 developed States. Further, the level of developing State representation in CESCR and ISO Technical Committee 224[125] should help to address such concerns in respect of the HRW. The existence of detailed normative and technical guidance on the HRW concept, contained in such documents as General Comment No. 15 and the 2007 ISO Guidelines, should go some way towards addressing concerns about the lack of clarity and consistency in respect of GAL norms. Indeed, several of the key areas of legal practice supporting and informing the HRW concept, such as human rights law and international investment law, are themselves 'established fields of normativity'[126] with established institutional structures and autonomous bodies of jurisprudential thought. In response to concerns about the 'juridification' of political processes, the areas of human rights law and international investment law are already substantially 'juridified', but the discretion of judicial and quasi-judicial decision-makers are largely constrained by highly-developed normative and technical guidance.

Notes

[1] B. Kingsbury, N. Krisch, R. B. Stewart and J. B. Wiener, 'Global Governance as Administration – National and Transnational Approaches to Global Administrative Law', (2005) 68/3&4 *Law and Contemporary Problems* 1, at 2.

[2] *Ibid.,* at 3.

[3] B. Kingsbury, 'The Concept of "Law" in Global Administrative Law', (2009) 20/1 *European Journal of International Law* 23, at 25.

[4] Available at www.asil.org/ilib/WaterReport2004.pdf.
[5] Supra, n. 3, at 25.
[6] See, for example, ICSID Case No. ARB/03/19, *Aguas Argentinas S.A., Suez, Sociedad General de Aguas de Barcelona S.A. and Vivendi Universal S.A. v. The Argentine Republic,* Order in Response to a Petition for Transparency and Participation as Amicus Curiae (19 May 2005); ICSID Case No. ARB/03/17, *Aguas Provinciales de Santa Fe S.A., Suez, Sociedad General de Aguas de Barcelona S.A. and InterAguas Servicios Integrales del Agua S.A. v. The Argentine Republic,* Order in Response to a Petition for Participation as *Amicus Curiae* (17 March 2006); ICSID Case No. ARB/03/19, *Suez, Sociedad General de Aguas de Barcelona S.A. and Vivendi Universal S.A. v. The Argentine Republic,* Order in Response to a Petition by Five Non-Governmental Organizations for Permission to Make an *Amicus Curiae* Submission (12 February 2007); ICSID Case No. ARB/05/22, *Biwater Gauff (Tanzania) Ltd. V. United Republic of Tanzania*, Procedural Order No. 5 (2 February 2007).
[7] Committee on Economic, Social and Cultural Rights, *General Comment No. 15, The Right to Water (Articles 11 and 12 of the International Covenant on Economic, Social and Cultural Rights)*, U.N. DOC. E/C.12/2002/11, 26 November 2002. Available at www.unchr.ch/html/menu2/6/gc15.doc.
[8] *Supra,* n. 1, at 2.
[9] *Ibid.,* at 4.
[10] *Ibid.,* at 5. (Original emphasis).
[11] *Ibid.*
[12] *Supra,* 3, at 23.
[13] *Ibid.*
[14] C. Harlow, 'Global Administrative Law: the Quest for Principles and Values', 17/1 (2006) *European Journal of International Law* 187-214, at 187.
[15] *Ibid.,* at 188.
[16] ISO 24510:2007 Activities relating to drinking water and wastewater services -- Guidelines for the assessment and for the improvement of the service to users; ISO 24511:2007 Activities relating to drinking water and wastewater services -- Guidelines for the management of wastewater utilities and for the assessment of wastewater services; ISO 24512:2007 Activities relating to drinking water and wastewater services -- Guidelines for the management of drinking water utilities and for the assessment of drinking water services. Available at www.iso.org/iso/iso_catalogue/catalogue_tc/catalogue_tc_browse.htm?commid=299764&published=on&includesc=true.
[17] *Supra,* 3, at 23.
[18] *Supra,* n. 3, at 32.
[19] *Ibid.,* at 32-33.
[20] *Ibid.*, at 34.
[21] *Supra,* n. 7.
[22] *Supra,* n. 16.
[23] *Supra,* n. 6.
[24] *Supra,* n. 14, at 188.

[25] *Ibid.*
[26] *Supra,* n. 3, at 33 (emphasis added).
[27] On the link between the human right to water and the human right to health, see General Comment No. 15, *supra,* n. 7, paras. 8, 10, 11, 12(b), 16(c) and (d), 21, 23, 28(b) and (e), 29, 44(a)(iii) and (b)(i). For example, para. 8 states that: 'Environmental hygiene, as an aspect of the right to health under article 12, paragraph 2*(b),* of the Covenant, encompasses taking steps on a non-discriminatory basis to prevent threats to health from unsafe and toxic water conditions.'
[28] See General Comment No. 15, *ibid.,* at paras. 12(c)(iv), 16(a), 24, 37(f), 48, 55 and 56.
[29] See Kingsbury, *supra*, n. 3, at 32-33.
[30] See, in particular, Harlow, *supra,* n. 14, at 207-214.
[31] Harlow, *ibid.,* at 213.
[32] *Ibid.,* at 207.
[33] *Ibid.,* at 209.
[34] *Ibid.,* at 211.
[35] *Supra,* n. 3, at 29.
[36] *Supra,* n. 14, at 208.
[37] *Ibid.*
[38] Kingsbury, *supra,* 3, at 25.
[39] Vogel, 'Administrative Law: International Aspects', in R. Bernhardt (ed.), *Encyclopedia of Public International Law* (1922) 22, at 23, quoted in Kingsbury, *ibid.*, at 23-24.
[40] UNGA Res. A/64/L.63/Rev.1 on The human right to water and sanitation (26 July 2010).
[41] UN HRC Res. A/HRC/15/L.14 on Human rights and access to safe drinking water and sanitation (24 September 2010).
[42] UNGA Res. 2200, 21 UN GAOR, 22nd Sess., Supp. 49, UN Doc. A/6316 (1967), (1967) 6 ILM 360. Adopted and opened for signature, ratification and accession by UN General Assembly Resolution 2200A (XXI) of 16 December 1966, the Covenant has now been ratified by 147 States. See further, J. Razzaque, 'Trading Water: The Human Factor', (2004) 13/1 *Review of European Community and International Environmental Law* 15, at 17.
[43] See further, P. H. Gleick, 'The human right to water', (1998) 1 *Water Policy* 487, at 492.
[44] S. McCaffrey, 'The Human Right to Water', in E. Brown Weiss, L. Boisson de Chazournes and N. Bernasconi-Osterwalder, *Fresh Water and International Economic Law* (2005, Oxford University Press) 93, at 94. Though it should be pointed out that, as a non-binding interpretation, General Comment No. 15 may be used to determine whether Stats have met their treaty obligations, see M. Williams, 'Privatization and the Human Right to Water: Challenges for the New Century', (2007) 28 *Michigan Journal of International Law* 469, at 475. See also, E. B. Bluemel, 'The Implications of Formulating a Human Right to Water', (2004) 31 *Ecology Law Quarterly* 957, at 972.

[45] *Ibid.,* at 103. He further points out that 'This is also true, *a fortiori*, of States that are not parties to the Covenant.' More positively, however, he also points out, at 94, that 'thus far States Parties to the Covenant have not objected to the interpretation contained in the General Comment'.
[46] *Supra,* n. 44, at 477. Williams also concludes that both CEDAW and CRC, *ibid.,* suggest that 'the right to water merits protection because of its connection to other rights'. See further, Bluemel, *ibid.*, at 963.
[47] Bluemel, *ibid.,* at 971.
[48] Williams, *supra,* n. 44, at 479.
[49] Bluemel, *supra,* n. 44, at 968-972; A. Hardberger, 'Life, Liberty and the Pursuit of Water: Evaluating Water as a Human Right and the Duties and Obligations it Creates', (2005) 4/2 *Northwestern Journal of International Human Rights* 331, at 360-362. See, Williams, *ibid.,* at 478. See also, Gleick, *supra,* n. 43, as an example of a commentator who argues, at 490 and 501, that an independent right to water exists in customary international law.
[50] See, for example, Hardberger, *ibid.,* who, while conceding that an independent human right to water is not considered customary international law, argues, at 361-362, that 'there should be a binding document that encompasses the ideas of General Comment 15'.
[51] See further, *infra.*
[52] D. J. Bederman, *International Law Frameworks* (2001), at 15, cited by Williams, *supra,* n. 44, at 478.
[53] *Ibid.*
[54] S. R. Tully, 'The Contribution of Human Rights to Freshwater Resource Management' (2004) *Yearbook of International Environmental Law* 101, at 103.
[55] *Supra,* n. 44, at 97.
[56] See Gleick, *supra,* n. 43, at 491.
[57] UNGA Resolution 217A (III), (New York, 10 December 1948).
[58] Articles 55 and 56 of the 1945 U.N. Charter provide, rather generally, that the U.N. shall promote 'respect for, and observance of, human rights and fundamental freedoms' and state that it is the duty of all members to promote these goals.
[59] See, for example, O. Schachter, 'International Law in Theory and Practice', (1982) 178 *Recueil des Cours* 9, at 340.
[60] McCaffrey, *supra,* n. 44, at 96. See further, S. McCaffrey, 'A Human Right to Water: Domestic and International Implications', (1992) 5 *Georgetown International Environmental Law Review* 1, at 8.
[61] UNGA Res. 2200, 21 UN GAOR, 22nd Sess., Supp. 52, UN Doc. A/6316 (1967), (1967) 6 ILM 368.
[62] McCaffrey, *supra,* n. 44, at 97. See also McCaffrey (1992), *supra,* n. 60, at 9-11.
[63] A. A. Cancado Trinidade, 'The Parallel Evolutions of International Human Rights Protection and of Environmental Protection and the Absence of Restrictions on the Exercise of Recognized Human Rights', (1991) 13 *Revista IIDH* 35, at 52, cited in McCaffrey, *supra,* n. 44, at 97.
[64] *Ibid.,* at 98.

[65] U.N. Doc. A/34/46 (1979). Entered into force 1981. See also, governmental declarations of intent to ensure universal access to safe drinking water and available sanitation through public distribution systems, contained in the 1995 Beijing Declaration and Platform for Action, UN Doc A/CONF.177/20 (1995) and UN Doc. A/CONF.177/20/Add.1 (1995), *Women and Health*, at paras. 92 and 106(x), cited in Tully, *supra,* n. 54, at 105.
[66] U.N. Doc. A/44/49 (1989). Entered into force 1990.
[67] Articles 20, 26, 29 and 46. (1949) 75 *UNTS* 135. Entered into force 1950.
[68] Articles 23, 55, 59, 85, 89 and 127. (1949) 75 *UNTS* 287. Entered into force 1950. See also, Additional Protocol to the Geneva Conventions Relating to the Protection of Victims of International Armed Conflicts (Protocol I), (1977) 1125 *UNTS* 3, at Articles 54 and 68-71 (entered into force 1978), and Additional Protocol to the Geneva Conventions Relating to the Protection of Victims of Non-International Armed Conflicts (Protocol II), (1977) 1125 *UNTS* 609, at Articles 5 and 18 (entered into force 1978).
[69] *Supra,* n. 43, at 494.
[70] *Supra,* n. 44, at 98.
[71] *Ibid.,* at 107.
[72] OAU Doc. CAB/LEG/24.9/49 (1990), entered into force 1999.
[73] (San Salvador, 17 November 1988). Available at www.oas.org/juridico/english/Treaties/a-52.html.
[74] *Supra,* n. 44, at 99.
[75] UNDP, *Integrating Human Rights with Sustainable Human Development* (UNDP, New York, 1998), at 173-174, cited in E. Filmer-Wilson, 'The Human-Rights-Based Approach to Development: the Right to Water', (2005) 23 *Netherlands Quarterly of Human Rights* 213, at 213.
[76] See, for example, the 1995 findings of the African Commission on Human and Peoples' Rights in respect of Zaire, African Commission on Human and Peoples' Rights, Comm. No. 25/89, 47/90, 56/91, 100/93, Decision taken at the 18th Ordinary Session, 1995; the 1997 findings of the Inter-American Commission on Human Rights in respect of Ecuador, Inter-American Commission on Human Rights, *Report on the Situation of Human Rights in Ecuador,* OEA/Ser.L/V/II.96, Doc. 10 rev. 1 (24 April 1997). See McCaffrey, *supra,* n. 44, at 99.
[77] Article 10(2) of the 1997 United Nations Convention on the Law of Non-Navigational Uses of International Watercourses, (1997) 36 *ILM* 719 (New York, 21 May 1997, not yet in force).
[78] See the oral report of the coordinator of the informal consultations on Article 10(2), U.N. Doc. A/C.6/51/SR.57 (1997), at 3. See further, A. Tanzi and M. Arcari, *The United Nations Convention on the Law of International Watercourses* (Kluwer Law International, The Hague / Boston, 2001), at 139.
[79] (1992) 31 *ILM* 1312.
[80] U.N. Doc. MP.WAT/AC.1/1991/1 (17 June 1999).
[81] Article 4(2)(a).
[82] Article 6(1)(a).
[83] 28 May 2002 (Mauritania, Mali, Senegal).

[84] McCaffrey, *supra*, n. 44, at 101, referring to Article 4(3) of the 2002 Charter.
[85] Article 14(1) provides that
'In determining an equitable and reasonable use, States shall first allocate waters to satisfy vital human needs'
Article 3(20) defines "vital human needs" to mean
'waters used for immediate human survival, including drinking, cooking, and sanitary needs, as well as water needed for the immediate sustenance of a household'.
[86] Article 17 provides, *inter alia*, that
'[e]very individual has a right of access to sufficient, safe, acceptable, physically accessible, and affordable water to meet that individual's vital human needs'.
[87] See, for example, General Comment No. 15, *supra,* n. 7, paras. 31, 34, 35 and 44(c)(vii).
[88] For example, section 27(1)(b) of the Constitution of the Republic of South Africa.
[89] For example, section 3(1) of the South African 1997 Water Services Act and Article 36 of the 1998 National Water Act.
[90] In respect of the Indian courts, see *Charan Lal Sahu v. Union of India*, AIR (1990) SC, at 1495; *F. K. Hussain v. Union of India*, AIR (1990) Kerala 321, at 340; *Subash Kumar v. State of Bihar* (1991) 1 SCC 598, at 608; *M. C. Mehta v. Union of India* (1998) 9 SCC 589, at 607.
[91] For example, Tully, *supra,* n. 54, is dismissive, at 120, of most of the examples commonly cited.
[92] Mar del Plata Declaration and Plan of Action, *Report of the UN Water Conference,* UN Publication No. E.77.II.A.12, Part One, Chapter 1 (1977). See General Comment No. 15, *supra,* n. 7, para. 4, note 5. The Mar del Plata Declaration stated that
'all peoples, whatever their stage of development and their social and economic conditions, have the right to have access to drinking water in quantities and of a quality equal to their basic needs'.
[93] *Report of the United Nations Conference on Environment and Development* (Rio de Janeiro, 3-14 June 1992), UN Doc. A/ CONF.151.26 (vol. II) (1992). Available at www.un.org/esa/sustdev/agenda21.htm.
[94] Dublin Statement on Water and Sustainable Development, International Conference on Water and the Environment: Development Issues for the Twenty-First Century, UN Doc. A/CONF.151/PC/112 (1992). Available at
www.wmo.ch/homs/documents/english/icwedece.htm Principle No. 4 states that
'it is vital to recognise first the basic right of all human beings to have access to clean water and sanitation at an affordable price.'
[95] For example, UN General Assembly Resolution 35/18 (1980) concerning a Proclamation of the International Drinking Water Supply and Sanitation Decade 1981-1990.
[96] CSD Decision 6/1 (1998) on Strategic Approaches to Freshwater Management, Official Records of the Economic and Social Council, Supplement No. 9, UN Doc. E/1998/29, Chapter 1, Section B, paras. 3, 10 and 19.

---

[97] For example, World Bank Operational Directive 4.30: Involuntary Resettlement, (June 1990), paras.9, 12 and 20 require that, where involuntary resettlement is unavoidable during the course of a project, access to water is to be considered under the detailed resettlement plan required to address and mitigate the impacts of the resettlement on the resettled and host populations.

[98] See generally, P. Cullet and A. Gowlland-Gaultieri, 'Local Communities and Water Investments', in E. Brown Weiss, L. Boisson de Chazournes and N. Bernasconi-Osterwalder, *Fresh Water and International Economic Law* (2005, Oxford University Press) 303.

[99] See, for example, General Comment No. 15, *supra,* n. 7, paras. 12(c)(iv), 16(a), 24, 37(f), 48, 55 and 56.

[100] See General Comment No. 15, para. 37(f). General Comment No. 15, para. 40, describes the core obligations set out in para. 37 as 'non-derogable'.

[101] See Cullet and Gowlland-Gualtieri *supra,* n. 98, at 313-315, citing Communication No. 155/96, *The Social and Economic Rights Action Center and the Center for Social and Economic Rights v. Nigeria*, African Commission on Human and Peoples' Rights, 30th Ordinary Session, (13-27 October 2001), at para. 53; Chapter X to the Second Report on the Situation of Human Rights in Peru, Inter-American Commission on Human Rights; Chapter IX to the Report on the Situation of Human Rights in Ecuador, Inter-American Commission on Human Rights, OEA/Ser.L/V/II.96, doc. 10 rev. 1 (Recommendations) (24 April 1997); Case 7615 (Brazil), Inter-American Commission on Human Rights, 1984-1985 Annual Report 24, OEA/Ser.L/V/II.66, doc. 10, rev. 1 (1985), the *Yanomami* case. See further, *Awas Tingni Mayagna (Sumo) Indigenous Community v. Nicaragua,* Judgment of 31 August 2001, Inter-American Court of Human Rights, (Ser. C), No. 79 (2001); *Guerra and Others v. Italy* (1998) 26 *European Human Rights Reports* 357; *Zander v. Sweden* (1993) 18 *European Human Rights Reports* 175.

[102] Convention on Access to Information, PublicParticipation in Decision-Making and Access to Justice in Environmental Matters, (Aarhus, 25 June 1998), 38 *ILM* 517 (1999).

[103] See, for example, Arts. 2(2), 2(6), 3(8) and 4(2) of the 1991 UNECE Convention on Environmental Impact Assessment in a Transboundary Context, (Espoo, 25 February 1991), 30 *ILM* 800 (1991).

[104] Agenda 21, (Rio de Janeiro, 13 June 1992), U.N. Doc. A/CONF.151/26 (Vols. I, II & III) (1992).

[105] See General Comment No. 15, paras. 13-16 and 37(b). On the emergence of the principle of non-discrimination in the context of international environmental law, see H. Smets, 'Le principe de non-discrimination en matière de protection de l'environnment', (2000) *Revue Europé du Droit de l'Environnment,* 1.

[106] See, for example, General Comment No. 15, paras. 16(d), and 37(b), (f) and (h).

[107] Convention concerning the Protection and Integration of Indigenous and Other Tribal and Semi-Tribal Populations in Independent Countries (26 June 1957); Convention concerning Indigenous and Tribal peoples in Independent Countries (27 June 1989), 28 *ILM* 1382 (1989).

[108] See further, B. Morgan, 'Turning Off the Tap: Urban Water Service Delivery and the Social Construction of Global Administrative Law', (2006) 17/1 *European Journal of International Law* 215, at 216.
[109] *Ibid.*
[110] *Ibid.,* at 217.
[111] See generally, S. R. Ratner, 'Corporations and Human Rights: A Theory of Legal Responsibility', (2001) 111 *Yale Law Journal* 443.
[112] See generally, S. D. Murphy, 'Taking Multinational Corporate Codes to the Next Level', (2005) 43 *Columbia Journal of Transnational Law* 389.
[113] See M. Williams, *supra,* n. 44, at 488-491, who points out that the corporate code of conduct of at least one large transnational water services company (RWE AG) expressly alludes to the UN Global Compact.
[114] See Organisation for Economic Co-operation & Development, OECD Guideline for Multinational Enterprises, June 21, 1976, available at www.oecd.org/dataoecd/56/36/1922428.pdf (revised 2000).
[115] See further, B. Morgan, 'The Regulatory Face of the Human Right to Water' (2004) 15/5 *Water Law* 179-186, at 181-182.
[116] Draft Declaration, Article 1.
[117] *Supra,* n. 16. On the background to this ISO initiative, see generally, Morgan, *supra*, n. 115, at 182-183. See also Morgan, *supra,* n. 108, at 224-227.
[118] U.N. Doc. A/HRC/8/5. See U.N. Doc. A/HRC/11/13, *ibid.,* at 3, para. 1.
[119] U.N. Doc. A/HRC/11/13, *ibid.,* at 3-4, paras. 3-5.
[120] *Ibid.,* at 6-7, para. 13.
[121] *Supra,* n. 6.
[122] ICSID Case No. ARB/03/19, *Aguas Argentinas S.A., Suez, Sociedad General de Aguas de Barcelona S.A. and Vivendi Universal S.A. v. The Argentine Republic,* Order in Response to a Petition for Transparency and Participation as Amicus Curiae (19 May 2005).
[123] See Harlow, *supra,* n. 14, at 207-214.
[124] See www.coalition-eau.org/spip.php?rubrique1.
[125] See, Morgan, *supra,* n. 115, at 182-183.
[126] See Kingsbury, s*upra,* n. 3, at 29.

CHAPTER THIRTEEN

KEEPING NATURE ALIVE: FROM MORAL MOTIVATIONS TO LEGAL IMPLICATIONS

KATHRYN KINTZELE

> "It would not be wrong to say that the love of nature, the desire for its preservation… is among those pristine and universal values which command international recognition."[1]

Introduction

Ethics form the foundation of law, from the very origins of jurisprudence, to the development of laws into constitutions and statutes, to its application in judicial opinions, case law and other means of enforcement. Ethics asks us, as an individual and as a society: what is just? What is good and bad, or right and wrong, and when does responsibility attach? Law, as a system of rules that guides behavior through appropriate institutions, includes positive and negative rights, and reciprocal duties. It describes and enforces ethics, what a particular society deems to be right and wrong. For purposes of this article, ethics will refer to the realm of public ethics, or the ethics of a society.

Through comparative law, the focus of this article will be ethical principles of environmental conservation as found in constitutions and judicial opinions from around the world, guided by the ethical principles of truth; democracy; justice and equity; and respect and love, supported by the work of the Biosphere Ethics Initiative.[2] It will begin with a comparative legal analysis of constitutions, the very framework of the nation-state (hereinafter, "state").[3] A constitution textualizes what is just and unjust, and offers protections for the vulnerable.

But ethics is much more than symbology and words, ethics is action. Therefore, this article will then give an overview of judicial opinions from states and international courts and show how those constitutions, and those ethical principles, are, or are not, being implemented. Some states carry more weight than others on decisions outside of their jurisdiction, but international principles of law are sources of law under many constitutions, as well as under the Statute of the International Court of Justice.

From these comparative analyses, this article will show that ethics that support a flourishing and sustainable future are already found in existing law and practice, and how they are continuing to expand and emerge in greater extent with each constitutional revision and judicial opinion.

The Right Framework: Ethics and Constitutions

Constitutions often carry a similar structure, regardless of the state: an opening, or preambulatory, aspirational statement, followed by numerated, operational clauses detailing the rights and duties of the government and the citizenry. Many of the clear ethical statements found in constitutions are often preambulatory, but many of the operative clauses show ethical considerations, as well. This section will highlight existing and missing ethical considerations as written in some of the most progressive constitutions from around the world, guided by the often overlapping ethical principles of truth; democracy; justice and equity; and respect and love.

Principles of Truth

There are powerful truths that seem evident to all, but for some reason are often difficult to textualize. The most relevant here are (1) humanity's utter dependence on nature; (2) ecological reconciliation, which asks humans to address our past harms to nature, and in the name of nature;[4] and (3) the precautionary principle, which requires humans to have humility and be honest with what we know and do not know. These are all simple statements, requiring only the recognition of reality, but they are each heavily consequential: they incur responsibility.

An important first step towards a flourishing future for humans and nature is for a state to acknowledge man's correct relationship with the rest of the living and non-living world. It should serve as the cornerstone for all further discussion of rights and duties, for without nature, humans, nor our governance systems, would exist. However obvious this statement seems, few states acknowledge it. In a world defined by capital, progress and growth, it is not difficult to imagine why so many states avoid humanity's own natural limits.

A few states, however, do touch upon humanity's dependence on nature. The Constitution of Ecuador states "Nature has the right to an integral restoration. This integral restoration is independent of the obligation on natural and juridical persons or the State to indemnify the people and the collectives that depend on the natural systems."[5] The Constitution of Cuba states, "The State protects the environment and natural resources of the country. It recognizes their close link with the sustainable economy and social development for making human life more sensible, and for ensuring survival, welfare and security of present and future generations."[6]

The Charter for the Environment, found within the Constitution of France, arguably has some of the strongest language to this point. The opening of the Charter states, "The French People, considering that natural resources and equilibriums have conditioned the emergence of mankind; the future and very existence of mankind are inextricably linked with its natural environment ..."[7]

The precautionary principle, on the other hand, is quickly emerging as a principle of international law. Bhutan states that, "Parliament may, in order to ensure sustainable use of natural resources, enact environmental legislation and implement environmental standards and instruments based on the precautionary principle, polluter pay principle, maintenance of intergenerational equity…"[8] France's Constitution goes even further,

> "When the occurrence of any damage, albeit unpredictable in the current state of scientific knowledge, may seriously and irreversibly harm the environment, public authorities shall, with due respect for the principle of precaution and the areas within their jurisdiction, ensure the implementation of procedures..."[9]

In this same vein of humility of what we know and do not know, and responding accordingly, states should recognize the danger in the commodification of life, the appropriation of life processes and the synthetic creation of new life forms being introduced into the biosphere. Several states speak directly to these concerns, including Ecuador, "The State will apply precaution and restriction measures in all the activities that can lead to the extinction of species, the destruction of the ecosystems or the permanent alteration of the natural cycles… The introduction of organisms and organic and inorganic material that can alter in a definitive way the national genetic patrimony is prohibited."[10]

As encompassing as the precautionary principle is, it should expand its reasoning to recognize that the application of scientific knowledge is not value-neutral. Neither the funding for scientific research, nor its application, is without bias. This should always be addressed when weighing scientific evidence.

Principles of Democracy

States should uphold decision-making that is participatory, inclusive, self-critical and democratic. They should maintain, promote and nurture bio-cultural diversity, with specific concern for indigenous knowledge and culture, and foster local and regional alliances which recognize the knowledge and understanding that each has to contribute. Democratic principles are inextricably linked to the conservation of nature.

Some states, like Montenegro, are straight and to the point, "Montenegro is a civil, democratic, ecological state with social justice, based on the rule of law."[11] Others, like the newly democratic state of Iraq, even require a constitutional oath by each member of the Council of Representatives,

> "I swear by God Almighty to carry out my legal duties and responsibilities with devotion and integrity and preserve the independence and sovereignty of Iraq, and safeguard the interests of its people, and ensure the safety of its land, sky, water, wealth, and federal democratic system, and I shall endeavor to protect public and private liberties, the independence of the judiciary, and pledge to implement legislation faithfully and neutrally. God is my witness."[12]

Democracy cannot exist without the participation of the citizenry. Colombia states, "The laws must guarantee the community's participation in the decisions that may affect the environment."[13] France requires "…a duty to participate in preserving and enhancing the environment," as well as gives everyone the right, "…in the conditions and to the extent provided for by law, to have access to information pertaining to the environment in the possession of public bodies and to participate in the public decision-taking process likely to affect the environment." The Constitution of Ukraine guarantees, "…the right of free access to information about the environmental situation… and also the right to disseminate such information," and that, "No one shall make such information secret."[14]

Respecting and promoting biocultural diversity is another key component of democracy. This is evident in the Constitution of Iraq, "The state shall promote cultural activities and institutions in a manner that befits the civilizational and cultural history of Iraq, and it shall seek to support indigenous Iraqi cultural orientations."[15] Kenya also promotes the protection of cultural values, "The State shall… ensure that social and cultural values traditionally applied by the communities of Kenya for the sustainable management of the environment and natural resources are observed…"[16] The Constitution of Papua New Guinea has as a National Goal and Directive, "…the conservation and replenishment, for the benefit of ourselves and posterity, of the environment and its sacred, scenic, and historical qualities."

Germany includes another aspect of democracy: building local, regional and international alliances, "…land distribution, regional planning and water conservation."[17] The Republic of China seeks to protect the people, culture and environment in its Frontier Regions,

> "The State shall actively undertake and foster the development of education, culture, communications, water conservancy, public health, and other economic and social enterprises among the various ethnic groups in the frontier regions. With respect to land utilization, the State shall, in the light of climatic conditions, nature of the soil, and the life and habits of the people, adopt measures for its protection..."[18]

France states, "This Charter shall inspire France's actions at both European and international levels,"[19] whereas Kenya requires that, "that the environmental standards enforced in the Republic are of the accepted international standards."[20] The important point here is that environmental problems and ecosystems are not constrained by political borders; a point that Colombia makes clear, "The state must also cooperate with other nations in the protection of the ecosystems in border areas."[21]

Principles of Justice and Equity

States should support universal human rights and efforts for social, economic and environmental justice, and recognize the power imbalances that are its obstacles. They should also advance just and sustainable forms of governance within their political and economic structures, including supporting efforts to recognize and implement the rights of nature. To ensure a flourishing a sustainable livelihood, states should also promote a universally fair manner of human development and consumption, sharing both the burdens and the benefits. Within these principles of justice and equity is also the principle of intergenerational equity, or the rights of future generations.

In the Constitution of Bhutan, "The State shall strive to create conditions that will enable the true and sustainable development of a good and compassionate society rooted in Buddhist ethos and universal human values."[22] Social justice is also a precursor to Montenegro's explicit declaration as a democratic and ecological state, "Stemming from the conviction that the state is responsible for the preservation of nature, sound environment, sustainable development, balanced development of all its regions and the establishment of social justice…"[23]

In a State's effort to advance just and sustainable forms of governance, it is crucial to enforce responsibility civilly and criminally. This can be through the polluter pays principle, as in Kenya, "…enforce the "polluter-pays" principle and ensure that the disposal and storage of environmentally undesirable substances, including hazardous waste and munitions, take into account the requirements for a clean, safe and healthy environment,"[24] or through criminal proceedings, as in Paraguay, "The law will define and establish sanctions for ecological crimes. Any damage to the environment will entail an obligation to restore and to pay for damages."[25] Brazil encompasses both, "Conduct and activities considered harmful to the environment shall subject the individual or corporate wrongdoers to penal and administrative sanctions, in addition to the obligation to repair the damages caused."[26] The polluter pays principle is a standing principle of international law, discussed further below. 'Ecological crimes', on the other hand, is a concept that is slowly gaining recognition and should be more strongly advocated.[27]

Only one state, Ecuador, has afforded rights to nature in their Constitution, "Nature or Pachamama, where life is reproduced and exists, has the right to exist, persist, maintain and regenerate its vital cycles, structure, functions and its processes in evolution… Every person, people, community or nationality, will be able to demand the recognitions of rights for nature before the public organisms."[28]This is of crucial substantive and procedural importance, as it shows respect to nature, as well as broadens the standing of parties who can bring claims before a judicial body.

Largely through a policy of sustainable development, several states seek to address a universally fair manner of human development and consumption, although few documents actually say the word 'consumption'. In Bhutan, they introduce an entirely new methodology for a just society, Gross National Happiness. "The Royal Government shall… Secure ecologically balanced sustainable development while promoting justifiable economic and social development" and, "The State shall strive to promote those circumstances that will enable the successful pursuit of Gross National Happiness."[29] The Constitution of South Africa gives everyone the right to, "…have the environment protected… through reasonable legislative and other measures that… secure ecologically sustainable development and use of natural resources while promoting justifiable economic and social development."[30]

Another principle that is quickly emerging across states is the recognition of the rights of future generations. Papua New Guinea promotes, "the conservation and replenishment, for the benefit of ourselves and posterity, of the environment and its sacred, scenic, and historical qualities."[31] Also, "We, the people of Kenya, respectful of the natural environment that is our heritage, and determined to sustain it for the benefit of future generations..." and "The national values, principles and goals include... recognition of the responsibilities of the State to future generations..."[32] Andorra ensures the rational use of the soil and resources "...so as to guarantee a befitting quality of life for all and, for the sake of coming generations..."[33]

Principles of Respect, Love and Integrity

Business as usual is not working, and states must be willing to think outside the box to address the issues of today. Although no constitutions currently use the word 'love', several, like Bhutan [with their Gross National Happiness] and Ecuador [with their Rights of Nature], are doing just this type of thinking. States should promote ecological solidarity between humans and nature, with the obligation to respect and the compassion of love as the basis for genuine care of living beings, places and people: love for the beauty and gift of the natural world with all of its living diversity; love for our places and our homes; and love for the people of today and tomorrow.

Bhutan strives, "...to ensure justice and tranquility and to enhance the unity, happiness and well being of the people for all time..." and has therefore adopted "Buddhism [as] the spiritual heritage of Bhutan, which promotes among others the principles and values of peace, non-violence, compassion and tolerance."[34] The Constitution of India states that it is a fundamental duty of every citizen "to protect and improve the natural environment including forests, lakes, rivers and wild life, and to have compassion for living creatures..."[35]

These principles of respect and love are also evidenced in Ecuador's rights of nature, "[The State] will promote respect towards all the elements that form an ecosystem."[36]The Philippines touches on the harmony of nature, "The State shall protect and advance the right of the people to a balanced and healthful ecology in accord with the rhythm and harmony of nature."[37]

Hand in hand with respect is integrity. States should value the ecological integrity of the biosphere and its diversity of interacting ecosystems and species, independently of other values that humans place on them. Kenya requires that the State, "respect the integrity of natural processes and ecological communities, including conservation of habitats and species."[38] Colombia connects integrity with areas of special importance, "It is the duty of the State to protect the diversity and integrity of the environment, to conserve the areas of special ecological importance, and to foster the education for the achievement."[39]

The Right Leadership: Ethics and the Judiciary

Courage and leadership from the judiciary is necessary to keeping nature alive. Even if constitutions include the most ethical of principles, a state cannot escape its duty to enforce those ethical principles. In most systems, the judiciary is seen as one of the key executors of the law, enforcing and even creating the rule of law upon claimed violations. This section will give the results of ethical principles found in judicial opinions from around the world. In order to see the current trends, it will review existing and emerging legal principles.

Article 38 of the Statute of the International Court of Justice states the sources of international law, including "the general principles of law, as recognized by civilized nations." [40] Laws become principles of international law when they are found in the laws of several nations. This section looks into cases from international and domestic tribunals,[41] and highlights existing and emerging ethical principles of international law related to truth, democracy, justice and equity, and respect and love.

Principles of Truth

Ethical principles of truth found in judicial opinions include recognition of humanity's dependence on nature; and the precautionary principle.

The Constitutional Court of Colombia acknowledged the utter dependence of mankind on the environment in 1995 in *Jose Cuesta Novoa and Milciades Ramirez Melo v. the Secretary of Public Health of Bogota*.[42] Citizen petitioners, in defense of their constitutional rights to life and a healthy environment, argued that the defendant failed to fulfill its duties of environmental protection in that they tolerated the performance of several corporations that did not comply with environmental legislation. The Court held that, "...such a right is understood as part of a set of basic circumstances that surround mankind and allows its biological and individual existence."

In 2009, the Supreme Court of Justice of Argentina confirmed the precautionary measures needed to protect the integrity of the forest of the Salta Province. In *Salas, Dino y otros con Salta, Provincia y Estado Nacional*,[43] the Court suspended all tree felling and clearing permits of approximately one million hectares that the Provincial Government had issued. The permits were given taking into account the environmental impact of each of them, but without study of the cumulative effect of them together. This decision recognizes that lack of information regarding the possible effects can lead to the suspension of activities, based on the possibility of danger of serious harm.

In 1992, the Supreme Court of Pakistan heard a human rights case and used both the precautionary principle and the principle of sustainable development in its decision. In *Shehla Zia and others v. WAPDA*, Human Rights Case No. 15-K, residents alleged that an electromagnetic field posed a health hazard from construction of a grid station.[44] The Court stated, "At present, scientific evidence regarding the possibility of adverse biological effects from exposure to power frequency fields, as well as the possibility of reducing or eliminating such effects, is inconclusive." They continued, "In such a situation, the precautionary principle should be applied. To stick to a particular plan on the basis of old studies or inconclusive research cannot be said to be a policy of prudence and precaution." They then referred to Principle 15 of the Rio Declaration and added, "In fact, a policy of sustainable development should be adopted."

Principles of Democracy

Ethical principles of democracy found in opinions focus on individual and social concerns, including the rights to participate, to be informed, and to have access to information and to justice. They also acknowledge local, regional, state and international responsibilities.

In 1988, the Supreme Court in India recognized the crucial link between the environment and culture. In *Rural Litigation and Entitlement Kendera v. State of U.P.*, there was unauthorized and illegal mining in the Dehra Dun area which adversely affected the ecology of the region and caused environmental damage.[45] The Court stated, "...the Dun Valley limestone is a gift of nature to mankind and forests provide the green belt and are a bequest of the past generations to the present," and continued, "The problem of forest preservation and protection was no more to be separated from the lifestyle of the tribal people."

A Federal Regional Tribunal in Porto Alegro, Brazil in 1992 acknowledged the aesthetic importance of natural sites and also addressed property rights. In *Pedro Korkowski v. Ministerio Publico*, there were construction efforts on private land, located near a UNESCO historical site and without proper permit.[46] The owner was ordered to demolish the works, which he responded were a violation of his property rights. The Court held that, "Historic, artistic and natural sites have an aesthetic function that is to be preserved," and most notably, "Property rights are no longer sacred, as they were during the 19th century, and new principles of environmental law favor social over private concerns."

In Argentina, in 2006, an Appeals Chamber heard *Calderon, Placido Horacio del Valle and others v. Guaymallén Municipality* over an alleged failure to submit an environmental impact assessment.[47] The Court held that, "The locus that art. 43 of the Constitution establishes for the affected should be assessed on a criterion of broad interpretation, especially in the protection of collective rights, as is the preservation of the environment."

In 2008, the Constitutional Court of Colombia ruled on the *Unconstitutionality against Law 1021 of the year 2006 'General Forestry Law.'*[48] The government and the official bank implemented a sweeping strategy in Congress that allowed the adoption of the rule ignoring fundamental individual and collective rights, in particular, consultation with indigenous communities and ethnic groups recognized in the ILO Convention 169, an international treaty signed by the country. The Constitutional Court accepted the arguments of the claimant community and the law was declared unconstitutional.

Principles of Justice and Equity

Ethical principles of justice and equity found in cases include the duty not to harm (including cause future harm); the polluter pays principle; civil liability; criminal culpability; sustainable development; and the rights of future generations.

Arguably, two of the longest-standing principles as related to the environment are the duty not to harm and the polluter-pays principles. Each state has a duty not to cause harm to another state. In addition, if harm is caused, whether by a state or entity, that state or entity must pay for causing the harm. The duty not to harm also includes the responsibility to prevent harm and the requirement of prior notification if it is suspected that an activity will cause harm.

In 1938, the International Court of Justice addressed both in the *Trail Smelter Arbitration* (US/Canada).[49] Here, high stacks were erected at a lead and zinc smelter at Trail, British Columbia. These stacks released heavy amounts of sulfur dioxide and caused damage to the neighboring United States. The Court stated, "Under the principles of international law, as well as the law of the United States, no state has the right to use or permit the use of its territory or the properties or persons therein to harm another state." It then continued, "…the Trail Smelter shall be required to refrain from causing any damage through fumes to the State of Washington," and required Canada to pay damages.

In 2001, the Federal Court of Australia heard *Booth v. Bosworth* addressed the implications of future harm to a species.[50] Here, an injunction was sought to restrain the killing of spectacled flying foxes on or near a private lychee orchard, situated near a World Heritage Site. The owner of the orchard had set up a set of 14 aerial electrical fences to kill the foxes above their orchard, in close proximity to the Wet Tropics Heritage Area, which is listed as a Property under the International Convention for the Protection of World Cultural and National Heritage. From expert opinion, the Federal Court concluded that with an average of 377 killings per night, the electrical fences would halve the population of spectacles foxes in less than five years, rendering it an endangered species. They also relied on the World Heritage Convention in that the foxes contribute to the heritage values of the Wet Tropics Heritage Area.

In a landmark decision in the United States, the Court of Appeals for the Second Circuit ruled that public nuisance law can be used to sue power companies based upon present and future injuries from global warming. In *State of Connecticut, et al. v. American Electric Power Company Inc., et al.* (2009), eight states of the US, New York City, and three nonprofit (NGO) land trusts that sought to preserve environmentally sensitive land all sued power companies that own and operate fossil fuel power plants in 20 states of the US.[51]

The Court decided that because the NGOs own property interests that can be harmed by weather events as a result of global warming, they have standing to sue. They determined that the standing of the NGO land trusts can be based on future injury because injuries are "already in process as a result of the ongoing emissions by defendants that contribute to increasing temperatures."

In April 2010, the International Court of Justice decided the *Case Concerning Pulp Mills on the River Uruguay* (Argentina/Uruguay).[52] Argentina claimed violation of the Statute of the River Uruguay by the development of two pulp mills on the river by Uruguay. The Court held that,

> "…the principle of prevention, as a customary rule, has its origins in the due diligence that is required of a State in its territory. It is 'every State's obligation not to allow knowingly its territory to be used for acts contrary to the rights of other States' (Corfu Channel (United Kingdom v. Albania), Merits, Judgment, I.C.J. Reports 1949, p. 22). A State is thus obliged to use all the means at its disposal in order to avoid activities which take place in its territory, or in any area under its jurisdiction, causing significant damage to the environment of another State. This obligation 'is now part of the corpus of international law relating to the environment' (Legality of the Threat or Use of Nuclear Weapons, Advisory Opinion, (1996)."

The Court also considered that an environmental impact assessment must be conducted prior to the implementation of a project. Moreover, once operations have started and, where necessary, throughout the life of the project, continuous monitoring of its effects on the environment shall be undertaken (the principle of continuity).

In 1993, the Supreme Court of the Philippines acknowledged the right of a group of citizens to sue on behalf of succeeding generations. In *Juan Antonio Oposa and others v. The Honorable Factoran*, 43 minors, represented by their parents, brought a case against the government to cancel timber licenses on the ground that destruction of natural resources affected their own rights, as well as generations unborn. They pleaded their constitutional right to a balanced and healthful ecology and that they were entitled to protection of the State in its role as *parens patriae*. The Court held that the petitioners have a right to sue on behalf of succeeding generations because,

> "…every generation has a responsibility to the next to preserve the rhythm and harmony of nature for the full enjoyment of a balanced and healthful ecology. Unless the rights to a balanced and healthful ecology and to health are mandated as State policies… the day would not be too far when all else would be lost not only for the present generation, but also for those to come – generations which stand to inherit nothing but parched earth incapable of sustaining life."

In 1996, the Supreme Court of India acknowledged that the principles of sustainable development, polluter pays and precaution were each now part of customary international law.[53] In *Vellore Citizens Welfare Forum v. Union of India*, the Court passed several orders after monitoring this petition for over eight years. They concluded that, "Although the leather industry is a major foreign exchange earner for India and provided employment, it does not mean that this industry has the right to destroy the ecology, degrade the environment or create health hazards… Sustainable development, and in particular the polluter pays and precautionary approaches, have become a part of customary international law."

The Court then instructed that, "if a polluter refuses to pay compensation, his industry will be closed, and the compensation recovered as arrears of land revenue." The Court then declared that the matter would be monitored by a Special Bench – a "Green Bench" of the Madras High Court.[54] It is important to note that on October 19, 2010, a National Green Tribunal was created to hear all environmental cases. [55]

In 1998, the Tribunal of Rio Grande do Sul in Brazil held defendants to be criminally liable for environmental damage. In *Ministerio Publico v. Orlando Linden and Hermes Gildo Masera*, the plaintiff alleged that the former and the present Municipal Prefects, while exercising their legal duties, had continuously exposed human beings, animals and vegetation to health hazards by waste deposits made without an environment license.[56] The Court held that the defendants are "…criminally liable for the resulting damage after considering that they did not do everything that they should or could have done to avoid environmental damage." They were sentenced for 2 years and 10 days and registered in the Book of Inculpated.

In 2009, the *Attorney General (Province of Pichincha) – TD* was heard before the Tenth Criminal Court of Pichincha, Ecuador.[57] The court examined a case of unauthorized collection of 741 tarantulas. The specimens were found in the luggage of a tourist-Quito International Airport. The defendant claimed that he bought the tarantulas for scientific research in his native country. Given the evidence of an environmental crime, in terms of section 437F of the Criminal Code of Ecuador, in the form of unauthorized collection of wildlife (a scientific research activity), the judge ruled for eight months in prison.

In 2008, the High Court of Paris gave its *Erika Judgment* concerning a spilled oil tanker.[58] The judge recognized the criminal responsibility of an Oil company for the offense of pollution. The company was penalized for its dangerous and imprudent behavior regarding the environment. On
In March 2010, the Paris Court of Appeal passed the *Erika Judgment* on civil and criminal liability.[59] On the criminal front, the Paris Court of Appeal found that company was reckless in implementing its vessel vetting process and ordered them to pay a fine of €375,000. However, on the civil aspects, the Court of Appeal decided that the company was not liable for the civil damages awarded to the plaintiffs.

Principles of Respect and Love

In 1997, an International Court of Justice case included some of the best rulings on all of the ethical principles above, including love. In the Separate Opinion of Justice Weeramantry of the International Court of Justice in the 1997 *Case Concerning the Gabcikovo-Nagymaros Project* (Hungary/Slovakia), Weeramantry addressed the rights of future generations, the principle of sustainable development, environmental protection as a human right, the multi-disciplinary nature of the field, the importance of biocultural diversity, the principle of contemporaneity, the principle of continuity (regarding monitoring Environmental Impact Assessments), trusteeship, compassion and love.[60]

Weeramantry stated, "The environment is not an abstraction but represents the living space, the quality of life and the very health of human beings, including generations unborn." He argued that, "The court must hold a balance between the environmental considerations and the developmental considerations. The principle that enables the court to do so is the principle of sustainable development," and declared it to be, "an integral part of modern international law," and "The [right to development] does not exist in the absolute sense, but is relative always to its tolerance by the environment."

Weeramantry continued, "The protection of the environment is a vital part of contemporary human rights doctrine, for it is a sina qua non for numerous human rights such as the right to health and the right to life itself." He then added that, "The discipline of international law needs to be multi-disciplinary," and "The court has a duty to draw upon the wisdom of the world's several civilizations… [It] cannot afford to be mono-cultural." He also introduced a new principle, the principle of contemporaneity, "The ethical and human rights aspects of environmental law bring it within the category of law so essential that we cannot apply to today's problems the problems of yesterday."

His opinion continued, "The first principle of modern environmental law is the principle of trusteeship of earth's resources." He then described how ancient irrigation systems "…were undertaken for the benefit of the country and out of compassion for all living creatures." He said, "Even birds and beasts have a right to freedom from fear. The notion of not causing harm to others is a central tenet of Buddhism… It was] the sacred duty of each generation to ensure that the [irrigation] system was kept in good repair and all able-bodied men were expected to take part." In his concluding remarks, Weeramantry stated, "It would not be wrong to say that the love of nature, the desire for its preservation… are among those pristine and universal values which command international recognition."

Conclusion

The environment is not a bubble, it is not one of three pillars of "economy, society and environment," but rather it is the very foundation of all life, or all economy, of all society. In this same vein, ethics is the foundation of all law, but what kind of ethics? The constitutions and judicial opinions of this article gave evidence to numerous existing and emerging ethical principles of biodiversity conservation, principles that support a flourishing future for humans and nature.

Constitutions outline the hopes and dreams of a government for their people (and some, for nature), with their corresponding rights and responsibilities. It is the guiding document for that state, and symbolizes the ethics of its people. Judicial opinions put the words and ethics of constitutions into direct action. To make the right decision, to act on the side of the just, and in the name of the vulnerable, is an act of courage and leadership. Several judges throughout the world are leading their colleagues and their states in ethical action, and these decisions are creating legal principles that are serving and will serve as legitimate rule of law for other judges in future decisions.

The ethical principles evidenced within should serve as a source of empowerment to lawmakers, judges, and the citizens, present and future, that these laws serve to protect. They are helping to build global solidarity for the future of life.

Notes

---

[1] Justice Christopher Weeramantry, International Court of Justice (1991-2000).

[2] The Biosphere Ethics Initiative (BEI) is a soft law program that seeks to highlight existing ethical principles in action, and incorporate those principles into law and policy. It is led by the Ethics Specialist Group of the IUCN Commission on Environmental Law, the Center for Humans and Nature, and the Paris Muséum national d'Histoire naturelle. The *Evolving Biosphere Ethic* and additional information can be found at www.iucn.org and www.humansandnature.org.

[3] This section cites constitutions from: Andorra, Bhutan, Brazil, Colombia, Cuba, France, Germany, Ecuador, India, Iraq, Kenya, Montenegro, Paraguay, Papua New Guinea, Philippines, Republic of China, South Africa, and Ukraine.

[4] Ecological reconciliation is a state and individual's responsibility to acknowledge the truths of humanity's past harms to nature, and in the name of nature. This acknowledgement is not to pass blame or guilt, but to instead allow humanity to be aware and free to move forward. This concept, however, has yet to find itself in any state constitution.

[5] República del Ecuador, 2008. Chapter: Rights for Nature, Art. 2.

[6] República de Cuba, 1992.
[7] Chartre de l'Environnement, République Française, 1958, amended 2007.
[8] Bhutan, 2008. Article 5. Environment. 4.
[9] Chartre de l'Environnement, République Française, 1958, amended 2007. Art 5.
[10] República del Ecuador, 2008. Art. 4.
[11] Montenegro, 2007. The State. Article 1. Montenegro is also the only state that declares itself, "an ecological state."
[12] Iraq, 2005. The Legislative Power. The Council of Representatives. Article 50.
[13] Républica de Colombia, 1991.
[14] Україна, Ukrayina, 1996. Article 50.
[15] Iraq, 2005. Section Two. Rights and Liberties. Chapter One. Second: Economic, Social and Cultural Liberties. Article 35.
[16] Jamhuri ya Kenya, Republic of Kenya, 2001 amended 2008. Chapter 8. 87.
[17] Bundesrepublik Deutschland, 1949, amended 2006. Article 75. 4.
[18] 中華民國, Jhonghuá Mingoú, Republic of China, 1947, amended 2005. Section 6 The Frontier Regions. Article 169.
[19] Chartre de l'Environnement, République Française, 1958, amended 2007.Art 10.
[20] Jamhuri ya Kenya, Republic of Kenya, 2001 amended 2008. Chapter 8. 87.
[21] Républica de Colombia, 1991.
[22] Bhutan, 2008. Article 9. Principles of State Policy. 20.
[23] Montenegro, 2007. The State. Article 1.
[24] Jamhuri ya Kenya, Republic of Kenya, 2001 amended 2008. (h)
[25] República del Paraguay, Teta Paraguái, 1992. Title II. Chapter I About Life and Environment. Section II. Article 7. (3).
[26] Républica Federativa do Brasil, 1988. 3.
[27] See generally, L. Westra, including *Ecoviolence and the Law: Supranormative Foundations of Ecocrime* (2004); and *Environmental Justice and the Rights of Ecological Refugees* (2009). Earthscan Publishers, UK.
[28] República del Ecuador, 2008. Chapter: Rights for Nature. Art. 1.
[29] Bhutan, 2008. Article 9. Principles of State Policy. 2.
[30] Republic of South Africa, 1996, amended 2007. 24 Environment (b)(iii).
[31] Independen Stet bilong Papua Niugini, The Independent State of Papua New Guinea, 1975. National Goals and Directive Principles. (2).
[32] Jamhuri ya Kenya, Republic of Kenya, 2001 amended 2008.
[33] Principat d'Andorra, Andorra, 1993.
[34] Bhutan, 2008. Article 3. Spiritual Heritage. 1.
[35] भारत गणराज्य, Bh¬arat Ganarajya, 1950. Part IVA. 51A. (g)
[36] República del Ecuador, 2008. Chapter: Rights for Nature. Art. 3.
[37] Republika ng Pilipinas, Republic of the Philippines, 1987. Article II Declaration of Principles and State Policies. Section 16.
[38] Jamhuri ya Kenya, Republic of Kenya, 2001 amended 2008. Chapter 8. 87. (a).
[39] Républica de Colombia, 1991.
[40] Statute of the International Court of Justice, United Nations. Article 38.

[41] This section cites opinions from Argentina, Australia, Brazil, Colombia, Ecuador, France, India, Pakistan, Peru, Philippines, South Africa, the United States, and the International Court of Justice.
[42] Colombia Constitutional Court. Jose Cuesta Novoa and Milciades Ramirez Melo v. the Secretary of Public Health of Bogota (1995).
[43] Argentina Supreme Court of Justice of the Nation. Salas, Dino y otros con Salta, Provincia y Estado Nacional (2009).
[44] Pakistan Supreme Court. Shehla Zia and others v. WAPDA. Human Rights Case No. 15-K (1992).
[45] India Supreme Court. Rural Litigation and Entitlement Kendera v. State of U.P. (1988).
[46] Federal Regional Tribunal, Porto Alegro, Brazil. Pedro Korkowski v. Ministerio Publico (1992).
[47] Argentina, Appeals Chamber 4a of Civil, Commercial, Mining, Peace and Tax Mendoza. Calderon, Placido Horacio del Valle and others v. Guaymallén Municipality (2006).
[48] Colombia Constitutional Court. Unconstitutionality against Law 1021 of the year 2006 “General Forestry Law”. (2008).
[49] International Court of Justice. Trail Smelter Arbitration (US/Canada) (1938/1941).
[50] Federal Court of Australia. Booth v. Bosworth (2001).
[51] United States Court of Appeals for the Second Circuit. State of Connecticut, et al. v. American Electric Power Company Inc., et al. (2009).
[52] International Court of Justice. Case Concerning Pulp Mills on the River Uruguay (Argentina/Uruguary)(2010).
[53] Customary international law is a source of international law. See supra note 41. It is a law that is evidenced by the custom of states.
[54] Supreme Court of India. Vellore Citizens Welfare Forum v. Union of India (1996).
[55] In October 2010, India launched the National Green Tribunal (NGT). “Launch of the National Green Tribunal.” Press Note. Ministry of Environment and Forests, Government of India. October 19, 2010.
“The National Green Tribunal marks the first time a tribunal exclusively dedicated to environmental issues has been set up… this is the first body of its kind that is required by its parent statute, to apply the polluter pays principle and the principle of sustainable development… The Ministry of Environment and Forests also intends to organise workshops in rural areas to educate people about issues of access and procedure related to the NGT...”
[56] Tribunal of Rio Grande do Sul, Brazil. Ministerio Publico v. Orlando Linden and Hermes Gildo Masera (1998).
[57] Tenth Criminal Court of Pichincha, Ecuador. Attorney General (Province of Pichincha) - TD (2009).
[58] High Court of Paris, France. The Erika Judgment. (2008).
[59] Paris Court of Appeal, France. The Erika Judgment. (2010).

[60] International Court of Justice. Case Concerning the Gabcikovo-Nagymaros Project (Hungary/Slovakia) (1997).

PART IV

ISSUES IN GLOBAL GOVERNANCE

INTRODUCTION

LAURA WESTRA

Kissinger, Timmer and Rees, in Chapter 14, argue that, once we accept the interdependence of all people beyond their state borders, the reality of the results of our expanding ecological footprint, beyond our immediate surroundings, becomes obvious. Thus, the "sustainability of any given region may well depend on the continued productivity and sustainability of other regions".

In this chapter, two examples are provided, both taken from events in the year 2010: the BP massive oil well blowout in the gulf of Mexico, and the unprecedented heat wave in Russia. For the former, not only have the ecosystems and the fisheries of the area been gravely affected, but the general sustainability of fisheries and endangered fish species globally have suffered. For the latter, Russia's loss of grain crops due to the high temperatures contributed to rising global grain prices, and to the reduced availability of grain staples to many other populations.

Hence, four interregional strands must be acknowledged: (1) the interregional pollution strand; (2) the interregional ecological footprint strand; (3) the interregional effects of local ecological change; (4) the policy-driven systemic interregional impact displacement strand. The study of these interwoven "strands" demonstrates the importance of considering interregional sustainability through a shift away from so-called "development".

This study of a further aspect of the ecological footprint is new, but the underlying message is based on a timeless concept: "Nature's bottom line is simple: humanity has no choice, but to learn to live more equitably within limits imposed by the ecosphere's regenerative capacity".

In Chapter 15, Sabina and Alex Lautensach discuss the interface between human and state security in the recent times (since the early 1990s). The security of states depends on that of regions, communities, families and individuals, and states themselves are no longer able to guarantee the security of any of these subsets. Increasingly, security, far more than the traditional absence of threats from violent conflict, now includes a relative safety from acute infectious diseases, minimum complements of safe fresh water and adequate nutrition, and a formal guarantee of basic human rights and dignity. As well, the category of present "security" now encompasses future generations, so that the growing presence of environmental disasters adds yet another dimension. This multiplicity of security aspects leads to a new "comprehensive definition" which should include "four pillars" of human security; that is, the traditional "military/strategic security of the state"; "economic security"; "health-related security", and "environmental security".

Michael Schröter addresses the problem of "combining liberalism with sustainability" in Chapter 16. He starts with a definition of the concept. Although "liberalism" has an illustrious history, dating back to the work of Immanuel Kant, the original "political liberalism" is no longer dominant, as an "atomistic", "economistic liberalism" reigns today instead.

Sustainability, on the other hand, is much "newer" as a concept, as its importance emerged within the time-aspects described in the 1972 Limits to Growth by Meadows et al. Today, sustainability also manifests two separate aspects: the strong ecological one, and its weaker sense of "environmental sustainability". While the latter gives parity to "three equivalent dimensions", the former asserts that the social and economic aspects of the concept must be secondary to its main ecological aspect. Humanity, as a whole, "enjoys a collective right to nature". Thus, he concludes, liberalism needs to be closely connected to sustainability in democratic societies.

In Chapter 17, the interface between liberalism and sustainability is clearly in evidence as Philippe Crabbé examines African countries under globalization. He considers the "natural capital" and "non-renewable resources" in Africa. Despite their resources, the effects of globalization (neo-liberalism) now includes "income inequality", "poverty, hunger,...and vulnerability" in general. In addition, unchecked climate change exacerbates all these problems as well as the grave health issues besetting the continent, especially in children from 0-5 years of age.

Crabbé notes that in 2009, Michael Spence and Paul Collier proposed an international convention, the Natural Resources Charter (2009); as well, several other documents were produced, with an emphasis on Social Corporate Responsibility, and the UN Priorities for Responsible Investments (2006) and Canada has an important role to play in these initiatives, as well as in the important "African Vision 2050" (October, 2008).

In Chapter 18, Vicky Karageorgu brings us back to the issue of water resources and water rights discussed in Chapter 12. Her focus is on the example of inter-basin water transfers in Greece (and the Achelloos river). Although there are several general Directives which address "innovative approaches for water management and protection", this document does not contain specific provisions for water transfers. In fact, the UNECE Convention on the use of Transboundary Waters and Lakes, and other EU documents do not set "the regulative contours of the principle of sustainable water use". In contrast, the newly revised German Federation Water Act of 2009, is the best example of an instrument that recognizes "sustainable water use as its most important guiding principle (Art.1)". The case of the Achelloos river has been aided somewhat by "progressive jurisprudence", but—she concludes—national and international legislators should adopt "a more sophisticated and integrated approach" regarding water management rules.

CHAPTER FOURTEEN

INTERREGIONAL INTERDEPENDENCE: A FRAMEWORK AND POLICY PRESCRIPTIONS FOR SUSTAINABILITY

MEIDAD KISSINGER, VANESSA TIMMER AND WILLIAM E. REES

Abstract

This chapter develops and analyzes a theoretical interregional approach to sustainability in an interconnected world. We make the case that achieving ecological sustainability – living equitably within the limits of supporting ecosystems – requires the recognition of our material interdependence and interconnectedness across regions and the attendant need for interregional policy to shape local resilience and global sustainability. Approaching sustainability conscious of interregional connections reveals that: 1) virtually every significant human population or country lives, in part, on energy/material flows to and from distant places elsewhere around the world; 2) production, consumption and policy decisions in any given locale have the potential to create unseen unsustainable burdens on connected productive ecosystems in distant locales; 3) ecological change in one region has the potential to jeopardize the sustainability of other regions; and 4) society in almost any region has interests in sustaining the vitality of ecosystems in other regions. We highlight a range of interregional ecological linkages and discuss the kinds of interregional feedback and policy responses they require. This reveals how an interregional perspective not only provides further insight into the social and ecological challenges global society is facing, but also reveals places for positive human intervention and leadership in the name of sustainability and resilience.

Introduction

Throughout human history, people have depended mostly on local ecosystems for resources and for waste assimilation. However, in the past two centuries, the expansion of the human enterprise has resulted in a historically unprecedented extension and thickening of the web of material connections among regions. This has facilitated continued population growth almost everywhere and enabled people to consume more than they did in the past, both in total and on a *per capita* basis (Meadows et al. 2004; MEA 2005; FAOSTAT 2010). It also means that billions of people have become dependent on resource supplies from all over the world for their well-being, even their survival (Princen 1997; 1999; Rees 1994; 2010; WTO 2006). Mutual interdependence, ecological impacts and permanent change have been elevated to the global scale (Daly 1996; Clark 2000; Haberl et al. 2007; Rockström et al. 2009; Rees 2010). In this highly interconnected and dynamic global system, cumulative or sudden ecological failure in one region threatens the human sustainability in other regions and the entire global system.

At present, despite increasing global connectedness, most environmental policies apply to a single spatial scale – local, national or global – and focus on narrow aspects of human well-being and ecosystems integrity. The main emphases are on the local pollution impacts of local production activities or on the cross-boundary effects of local waste discharges on neighbouring countries. Even issues acknowledged to be global (e.g., climate change) are framed so that corrective action is essentially voluntary and fragmented among individual states. However, in an ecologically full world, one approaching or beyond biophysical limits (Daly 1991; Meadows et al. 2004); a world in which activity in any one region can impact other regions and even the entire ecosphere (Rees 2010; Kissinger and Rees 2010); a world in which resource scarcity and the consequences of ecological changes are becoming major security issues (Pirages 2005), disjointed single-scale approaches to assessing, quantifying and acting towards sustainability are futile. It has become essential to incorporate the interregional across-scale dimension of sustainability into both national and global policies for sustainability.

In recent years various researchers have emphasized the need to examine cross-scale linkages among complex nested social-ecological systems (MEA 2005; Cash *et al.* 2006; Kissinger and Rees 2010). This chapter starts from the premise that the sustainability of any given region may well depend on the continued productivity and sustainability of other regions. We develop a classification of interregional impacts central to sustainability. We then make the case that achieving ecological sustainability – living equitably within the limits of our supporting ecosystems – requires the recognition of ecological connectedness and interdependence across regions and the development of corresponding trans-regional policies that enhance both local resilience and global sustainability.

The year 2010 provides several illustrations of the interregional entanglement of human ecological relationships thus highlighting the emergent complexity of global sustainability. Two examples are British Petroleum's massive oil-well blowout in the Gulf of Mexico, and the unprecedented heat wave in Russia. BP's blowout will doubtlessly influence the gulf region's eco-systems for years to come but it is also affecting ecosystems and human well-being elsewhere. Consider the direct impact of oil- and dispersant- polluted water that has made its way via the Gulf Stream to distant ecosystems in the Caribbean and Atlantic coast. Moreover, since the Gulf of Mexico is the spawning and feeding ground for a variety of fish species, including North Atlantic Bluefin Tuna, the oil spill may have an indirect impact on several fisheries along the North American Atlantic coast, affecting the well-being of communities that rely on fishing for their livelihoods. Because the Gulf is one of the U.S.'s major seafood producing regions, the blowout significantly reduced production from the gulf, and increased seafood prices. However, this has not significantly reduced U.S or global demand so exploitation of marine ecosystems in other (already ecological sensitive) regions of the world has increased.

The summer 2010 drought and high temperature in large parts of Russia greatly lowered the quality of life of millions of Russians. Indeed, death rates shot up with spreading forest fires and polluted air. However, the impacts of Russia's extreme weather are also being felt world-wide. Consider the CO_2 released to the atmosphere from the fires and now contributing to accelerating global climate change (in a positive feed-back loop – climate change likely contributed to the extreme drought in the first place). Consider also the loss of a third of Russia's grain crop and its impact on millions of people elsewhere in the world who depend, in part, on that country's agricultural bounty. To assure domestic supplies, the Russian government temporarily banned grain exports, a decision that has contributed to rising global grain prices and to decreasing poorer populations' access to a basic staple. Note also that the loss of Russian output has increased the demand on other grain producing regions, some of which are already suffering from environmental constraints (e.g., flooding in the Canadian prairies and drought in Australia,). The cumulative effects of all such ecological change should be of concern to people and governments everywhere.

The purpose of the paper is to advance a theoretical interregional approach to sustainability in today's interconnected world and to discuss its implications for local and global governance. We outline four strands of interregional thinking, and discuss types of interregional feedback and policy needed to address mismatches between global change and 'management' responses. The analysis shows how the interregional perspective not only provides novel insight into the social and ecological challenges facing the global community, but also reveals loci for positive human intervention to enhance resilience and sustainability.

Defining and Analyzing Interregional Sustainability

By 'interregional' we refer to the ecological relationships and linkages among geographical entities that are widespread, often international but less than global. Thus, regions can refer to administrative units such as the European Union, nation-states and metropolitan regions; to economic regions within or shared between countries; and to bioregions such as a drainage basins or ecotypes. Approaching sustainability conscious of interregional connections reveals that: 1) virtually every significant human population or country lives, in part, on energy/material flows to and from distant locales elsewhere the world; 2) ecological change in one region may jeopardize the sustainability of other regions; 3) production, consumption and related policy decisions in any geographic locale have the potential to create unseen unsustainable burdens on productive ecosystems in other regions and; 4) people in most countries or regions today have direct and indirect interests in sustaining the vitality of ecosystems in other regions.

We distinguish among four strands of thinking about relevant interregional biophysical relationships: interregional pollution, interregional extension of ecological footprints, interregional impacts of local ecological change, and policy-driven systemic displacement of activities or impacts among regions. It is important to note that the four strands operate in complex, dynamic and interconnected ways and can be present simultaneously.

(1) *The interregional pollution strand* focuses on the impacts of unintended cross-boundary waste emissions and on the deliberate transfer of waste from one region to be processed and/or disposed of in another region. It recognizes that economic production in one location of the world imposes assimilation burdens on ecosystems and communities in distant

locales. In short, modes of production, types of energy and material used, and trans-boundary wastes flows in region 'A' can negatively affect the sustainability—the health of ecosystems and communities—in distant regions 'B'… 'Z'.

Consider transfrontier pollution, long-range transboundary pollution, and global pollution (Schleicher 1992; Okowa 2000; Kasperson and Kasperson 2001; Mason 2005). Transfrontier pollution (i.e., air or water pollution transferred from one country to its neighbor(s)) has long been recognized as a factor that impairs relations between neighboring countries and has thus received considerable attention from authorities and researchers (Okowa 2000). There are two sub-categories (Kasperson and Kasperson 2001): near-border impact risks and point source transboundary risks which are not necessarily located close to the border.

Long-range transboundary waste transfers from one or several countries (e.g. acid rain) can cause pollution damages in other, not necessarily neighboring, countries (Kasperson and Kasperson 2001). This kind of interregional pollution usually involves industrial activities or transportation but also includes impacts resulting from the shipment of hazardous waste betweens regions. Consider, for example, the rise in e-waste traffic from North America to China.

Global pollution refers to discharges associated with human activities in one or many regions that negatively affect global-scale life-support systems so that the damages are essentially universal. This type of interregional pollution or its effects (e.g., stratospheric ozone depletion) degrades both national assets and the global commons.

As noted, problems associated with interregional pollution are well recognized. (Kasperson and Kasperson 2001; Mason 2005). Their impacts have led to both bilateral and more widespread international environmental agreements (e.g., Montreal accord on ozone depletion) which aim to minimize both polluting activities and their impacts (Mason 2005; Clapp and Dauvergne 2005).

(2) *The interregional ecological footprint strand* reveals that *consumption-driven demand* for imported goods and materials imposes significant and often unrecognized (by importers) burdens on productive ecosystems in exporting regions. In particular, densely populated high-income countries are extending their 'ecological footprints' into the global commons and other countries all over the world (Rees 2006, 2010; WWF 2010; Kissinger and Gottlieb 2010). As globalization and international trade intensify, and incomes in emerging economies rise, such interregional pressure can only increase. Ecological footprint analysis forces recognition that the sustainability of people living in import-

dependent regions is increasingly tied to the continued productivity and sustainability of ecosystems in supporting regions, wherever on earth the latter may be located. Thus, both irresponsible consumption and lax production methods can jeopardize the sustainability of both consuming (importing) and producing (exporting) regions (Kissinger & Rees 2010).

Globalization and trade are generally understood as positive factors that shrink the world (Friedman 2002; Rees 2010). However, in a globalizing world, commodity chains actually grow longer, more complex, and more deeply transnational, thus *increasing* the distance between cause and effect. Importing populations are, therefore, effectively blind to the distant negative ecological effects of resource exploitation and production created by their resource demands – out of sight, out of mind (Rees 1994, 2010; Princen 1997; Conca 2002; Dauvergne 2008; Kissinger and Rees 2009). By blocking consumers' knowledge of production impacts, globalization likely fosters over-consumption (Princen 1997; Conca 2001). Consumers lack the information and direct negative feedback that might otherwise induce them to behave more 'sustainably' (Rees 1994; Princen 1997; Conca 2002). Formal recognition of this dimension of interregional sustainability goes back several decades (Borgstrom 1972; Rees 1992); however, the issue has only recently begun to attract serious policy attention stimulated by a recent sharp increase in formal acquisitions by rich countries of 'surplus' land in poor countries (Cotula et al. 2009).

(3) Analyses of *the interregional effects of local ecological change* show that local production and consumption can result in on-site ecological degradation that also risks the sustainability of distant regions. Ecological change can be a result of both indirect and direct drivers (MEA 2003). Direct drivers include land use and cover change; species introduction and removal; technology adaptation; physical inputs (e.g., chemicals); and local resource consumption (MEA 2005). Indirect drivers include demographic; economic; sociopolitical; technological; and cultural/religious factors that impact the direct drivers. Some drivers are influenced by international and global processes while others mainly reflect local circumstances. Most research attention has focused on production-related damage stimulated by expanding global markets. However, in an interconnected world, even when the extraction of resources is mainly for domestic consumption, 'local' ecological changes (e.g., soil erosion, falling water tables, significant pollution events, species extirpations, habitat losses, etc.) can contribute to the loss of human wellbeing elsewhere.

We can explore the interregional interest in maintaining the ecological sustainability and reliability of distant 'producer' regions on three levels:

i) If region 'A' depends on region 'B' for certain food staples it is in 'A's interest, and a requirement for its own sustainability, that region 'B' is managed so as to be able to continue supplying those products. Consider the effect of recent U.S. biofuels policy (subsidies to encourage the use of maize to produce fuel ethanol) on its capacity to supply export markets for maize and other grains. In 2007, food maize in Mexico became relatively scarce as US growers redirected their output to ethanol production. Mexican prices rose steeply causing food riots and underscoring Mexico's direct interest in U.S agricultural and trade policy. Similarly, the 2010 heat wave in Russia severely reduced grain production inducing the government to ban exports. The ban destabilized global markets raising grain prices and instantly increasing the numbers of people 'elsewhere' unable to purchase food. This should raise concern about the potential interregional impacts of water mismanagement and increasing local demand for water in the U.S. Southwest and in California, for example. Combined with climate change, these factors now threaten the viability of California's cropland, the major source of North America's table vegetables.

ii) Local ecological degradation can reduce the supply of globally significant ecosystem 'services' (e.g., the carbon sink function, biodiversity preservation) or of critical natural capital (e.g., ozone layer, productive soils). For example, forest fires (notably in Brazil, Indonesia Canada and Russia) contribute to greenhouse gas accumulations, global climate change and biodiversity loss; deforestation for any economic purpose anywhere also contributes to biodiversity loss which is ultimately a global issue; the local use of toxic chemicals can lead to dangerous accumulations in distant food chains, putting predatory birds and mammals, including humans, at risk (e.g., toxic pesticide and industrial residues from 'the South,' particularly Asia, degrade the 'country food' diets of Canadian Inuit). Slowing global ecological change anywhere is rapidly becoming an interest of people almost everywhere.

iii) In an increasingly interconnected world, large populations rely on political stability in both producer regions and other distant places where conflict might interfere with customary supply lines. Large-scale conflicts can disrupt either the production or shipment of key supplies to dependent regions. All trade-dependent regions and countries therefore have a growing interest in maintaining geopolitical stability.

(4) *Policy-driven systemic interregional impact displacement* considers linkages between economic policies and activities in one region and large-scale ecological change in other regions. Anticipating such effects requires adopting a whole systems perspective. Sometimes negative impacts can be the unintended consequences of what appear to be ecologically sound decisions. For example: (1) Schutz et al. (2004:36) showed that the introduction of catalytic converters to the western European auto fleet in the mid 1980's had the desirable effect of decreasing transportation source

air emissions and improving health in European cities but it also increased toxic emissions in Siberia where one of the major components of the catalyst (the metal palladium) is extracted. Some of those emissions extend as far as Alaska and Canada. (2) Mayer et al. (2005) showed how forest conservation policies in China and Finland have contributed to increased rates of wood imports from Russia resulting in increasing deforestation in that country. (3) Barker et al. (2007) and Reinaud (2008) discuss ‘carbon leakage’, the increase in carbon dioxide emissions in one country as the result of an emissions reduction by a second country with stricter climate policy.

More broadly, consider how the migration of ‘dirty’ industrial enterprises from Europe and North America to developing countries with low wages and lax environmental policies has redistributed global emissions of many pollutants. For example, a third of China’s carbon emissions and air pollution can be attributed to manufacturing for export to the same high-income consumer countries that cleared their own air by out-sourcing such dirty manufacturing. While such linkages are not obvious at first, thinking in terms of interregional displacement brings greater understanding of the complexity of ecological change and the linkages between action in one part of the world and impacts on another.

Policy Responses

The idea of international cooperation and agreements to reduce cross boundary and global pollution is not new. Consider such bilateral and multilateral pollution prevention agreements in place such as between the US and Canada that limit the movement of acid rain pollutants, the China-Russia bilateral agreement to control transboundary water pollution of the Amur-Heilong River, and the UN Economic Commission for Europe Water Convention. There are also global treaties such as the Montreal accord to reduce major stratospheric ozone–depleting chemicals, the Basel Convention on Hazardous Wastes, and the (albeit frail) agreements focusing on climate change in the last two decades. All acknowledge to some degree the complexity of cross boundary and global issues and recognize that solutions (the development of policy instruments that apply beyond domestic boundaries) necessarily require intensive international cooperation.

That said, the world still has a long way to go in the domain of transnational governance, particularly on aspects of ecological interdependence discussed here that are poorly documented and thus under-represented in policy decisions. Addressing this fundamental disconnect requires an understanding of complex social-ecological systems and underscores the need to explore entirely novel institutional responses. The following sections illustrate two domains of potential policy response to systemic interregional connections. The first, and to date the most common approach, involves one self-interested region striving for sustainability either knowingly or unknowingly at the expense of other regions. The second category explicitly acknowledges interregional dependence and involves cooperative engagement among regions in the development of governance institutions that formalize their interdependence and express their mutual interest in sustaining the relationship.

The Self-Interested Approach to Interregional Policy

Military power – At the extreme end of the self-interested policy spectrum is one country's use of military force to occupy another nation's territory. A country with a large interregional ecological footprint may, in future, resort to violence to ensure a continuing flow of vital resources. Human history is a sorry litany of invasions by one country by another in part to satisfy the material needs and wants of the conquering power.

Economic 'land grabs' – In the 21st century, some nations are achieving through commerce what used to require territorial occupation (Rees 2010). Economic occupation is replacing physical conquest. As a means to gain food and energy security as well as financial profit, the governments and corporations of wealthier, food-insecure nations have begun acquiring large tracts of foreign land in mostly poor, developing countries (Daniel and Mittal 2009; IFPRI 2009). The International Food Policy Research Institute (IFPRI) claims that foreign investors sought or secured between 37 million to 49 million acres (15 to 20 million ha) of farmland in developing countries between 2006 and mid-2009 (The Economist 2009). Cotula *et al.* (2009) documented a total of 6,156,930 acres (2,492,684 hectares) of approved land acquisitions from 2004 to early 2009 in just five African countries, Ethiopia, Ghana, Madagascar, Mali and Sudan.

Both military and economic takeovers can appear to enhance the short-term sustainability of the acquisitor populations. However, these strategies are shortsighted and even the legal arrangements ignore major ethical and practical circumstances. Tensions can only increase, particularly in times of stress (e.g., climate change, water scarcity) if country 'A' has acquired the right to exploit country 'B's natural capital, and thus prevents host populations from using domestic resources for their own purposes. Future generations that have had no say in the agreement are particularly likely to rebel if their development ambitions or personal security are compromised by the direct foreign control of domestic assets. Such arrangements also ignore the possibility that, once immediate shortages are alleviated (negative feedback is removed), the population of, or per capita demand by, the dependent country may continue growing until a new stress point is reached. Nothing will have been gained, but now the population's dependence of on foreign sources has increased, possibly irreversibly. These concerns highlight the fact that 'land-grab' agreements implicitly assume that the future will be a smoothly reversible extension of the past and that additional such arrangements can be made without limit. Neither assumption is remotely valid.

The 'Mutual Interest' Approach to Interregional Policy

At present any country has the right to exploit its natural resources for its own economic and social well-being. As long as domestic activity does not generate cross boundary pollution or jeopardize international agreements, other nations' ability to intervene in maintaining ecological sustainability is constrained. This historical approach to the management of natural capital is no longer adequate in the context of deepening ecological interdependence discussed above. Arguably, various forms of 'resource management partnerships' have become essential for global sustainability. This can include measures ranging from addendums to formal sustainable trade agreements that ensure sound management of essential natural capital, through knowledge transfer for sustainable resource management, compensation payments for ecological services, and aid.

Sustainable formal trade agreements – At the moment trade is determined by market forces and the fluctuating price of commodities. An alternative might be a long term sustainable trade treaty in which the producer / exporter regions would be guaranteed a long term market for a certain volume of a commodity in exchange for adequate price support to help maintain the sustainability of its natural capital (e.g., soil, water).[1] Consumers / importing regions might also negotiate the right to require sustainable modes of production and the right to monitor the condition of the relevant productive natural capital to ensure that management practices (involving chemical use or soil management, for example) satisfy their long-term needs. Such arrangements would serve the same security function as land acquisition but avoid the tensions associated with the pseudo-occupation of the exporter's territory by the dependent partner. They would not be a barrier to the risk of exogenous factors such as climate change, but could contain safeguards such prescribed limits to export quantities that might encourage the importer country to engage in a program of population control, for example.

Several steps toward interregional sustainability have been taken in recent years in novel push-pull cooperative strategies, although none attain the long term treaty approach described above. For example, consider the increasing development of trade in sustainability-certified products such as forest products that require producers to adopt improved long-term management strategies in exchange for higher prices as necessary, and fair trade in various agricultural commodities to ensure that an adequate share of the consumer price goes to the primary producers, often peasant farmers. Another example is the EU regulation around pesticides residues in imported food products (EU 2005). This is intended to ensure that the amount and kinds of pesticides used in exporting regions conform to EU concerns about food safety (consumer health).

Sustainable knowledge transfer – Another form of mutual resource management could be the transfer of knowledge and technology from an importing to an exporting country to minimize the ecological consequences of cropping or manufacturing activities. Again, if region ‘A’ requires resources or products from region ‘B’, then ‘A’ has an interest in ensuring that ‘B’ is operating sustainably.

Such knowledge transfers could be part of a formal sustainable trade accord, but in an ecologically full world they should not necessarily be limited to agreements between trading partners. It is in everyone’s long-term interest to share sustainable agricultural technologies, including traditional ecological knowledge, or to engage in industrial technology transfers that will use less energy and resources and inject fewer dangerous emissions into the global commons. It may be necessary for the international community to establish a resource pool to provide reasonable compensation to license/patent holders for knowledge transfers undertaken for the common good in order to overcome ‘intellectual property right’ barriers that impede the free-flow of knowledge.

Compensation payments – An emerging form of resource management for the common good is compensation payments, usually by governments on behalf of the wider community, to individuals who agree to maintain privately-owned natural capital assets for their life-support functions instead of ‘developing’ them for their marketable commodities. So far, these agreements involve payments to private land owners, mostly in the tropics, to maintain their land in its natural state (Miranda et al. 2006; Russo and Candela 2006).

The purely ecological benefits of such programs are self-evident but compensation is also *economically* justified if the money value of the public good provided by a stand of forest, for example (e.g., carbon assimilation, water purification, runoff and flood-control), exceeds the private gain that would be realized if the forest were harvested. Although still small in scale, such agreements acknowledge the value of ecological services produced by distant ecosystems and show that once people are sufficiently informed, society is willing to pay to preserve the relevant ecosystems. Precedents in this realm include historical 'debt-for-nature' swaps. Recently, international climate negotiations agreed on a global strategy to reduce emissions from deforestation and forest degradation (REDD) by having developed countries pay developing countries to not cut down their forests.

Economic / Market mechanism – Unaccounted ecological externalities are symptomatic of unsustainable markets. The ecological costs of production should be included in the price of both domestic and trade goods in the interest of market efficiency and to generate the producer surpluses necessary for resource managers to maintain the health of natural capital stocks. In the absence of either direct biophysical or economic feedback, neither the corporate sector, governments nor individual consumers have an incentive to develop more sustainable modes of production, to implement corrective tax policies or to reduce consumption. In short, the prevailing system of costs, prices, and market incentives fails absolutely to reflect critical ecological scarcity, determine the appropriate levels of natural capital or induce sustainable patterns of production and consumption (Rees 1995). The world community is suffering from an extreme form of market failure (Daly 1999; Norgaard and Liu 2007).

There are ways to correct for this failure. Given the 'common pool' nature of the global sustainability problem, new international regulatory mechanisms and institutions will be required to oversee the process. The new international agencies are necessary to 'internalize' previously ignored external costs and to supervise the implementation of commercial regulations designed, for the first time, with ecological sustainability in mind. In this context, we will need ecological analyses to determine the state of critical ecosystems and identify the economic processes and products most responsible for negative ecological change.

It might subsequently be necessary to place an absolute cap on exploitation of critical resources. This would insure the sustainable harvest of renewable resources and the controlled extraction of non renewable resources in a way that matches the pace at which substitutions can be made. The allowable harvest / extraction could then be auctioned by governments or designated international agencies to private sector processing and manufacturing firms in the form, for example, of transferable quotas. Revenues so generated would fund necessary regulatory activities such as monitoring exploitation rates and ecosystem restoration.

Note that ecological monitoring is also essential to support implementation of nationally-imposed resource depletion taxes, pollution charges and export taxes. In these cases too, revenues generated would be available for monitoring/enforcement and damage repair. The resultant higher costs would be passed on to consumers, thus providing needed negative feedback on unnecessary consumption. Such 'true cost pricing' would stimulate both conservation and, by improving the relative position of competing products or processes, stimulate the development of alternative more sustainable products and technologies.

Summary and Conclusions

We have argued that sustainability and human security require rigorous analyses of the interregional dimensions of the growing ecological interdependence among nations. By articulating the multiple ways in which countries and regions on a crowded planet are linked through production and consumption, ecological change, and the unintended consequences of environmental policy decisions, we devise a novel framework to understand complex social-ecological systems failures and shape more effective responses. In the context of accelerating global change, there is an urgent need for the world community to adopt a whole-Earth systems perspective better to reflect countries' mutual interest in developing new transnational institutions for developmental policy- and decision-making across scales.

More fundamentally, achieving interregional sustainability requires a shift away from so-called 'development' based on perpetual material growth stimulated by globalization and expanding trade. There is no longer any question that this is an ecologically full world, in which we have reached or breached safe limits to the scale of material economic activity (Daly 1991, Rees 2006, WWF 2010, Rockström *et al.* 2009). By staying our present course, we tempt ecological implosion, including the possibility that we tip critical systems (e.g., global climate) into a new regime hostile to human prospects (and those of countless other species).

How do we retreat from the edge of the abyss? We argue that to reduce exploitation pressure and waste accumulation in natural systems the world must recognize biophysical limits to quantitative growth and focus instead on qualitative improvement. Rather than responding to resource scarcity by subsidizing the search for new supplies, governments should place primary emphasis on efficiency, demand-side management and on investing in the restoration of vital natural capital stocks. Policies that increase regional self-reliance for at least self-producing resources and waste assimilation represent a complementary step towards sustainability. Populations that live geographically close to their supportive ecosystems will see or experience indications of over-exploitation and take steps necessary to reverse the problem. Every eco-region should minimize its dependence on potentially unreliable imports without cutting off the options of trade as required. An appropriate maxim might be, 'trade if necessary but not necessarily trade'.

Nature's bottom line is simple: humanity has no choice but to learn to live more equitably within limits imposed by the ecosphere's regenerative capacity. Operating sustainably within such limits requires a long-term perspective and an adaptive, whole systems approach. In particular, both regional and global development planning must take into account biophysical limits and *de facto* interregional ecological interdependence.

References

Barker, T., S. Junankar, H. Pollitt & P. Summerton. 2007. Carbon leakage from unilateral tax reforms in Europe 1995-2005. Energy Policy 35(12): 6281-6292.

Borgstrom, G. 1972. The Hungry Planet – the modern world at the edge of famine. The Macmillan Company, New York.

Cash, D.W., W. Adger, F. Berkes, P. Garden, L. Lebel, P. Olsson, L. Pritchard, & O. Young. 2006. Scale and cross-scale dynamics: governance and information in a multilevel world. Ecology and Society 11(2): 8. Available at: www.ecologyandsociety.org/vol11/iss2/art8/ (accessed 3 Sept 2009).

Clapp, J., & P. Dauvergne. 2005. Paths to a Green World – the political economy of the global environment. MIT press, Cambridge.

Clark, W. C. 2000. Environmental Globalization. In J. Nye & J. D. Donahue, eds. Governance in a Globalizing World. Brookings Press, Washington.

Conca, K. 2001. Consumption and environment in a globalized world. Global Environmental Politics 1(3): 53-71.

—. 2002. Consumption and environment in a global economy. In Princen, T., Maniates, M., and Conca, K. (Eds.) Confronting consumption. MIT press, Cambridge.

Cotula, L., S.Vermeulen, R. Leonard, & J. Keeley. 2009, Land Grab or Development Opportunity? Agricultural Investment and International Land Deals in Africa. London/Rome: IIED/FAO/IFAD. ISBN: 978-1-84369-741-1.

Daly, H.E.1991. From empty- world economics to full-world economics, recognizing the historical turning point in economic development. In: R. Goodland, H.E. Daly, S. El Sarafy, B.Von Droste, eds. Environmentally Sustainable Economic Development: Building on Brundtland. United Nations Educational, Scientific and Cultural Organization, Paris.

Daly, H.E. 1996. Beyond Growth – the economics of sustainable development. Beacon press, Boston.

—. 1999. Uneconomic Growth in Theory and in Fact. The First Annual Feasta Lecture Trinity College, Dublin (26th April, 1999). Available at www.feasta.org/documents/feastareview/daly.htm (Accessed 28 July 2010).

Daniel, S., & A.Mittal. 2009. The great land grab. Rush for world's farmland threatens food security for the poor. The Oakland Institute, Oakland, USA.

FAOSTAT. 2010. Available at: http://faostat.fao.org/default.aspx.

Friedman, T. 2002. Technologic in state of discord. A debate between Thomas Friedman and Robert Kaplan. Foreign Policy. 129: 64-70.

Haberl, H., K. H. Erb, F. Krausmann, V. Gaube, A. Bondeau, C. Plutzar, S. Gingrich, W. Lucht, and M. Fischer-Kowalski. 2007. Quantifying and mapping the human appropriation of net primary production in earth's terrestrial ecosystems. Proceedings of the National Academy of Sciences. Available at: www.pnas.org/content/104/31/12942.full (Accessed 3 September 2010).

IFPRI. 2009. "Land grabbing" by foreign investors in developing countries: risks and opportunities. International Food Policy Research Institute, Washington.

Kasperson, R, E. & J. X. Kasperson, eds. 2001. Global Environmental Risk. Earthscan, London.

Kissinger, M. & D. Gottlieb. 2010. Place oriented ecological footprint analysis: The case of Israel's grain supply. Ecological Economics 69, 1639-1645.

Kissinger M. & W.E. Rees. 2009. Footprints on the prairies: Degradation and sustainability of Canadian agricultural land in a globalizing world. Ecological Economics 68 (8-9): 2309 - 2315.

Kissinger, M. & W.E. Rees. 2010. Importing terrestrial biocapacity: The U.S. case and global implications. Land Use Policy 27: 589–599.

Mason, M. 2005. The New Accountability – environmental responsibility across borders Earthscan, London.
Mayer, A.L., P.E. Kauppi, P. K. Angelstam, Y. Zhang, P.M. Tikka. 2005. Importing timber, exporting ecological impact. Science 308: 359-360.
Meadows, D., Randers, J., & D. Meadows. 2004. Limits to Growth – The 30 Year Update. Chelsea Green publishing, Vermont.
MEA. 2005. Millennium Ecosystem Assessment – Ecosystems and Human Well-being: Synthesis. Island Press, Washington. Available at: www.maweb.org/documents/document.356.aspx.pdf (Accessed 11 Oct 2010).
Miranda, M., C. Dieperink, & P. Glasbergen. 2006. Voluntary agreements in water- shed protection experiences from Costa Rica. Environment Development and Sustainability 9 (1): 1–19.
Norgaard, B. R. & L. Xuemei. 2007. Market governance failure. Ecological Economics 60: 634-641.
Okowa, P. 2000. State responsibility for transboundary air pollution in international law. Oxford University Press, Oxford.
Pirages, D. 2005. From Limits to growth to ecological security. In D.Pirages & K. Cousins, eds, From Resource Scarcity to Ecological Security – exploring new limits to growth.(the MIT press, Cambridge, MA).
Princen, T. 1997. The shading and distancing of commerce: when internalization is not enough. Ecological Economics 20 (3): 235–253.
—. 1999. Consumption and environment: some conceptual issues. Ecological Economics 31: 347–363.
Rees, W. E. 1992. Ecological Footprints and Appropriated Carrying Capacity: What Urban Economics Leaves Out. Environment and Urbanization 4 (2): 121-130.
—. 1994. Pressing global limits: trade as the appropriation of carrying capacity. In: Schrecker, T., Dalgleish, J. (Eds.), Growth, Trade and Environmental Values. Westminster Institute for Ethics and Human Values, London.
—. 1995. Achieving sustainability: Reform or transformation? Journal of Planning Literature 9 (4): 343-361.
—. 2004. Waking the sleepwalkers – globalization and sustainability, in W. Chesworth, M. Moss, & V. Thomas, eds. The Human Ecological Footprint . University of Guelph, Guelph, ON.
—. 2006. Ecological footprints and bio-capacity: essential elements in sustainability assessment. In: Jo Dewulf & H.Van Langenhove, eds. Renewables based Technology: Sustainability Assessment. John Wiley and Sons, Chichester.

—. 2008. Human Nature, Eco-Footprints and Environmental Injustice. Local Environment – The International Journal of Justice and Sustainability 13 (8): 685 – 701.

—. 2010. Globalization and Extended Eco-footprints: Neo-colonialism and (Un)sustainability. Chapter 24 in J.R. Engel, L. Westra & K. Bosselmann, eds. *Democracy, Ecological Integrity and International Law*. Newcastle: Cambridge Scholars Publishing.

Reinaud, J. 2008. Climate Policy and Carbon Leakage – Impacts of the European Emissions Trading Scheme on Aluminium. IEA information paper. International Energy Agency, Paris.

Rockström J. et al. 2009. A safe operating space for humanity. Nature. 461(7263): 472-5. Available at: http://dx.doi.org/10.1038/461472a (Accessed 3 September 2010).

Russo, R.O., & G.Candela. 2006. Payment of environmental services in Costa Rica: evaluating impact and possibilities. Tierra Tropical 2 (1), 1–13.

Schleicher, K., ed. 1992. Pollution Knows No Frontiers. Paragon House, New York.

Schutz, H., S. Moll, & S. Bringezu. 2004. Globalization and Shifting Environmental Burden. Wuppertal institute, Germany.

WTO. 2006. International trade statistics 2005. World Trade Organization, Geneva, Switzerland.

WWF. 2010. Living Planet Report. World Wild Fund for Nature. Gland, Switzerland.

Notes

[1] In some respects, this would resemble the supply management strategies employed in Canada (and elsewhere) that have been so successful in supporting family farming operations in various parts of the country. Farmers are allocated or purchase tradable quotas for certain products (e.g., eggs, chickens and dairy products), guaranteeing them stable market conditions and a predictable income for planning purposes.

CHAPTER FIFTEEN

IRRECONCILABLE DIFFERENCES? THE TENSION BETWEEN HUMAN SECURITY AND HUMAN RIGHTS

SABINA W.B. LAUTENSACH
AND ALEXANDER K. LAUTENSACH

Chapter Summary

The current global environmental crises are generating catastrophic events at an increasing rate and of increasing severity, threatening the survival and security of millions. Aid and disaster relief is given by the UN and its branches, by governments and by NGOs who regard it as their duty to address such large-scale threats. The justifications of that duty rely on common interpretations of human security and on various definitions of survival that can create moral dilemmas. More comprehensive concepts of human security, such as the Four Pillar Model, can help to avoid such contradictions but that approach leads to clashes with common interpretations of human rights.

Consequently, under the imperative of acceptable sustainable survival for humanity, human rights need to be re-examined. This re-examination goes beyond the traditional compromise between security considerations and human rights because under the current unprecedented threats some so-called 'rights' have become ungrantable. Even the moral extensionism that gave rise to concepts of sustainability as in the Brundtland document cannot resolve this problem, nor can it entirely reconcile rights with duties. We propose that only a rigorous enforcement of impact limits can accomplish such a reconciliation. However, such an approach will have grave consequences for the future of democratic societies.

Introduction

Human security as a concept first surfaced in the early 1990s when it became increasingly clear that the end of the cold war would not be accompanied by an end to armed conflict but that instead the nature of violent conflict was changing, away from traditional interstate war towards intrastate conflicts fuelled by ethnic, religious or ideological divisions (Hampson et al 2002). The discourse about security became enriched with the new insight that states are not the only entities whose security ought to concern us. Regions, communities, families and individuals can only feel secure if they have reason to believe that their continued functioning is not going to be threatened at every turn. Furthermore, the security of the state largely depends on the security of regions, communities, families and individuals. And occasionally states fail to fulfil their obligations as security guarantors, even to the point of threatening the very security of their own citizens (ICISS 2001: 15). It was realised that a primary requirement for human security was not merely the absence of war but the absence of structural and personal violence (Galtung 1969). This realisation informed a shift in perspective from the state as the subject and object of security policy to the human individual as the centre of security considerations – from state security to human security (Griffin 1995). And since human beings, unlike states, are capable of sensations and emotions, human security was recognised as partly contingent on those particular states of mind that we tend to associate with human well-being.

It follows that human security depends on variables that extend beyond what has traditionally been regarded as the political arena. The absence of violent conflict is only one of many determinants of human security; they include a relative safety from acute infectious disease, minimum complements of safe fresh water and adequate nutrition, and a formal guarantee for basic human rights and dignity. Concern for security also became extended further into the future. It has become acceptable to express concern with the future well-being of one's children, and, from middle age onward, with the well-being of their children, and so on. This long-term humanitarian concern has gradually come to inform the agenda of human security, as indicated by some common definitions of sustainability (WCED 1987; UN Millennium Project 2005).

With those concerns in mind, development agencies operating under national, supra-national or non-governmental umbrellas have adopted these extensions of the security concept into environmental and ethical dimensions. This reconceptualisation is evident in several key policy documents of the United Nations. In the Secretary General's Millennium

Report the UN's security agenda are defined as 'freedom from fear', and their development agenda as 'freedom from want' (United Nations 2000). The UN's guiding principles on security are paraphrased in negative terms as freedom from a condition that is evidently undesirable. Similarly, Alkire (2002: 2) defined the objective of human security as 'to safeguard the vital core of all human lives from critical pervasive threats, and to do so without impeding long-term human flourishing'. Elsewhere (Lautensach 2006) we suggested that those definitions are unhelpful because of the logical difficulties with negative definitions, the subjectivity and relativity of the criteria involved, and the neglect of systemic limits.

In order to arrive at a definition of human security that might realistically allow us to promote it in specific contexts, it is helpful to first examine what sources of insecurity might threaten the global citizen. Because of the subjective nature of human security such an examination must involve consultation with the people in question. Multinational opinion surveys[1] point towards criminal violence, armed conflicts (civil or international), terrorism, infectious disease, and 'natural disasters' as the events that people are most concerned about. The latter include extreme weather events, climatic aberrations, pest invasions, famines, floods, land slides, earthquakes and volcanism, and meteorite impacts. Other sources of insecurity include economic collapses, personal bankruptcy, personal accidents with traumatic health effects, and chronic health problems. Of course all of those factors potentially give rise to acute wants and needs in the individual. But by focussing on those sources of insecurity we eliminate some of the subjectivity and heterogeneity of the abovementioned 'freedoms' while gaining the advantage of focussing on more clearly defined targets. This can better facilitate proactive and preventive planning of security policies and it enables us to enlist a host of descriptive-analytical sciences in our planning efforts.

A survey of the sources of insecurity suggests that a comprehensive definition of human security needs to include four broad areas which we refer to as the 'four pillars' of human security (Lautensach 2006). They include the traditional area of military/strategic security of the state; economic security, particularly the contribution made by heterodox models of sustainable economies; health-related security, informed by epidemiology and the complex determinants of community health and health care priorities; and environmental security that seeks to maintain the complex interactions between human populations and their ecological support structures, namely the source and sink functions of their host ecosystems. Environmental security has also been defined as security from 'critical adverse effects caused directly or indirectly by environmental change' (Barnett 2007: 5). While all four pillars influence each other, environmental security forms the base on which the other three in many respects are founded (Lautensach & Lautensach 2010).

This dependence of all aspects of human security on environmental security becomes most evident when we examine its prospects over the long term, i.e. its sustainability. In the most elementary sense, sustainability just means survival. However, human populations can survive in different modes. Recognising that the survival of societies is contingent on collective choices, Potter (1985) distinguished five distinct modes of human survival – mere, miserable, unjust, idealistic, and acceptable. Given the central importance of human well-being and of principles of justice in our formulation of human security we can conclude that sustainable human security on a global scale is identical with the *acceptable* survival of humanity. We will return to the issue of such collective choices once we have outlined the ecological basis of sustainability.

Human Security and Sustainability

Sustainability is described by the balance between supporting the quality of life for a human population and the continued functioning of its environmental support structures, namely ecosystems.[2] Ecosystems consist of local communities of species and their physical environment. They serve as sources of food, fresh water, raw materials and energy, and they recycle the population's wastes back into biomass. Complex ecosystems that are rich in species (occurring especially in the tropics) tend to be more resilient to disturbances, whereas ecosystems that consist only of a few species tend to be more fragile.

Human populations, like all other animal populations, obtain their sustenance from ecosystems which provide the abovementioned "ecosystem services" (Vitousek et al 1997). Human populations are special in that they employ technology to maximize the benefits of those ecosystem services. But regardless of this technological windfall, the capacities of local ecosystems remain limited. Generally, the environmental impact I of a human population on local ecosystems is described by the $I=PAT$ formula, where P means population size, A stands for the affluence or economic means per capita, and T represents the technological impact per capita (Ehrlich & Holdren 1971; York *et al.* 2003). The maximum sustainable impact, also referred to as carrying capacity (Curry 2006: 126) is thus described as the product of the three variables; it can be reached by small populations with a high-impact lifestyle or by larger populations where each individual demands less in terms of support services. When a population exceeds the maximum sustainable impact it enters into overshoot, where the services of the local ecosystems are being overtaxed and, depending on the system's fragility, may undergo irreversible structural changes (Catton 1980; McMichael 2001; Wackernagel *et al.* 2002; Meadows *et al.* 2004). Inevitably the consequence for the population is such that various biological regulatory mechanisms lead to a decrease in population size, below the system's carrying capacity. Numerous precedents from animal populations have allowed ecologists to characterize and predict those dynamics with impressive accuracy.

The population's environmental impact can also be expressed in terms of the area of productive land required to support the population's life style. This is referred to as that population's ecological footprint (Wackernagel & Rees 1996). A population whose footprint exceeds the amount of accessible land is clearly in local overshoot. This may not always have immediate negative consequences for people's security as they may obtain the shortfall from other regions that are either underpopulated, defenceless, or otherwise disempowered. That practice is, however, often unjust and supports unsustainable patterns of consumption.

To summarise this sequence of causation, unsustainable practices sooner or later lead a population into overshoot, which in turn erodes environmental support structures and decreases their capacity to deliver resources and to accept wastes in the future. This loss of ecological integrity means that environmental security of the population is threatened which can manifest itself in shortages of food, energy, or of other commodities, or in elevated levels of pollution. Such changes invariably compromise population health and lead to economic decline, civil disorder, and vulnerability to external enemies (Homer-Dixon 1999). Evidence is provided by the historical precedent of cultures that disappeared as a result of that sequence of effects (Diamond 2005). The upshot is that whatever safeguards may be in place to protect the economic security of a population, its public health, its national security, and the rule of law – they seem of little help in the long term unless sustainability and environmental security are guaranteed. This resonates with Barnett's (2007) finding of a mutual dependence between environmental security and peace.

The spectrum of possible combinations of different population sizes, consumption levels, and technological impacts illustrates again the multiplicity of choices by which a society determines its mode of survival. Once its environmental impact approaches the sustainable maximum (or its footprint approaches the available maximum bioproductive area) those modes become a function of population size, with *mere* survival becoming most likely for a sufficiently large population and *acceptable* survival remaining an option only for relatively small population sizes. Thus we can establish a causal connection between population size and human security. This raises some important questions about the status of human rights, which, as we shall see, leads to some significant political ramifications.

Human Security Conflicts with Human Rights and Civil Liberties

As outlined above, many popular conceptualisations of human security are founded on human rights. These rights developed historically in three stages, each based on agreement among human communities on a set of basic needs (Fien 1993). The first generation of human rights was civil and political in nature and was based on the cardinal value of freedom. The second generation are economic and social human rights, based on the cardinal value of human equality. At this stage a third generation of rights is being formulated, as illustrated by the UN and numerous major NGOs focusing on the right of every global citizen to enjoy freedom from fear and freedom from wants (Annan 2005). Thus they add to the list of human rights specified in the UN Charter of Rights and Freedoms (United Nations 1948)[3] the rights to certain 'freedoms'. Those freedoms are distinct from civil liberties and from the negative rights specified in the Charter which both pertain to the functioning of civil society. They extend on the rights specified in Article 25 of the Charter which refers to "the right to a standard of living adequate for the health and well-being of himself and of his family, including food, clothing, housing and medical care and necessary social services ..". In terms of environmental security those freedoms amount to certain quality attributes pertaining to environmental support systems, sometimes referred to as 'environmental rights'. The UN's Millennium Development Goals (United Nations 2009) refer to them as the right to clean air, safe potable water, adequate nutrition, shelter, the safe processing of wastes, and adequate health care. The fashionable slogan of 'ending poverty' is in part based on the rigorous application of the belief that every human being has a moral claim to a certain standard of living, while usually failing to define what kind of poverty it refers to (Shiva 1998).

In our examination of environmental security we established that the purification of air and water, the provision of foods and shelter, and the processing of wastes are directly contributed by the local ecosystems. It follows that the sustainable provision of those services depends on the biological integrity of those ecosystems (Westra 2005; Karr 2006). Accordingly, it would make sense for human communities to claim the right that 'their' ecosystems not be harmed or diminished in their capacities. The fact that such a claim is not usually made, and that instead the demands pertain exclusively to the human recipients of those services represents both a grave logical fallacy and a strategic error in judgment by human rights advocates. The integrity of an ecosystem can, given sufficient care, experience and motivation, be maintained sustainably, barring any major external threats such as climate change. Among the conditions of such a policy would be that the total environmental impact of the community does not exceed the sustainable maximum, i.e. the ecosystem's carrying capacity. In contrast, claiming that the individual community member has a right to a certain quality of service makes no sense because, as we have seen, no-one has the power to grant such a demand, not even the most absolute dictator, once the population's impact has exceeded that threshold. Thus, those third generation 'environmental rights' belong in a different category from the other human rights, the category of ungrantable 'rights'.[4] Being grantable, however, is an essential property of any right (Rawls 1971). Therefore, a right that cannot be granted is no right at all (hence our use of inverted commas), and it makes no sense to promise or to claim it. While the goal of ending poverty may have been realistic when the world population measured two billion or so, it was reduced to an unscientific pipe dream after we entered global overshoot in the mid-1980s (Wackernagel et al. 1997; 2002).

The strategic error in judgment associated with claiming ungrantable 'rights' derives from the effect that such a claim diminishes the status of other rights to which realistic and legitimate claims could be made. For example, if the UN's Human Rights Council added to the list of human rights the right to own a circus, clearly ungrantable, the entire list would as a result acquire a less serious, less binding, and more conditional appearance. This would be regarded as a disservice by most people who harbour genuine concern for human rights and their enforcement. Including ungrantable 'rights' among the list is likely to diminish the sense of urgency with which all human rights ought to be respected by the powerful worldwide.

This problem does of course not diminish the need for guarantees that promote the environmental security of communities, especially when it comes to the world's disempowered. As we suggested above, the concept of ecosystem integrity could be instrumental in formulating policy guidelines that would go a long way towards such guarantees. This approach might also initiate a more balanced manner of moral reasoning, away from the heavy emphasis on deontological rights-based arguments and including arguments from utilitarian and virtue ethics. Arguments based on rights and duties often do not go far enough to promote human security in the form of effective policies and legislation because they often do not specify codes of behaviour clearly enough. Grantable rights depend primarily on human behaviour whereas ungrantable 'rights' depend heavily on environmental resources. In the light of this distinction, the question arises whether the latter are of any use at all.

From Environmental 'Rights' to Environmental Demands

The preceding discussion emphasised the need to use rights-based arguments prudently when debating human security, and to invoke grantable rights in different ways and on different occasions compared to those kinds of demands that might be construed as invoking ungrantable 'rights'. Heretofore we will refer to the latter as 'environmental demands'. We do not mean to insinuate that environmental demands have no place in the human security debate – on the contrary: The qualities of air and water, of nutrition, of shelter, of the ways of recycling wastes, and the status of public health are still among the best indicators that help us assess the status of environmental security of a community. And they can help bolster some rights-based arguments, namely in connection with the right to justice. We perceive at least three distinct uses for environmental demands in the promotion of human security.

First, in the case of a community or region that has not yet reached the maximum sustainable ecological impact environmental demands can highlight situations of injustice and environmental harm. Based on claims for distributive justice and on evidence of ecological deterioration they can inform policies that would elevate the minimum standard of living in the community and the sustainability of its ways of living, respectively. In this case, "ending poverty" is a noble and achievable goal.

Secondly, in situations where the maximum sustainable impact has been exceeded environmental demands serve to highlight that very circumstance. No other physical observation illustrates the fact of ecological overshoot more clearly than the widespread squalor resulting from polluted air and water, famine, and poor public health (McMichael 2001). Also, comparisons of national footprints with available bioproductive areas (e.g. by Wackernagel et al. 1997; GFN 2009) have made it abundantly clear that overshoot is now the norm for many regions and historical reality for the globe since the mid-1980s (MAB 2005; Wackernagel et al. 2002). Yet the worldwide efforts by powerful groups to ensure that information about overshoot is delayed, distorted, ignored or denied require concerted efforts to disseminate this information and to educate the public (Rees 2004). The language of environmental demands is one that everyone understands, even if they remain largely unmet.

Thirdly, environmental demands are the mainstay of the discourse on environmental justice. An effective way to illustrate the widening gap between global rich and poor is to compare the environmental indicators that shape their respective lives, in addition to the often invoked data on *per capita* consumption. Regardless of the extent of overshoot, such comparisons highlight the injustice inherent in the global economic order, its trading schemes, and its underlying maldistribution of political power. [5]

Insisting on environmental demands in those contexts represents of course only the first step in an argument that necessarily leads to a discussion of Potter's (1985) five modes of survival, and of measures that might lead us from the current prevailing mixture of mere, miserable, and irresponsible modes towards acceptable alternatives. However, that transition confronts us with another problem where many of the grantable human rights–the ones that refer mostly to human behaviour in a civil society–conflict with environmental security.

Reconciling Human Security with Human Rights and Civil Liberties

History abounds with well-intentioned efforts by powerful rulers to enforce measures for the 'common good', which arguably required that the individual rights and liberties of their subjects be curtailed. At times it is difficult to assess whether omitting such specific limitations would have necessarily led to preferable outcomes. Yet human rights were often violated in the course of those measures. Article 29 of the Charter serves the same purpose, albeit not in a dictatorial way. In principle, every law that is passed represents a compromise between benefits to society and sacrifices to individual autonomy. This basic dilemma is amplified many times in scenarios of transition towards sustainability.

The imperative of attaining sustainability rests either on intergenerational justice or on a more ecocentric motive to spare the Earth further ecological devastation from the hands of humanity. This moral extensionism is evident in such mainstream definitions of sustainability as in the Brundtland document (WCED 1987). However, even such a purely anthropocentric application of intergenerational justice requires that the welfare of future generations be weighed against the rights and liberties of existing humans. The requisite drastic changes in lifestyle choices towards greater efficiency, reduced consumption, adaptation to global changes, and organisational reform, probably accompanied by a reduction in the global population, would necessitate either an unprecedented amount of consensus or a draconian repression of individual autonomy (Bowers 1993; Daly & Cobb 1994). Neither option sits well with the humanistically inclined advocates of intergenerational justice. Of course, denial is always an option: In 2005, then Secretary-General Kofi Annan (2005: 1) stated in his report 'In Larger Freedom' advocating development, security and human rights for all, "The world must advance the causes of security, development and human rights together, otherwise none will succeed. Humanity will not enjoy security without development, it will not enjoy development without security, and it will not enjoy either without respect for human rights."

There is good reason to assume that only an ecocentric ethic can ultimately provide the moral basis for sustainable behaviour (Curry 2006; Lautensach 2009). To ecocentrists, even of the light green persuasion, many human rights and liberties carry less moral weight than the survival of other species and the existence and integrity of entire ecosystems and the biosphere. For them, this dilemma presents less of a challenge. This has given cause for occasional charges of 'ecofascism' from rights-oriented circles (Regan 1983: 362).

To the determined anthropocentrist, however, the only recourse is to weigh the extent to which human rights and liberties will have to be curtailed if the transition is to be accomplished, against the extent to which rights and liberties will be lost amidst anarchy, chaos, famine, disease, and incessant warfare over diminishing resources, if sustainability is not achieved (Homer-Dixon 1999). To anyone who values the future welfare of our children and grandchildren the mere comparison of benefits and harms the dilemma might appear less than paralysing. Nevertheless, many will be shocked by the realisation that, whichever choice is taken, democracy as we have come to know it in affluent countries will not last.

References

Alkire, Sabina. 2002. 'Conceptual framework for the Commission on Human Security'. www.humansecurity-chs.org/doc/frame.html.

Annan, Kofi. *In Larger Freedom: Towards Development, Security, and Human Rights for All.* Executive Summary. New York: United Nations, 2005. www.un.org/largerfreedom/executivesummary.pdf.

Barnett, J. 2007. Environmental Security and Peace. *Journal of Human Security* 3 (1): 4-16.

Bowers, Chet A. 1993. *Education, cultural myths, and the ecological crisis: Toward deep changes.* Albany, NY: State University of New York Press.
Catton, W.R. Jr. 1980. *Overshoot: The ecological basis of revolutionary change*. Urbana: University of Illinois Press.
Chambers, N. Simmons C. and M. Wackernagel. 2000. *Sharing nature's interest: ecological footprints as an indicator of sustainability*. London: Earthscan Publ. Ltd.
Curry, P. 2006. *Ecological Ethics: An Introduction*. Cambridge, UK: Polity Press.
Daly, H.E., and J.B. Cobb. 1994. *For the common good: Redirecting the economy toward community, the environment, and a sustainable future*. Boston: Beacon Press, 2nd ed.
Diamond, Jared. 2005. *Collapse: How societies choose to fail or succeed.* London: Viking Penguin.
Ehrlich, Paul R. and J. Holdren. 1971. "The Impact of Population Growth". *Science* 171: 1212-1217.
Fien, John. 1993. *Education for the environment: Curriculum theorising and environmental education.* Geelong, Victoria, Australia: Deakin University Press.
Galtung, Johan. 1969. 'Violence, peace, and peace research'. *Journal of Peace Research* 6 (3): 167-191.
GFN (Global Footprint Network). 2009. *Ecological Footprint Atlas.* www.footprintnetwork.org.
Griffin, Keith. 1995. 'Global prospects for development and human security'. *Canadian Journal of Development Studies* XVI (3): 359-370.
Hampson, Fen O., Jean Daudelin, John Hay, Todd Martin and Holly Reid. 2002. *Human Security and World Disorder*. Oxford University Press.
Homer-Dixon, Thomas. 1999. *Environment, Scarcity, and Violence.* Princeton: Princeton University Press.
ICISS (International Commission on Intervention and State Sovereignty). 2001. *The Responsibility to Protect*. Ottawa: International Development Research Centre. www.idrc.ca.
Karr, James B. 2006. Measuring Biological Condition, Protecting Biological Integrity. In *Companion to Principles of Conservation Biology*, ed. M.J. Groom, G.K. Meffe & C.R. Carroll. Sunderland, MA: Sinauer Associates. www.sinauer.com/groom/article.php?id=23.
Lautensach, A.K. 2006. Expanding Human Security. *Journal of Human Security* 2 (3): 5-14.

—. 2009. "The ethical basis for sustainable human security: A place for anthropocentrism?" *Journal of Bioethical Inquiry* 6 (4): 437-455.

Lautensach, A. & S. W. Lautensach. 2010. "Prioritising the Variables Affecting Human Security in South-East Asia" (keynote presented at the Conference of the South East Asia Society, Vienna, 28-29 May).

Lemons, John. 1996. "Afterword: University education in sustainable development and environmental protection". In *Earth summit ethics: Toward a reconstructive postmodern philosophy of environmental education*, ed. Callicott, J. Baird and Fernando J.R. da Rocha, 193-217. Albany New York: State University of New York Press.

McMichael, A.J. 2001. *Human frontiers, environments and disease : past patterns, uncertain futures*. Cambridge : Cambridge University Press.

Meadows, D., Randers, J. & D. Meadows. 2004. *Limits to Growth: The 30-Year Update*. White River Junction, VT: Chelsea Green Publishing Co.

MAB (Millennium Assessment Board). 2005. *Millennium Ecosystems and Human Wellbeing*. Washington DC: Island Press. www.millenniumassessment.org//en/Products.Synthesis.aspx.

Potter, Van Rensselaer. 1985. A response to Clements' environmental bioethics: A call for controlled human fertility in a healthy ecosystem. *Perspectives in Biology and Medicine* 28(3) (Spring): 426-433.

Rawls, John. 1971. *A theory of justice*. Cambridge, Mass.: Harvard University Press.

Rees, William. 2004. "Waking the Sleepwalkers – Globalisation and Sustainability: Conflict or Convergence". In *The Human Ecological Footprint*. Guelph, Canada: University of Guelph, 1-34.

Regan, Tom. 1983. *The Case for Animal Rights*. Berkeley, CA: University of California.

Shiva, Vandana. 1998. "Development, Ecology and Women". In *Applied Ethics: A Multicultural Approach*, ed. L. May, S. Collins-Chobanian & K. Wong. Upper Saddle River, NJ: Prentice Hall, 170-180.

UN Millennium Project. 2005. Investing in development: A practical plan to achieve the Millennium Development Goals. New York: United Nations.

United Nations. 1948. The Universal Declaration of Human Rights. www.un.org/en/documents/udhr/index.shtml.

—. 2000. We the Peoples: The role of the United Nations in the 21st century. New York: United Nations. www.un.org/millennium/sg/report/ key.htm.

—. 2009. End Poverty 2015: Millennium Development Goals. New York: United Nations www.un.org/millenniumgoals/.

Vitousek, P.M. Mooney, H.A. Lubchenko, J. and J.M. Melillo. 1997. Human domination of Earth's ecosystems. *Science* 277 (25 July): 494-499.

WCED (World Commission on Environment and Development). 1987. *Our Common Future: The Brundtland Report*. Oxford: Oxford University Press.

Wackernagel, M., Schulz, N.B., Deumling, D., Linares, A.C., Jenkins, M., Kapos, V., Monfreda, C., Loh, J., Myers, N., Norgaards, R., & J. Randers. 2002. Tracking the ecological overshoot of the human economy. *Proceedings of the National Academy of Sciences* (USA) 99: 9266-9271. . www.pnas.org/cgi/reprint/99/14/9266.pdf.

Wackernagel, M., L. Onisto, A. Callejas Linares, I..S. López Falfán, J. Méndez García, A.I. Suárez Guerrero, Ma. Guadalupe Suárez Guerrero. 1997. Ecological Footprints of Nations: How Much Nature Do They Use? -- How Much Nature Do They Have? Xalapa, Mexico: Centre for Sustainability Studies ; Costa Rica : Earth Council. www.ecouncil.ac.cr/rio/focus/ report/english/footprint/.

Wackernagel, Mathis and William Rees. 1996. *Our ecological footprint: Reducing human impact on the earth*. Oxford: John Carpenter.

Westra, Laura. 2005. "Ecological Integrity". In *Encyclopedia of Science, Technology and Ethics*, ed. C. Mitcham. Vol. 2 D-K. Detroit, MI: McMillan Reference USA, 574-578.

York, R., Rosa, E.A., and T. Dietz. 2003. STIRPAT, IPAT and ImPACT: analytic tools for unpacking the driving forces of environmental impacts. *Ecological Economics* 46(3) (Oct): 351-365.

Notes

[1] For example, one survey conducted in the UK identified crime and health as the greatest concerns. www.angus-reid.com/polls/index.cfm/fuseaction/viewItem/ item ID/8084.

[2] We use the term sustainability only in its original environmental meaning and do not refer to other, secondary interpretations such as cultural or social sustainability. Lemons (1996: 198) defined sustainability as "the continued satisfaction of basic human physical needs, such as food, water, shelter, and of higher-level social and cultural needs, such as security, freedom, education, employment, and recreation", along with "continued productivity and functioning of ecosystems". We regard the popular 'Brundtland' definition of sustainability (WCED 1987) to be quite useless because of its lack of conciseness, inattention to metaethical considerations and its neglect of fundamental ecological limitations. A more useful definition, attributed to Steve Goldfinger (Chambers *et al*, 2000: 2), states that a sustainable community

is one that converts resources into waste no faster than ecological support structures can convert the waste back into resources.

[3] In thirty articles, the Charter specifies the human right to life, liberty, and security of person; to freedom from discrimination by race, creed, gender, and equality before the law and due process; to a fair and public hearing in case of criminal accusations; to be presumed innocent until proven guilty; to free association and nationality, to freedom of movement; to own property; to freedom of expression and of religion; to democratic choice of representation; to respect for human dignity; to work and to equal pay as well as to leisure; and to a basic humanistic education. Negative rights include freedom from inhuman or degrading treatment or punishment; from arbitrary arrest, detention, or exile. Article 29 recognises appropriate duties and limitations of the rights of the individual for the common good but does not mention ecological limits.

[4] Ungrantable 'rights' do not necessarily contravene Rawls' (1971) basic postulate that a person exercising a right must not impinge on the opportunities of other people exercising the same right. In the case of 'environmental rights', their grantability depends on the total number of people claiming them. Thus they violate Rawls' postulate only if their collective exercise results in the carrying capacity being exceeded. But their property of being ungrantable remains true under all conditions, in the sense that they cannot be guaranteed even by a ruler with the highest political power and the best intentions.

[5] A thought experiment can illustrate the fact that environmental demands can, in the long run, only satisfy considerations of justice, not considerations of rights. In a highly inequitable society that has reached its maximum ecological impact only an elite minority can enjoy the full benefits of abundant ecosystem services while the majority survive (or not) in relative misery. A benign dictator set on reforming this society along the environmental demands made by the masses could distribute those benefits more equitably. This would improve the situation in terms of distributive justice but it cannot bring about acceptable living conditions for all, thus leaving many environmental demands unmet. Furthermore, neither total utility nor average utility would increase, even if the population has stopped growing.

CHAPTER SIXTEEN

SUSTAINABILITY AND LIBERALISM – TWO CONFLICTING PRINCIPLES?

MICHAEL W. SCHRÖTER*

The accusation of eco-dictatorship was leveled as early as the 1980s[1] to reveal the "real purpose" of environmental-activists, and to compromise their political suggestions: Not to save nature but to install a political system far away from individual liberty in which an elite would dominate for the sake of a higher good. In actual fact, they were pursuing the establishment of a new political and economic order. This rhetoric continues to this day.[2] On the other hand, there have always been authors predicting the loss of legitimacy if a state does not react appropriately to the ecological crisis.[3] There can be no doubt, though, that the task of environmental protection is rooted in the fundamental source of legitimacy of the state: to secure the basis of society and its members. Even conservative state ideology starting with Hobbes' *Leviathan* affirms this.[4] Looking at these two contrary positions we must concede a nearly paradoxical structure of discourse: While the danger of an eco-dictatorship looms in the event the state becomes too protective, the state will at the same time lose its legitimacy if it protects too little.[5] Is there a chance of reconciling these extremes?

I will take up the discourse of "Liberalism and Sustainability" starting with defining the two terms, followed by discussing their relation. If a positive relation can be ascertained the question of a legal modus to institutionalize the relation will be eventually raised. But before getting on with this discourse I would recall that it has not begun recently but at the latest around the time of the Enlightenment.[6] A quote of Friedrich Schiller from 1795 demonstrates this:

""They (the objects of nature; MS) *are* what we *have been*, what we are *going to be again*. We have been nature like them and our culture shall lead us on the path of liberty and reason back to nature. Thus they are also representations of our lost childhood which will remain our greatest treasure; and fulfill us with certain nostalgia. At the same time they are representations of our highest perfection in ideal, that's why they put us in sublime quietness.""[7]

1. Liberalism and Sustainability

In tackling the effort to define "Liberalism" and "Sustainability" in the brevity appropriate for this paper, the purpose cannot be to find comprehensive, sound and all-convincing descriptions but to determine the special functions these terms serve in the discourse. The focus will therefore be on basic intuitions in which all meanings and interpretations are rooted.

1.1 Liberalism

Liberalism is not a philosophical construct but a political doctrine. Nevertheless it justifies itself by philosophical ideas and thoughts. One of these was stated by Immanuel Kant. Probably inspired by Jean-Jacques Rousseau in this regard, Kant acknowledged only one right as innate: freedom.[8] Kant differentiated between an inner and an outer freedom. The inner freedom means the capability to govern the will with self-given laws of reason. The outer freedom is the independence from suffering the enforcement of someone else's arbitrariness.[9] Only the outer freedom forms the innate right of every human. In this, Kant is conforming to John Locke and the English tradition of liberalism that has its characteristic utterance in John Stuart Mill's harm-principle: People can do anything they like as long as they do not harm anybody else.[10] This harm can be of a physical or moral nature. And this possibility of harm, and the need to be protected against it, is the foundation of the legitimacy of the state. Again there is a strong parallel to Kant for whom the outer freedom constitutes the state *a priori*: no person shall be forced to happiness but may pursue their happiness as they please.[11]

Those sketched philosophical thoughts underpinned a fundamental political change in the relation between the individual and the state: It was not the state anymore but the individual which became almighty, only tamable by the reasonable interest of the others. The argumentative basic pattern for the individual to justify freedom from the state and others became: Because I am there! This new rationality of discourse forced and forces everyone, but especially the state, to find good reasons why "there-being" is not a sufficient reason for a way of action for the individual. To this extent, liberalism can be understood as a political doctrine shaping a state where the individual is free to act as he/she wants as long as there are no good reasons presented and accepted why he/she should not.[12]

Liberalism marks one apex of what can be understood as a triangle of political doctrines.[13] The other two apices are occupied by anarchism and authoritarianism:

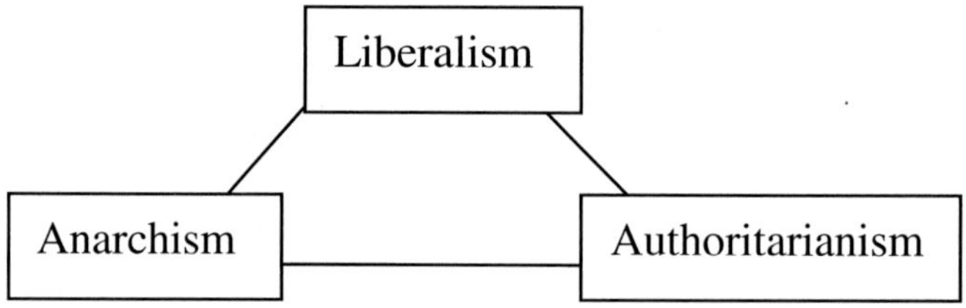

While anarchism shares with liberalism the conviction of individual freedom,[14] it denies the necessity of the state. Authoritarianism regularly gives the state or the forming community a higher importance than the individual. Of course, between these ideal positions there exist many possible combinations, and these can be found in the discussions through the decades up to the present day. But in these discussions, liberalism as a pattern of argument carries greater weight, at least in the Western world, than the two other positions. Why is this?

The obvious answer is a historical one: Armed with the arguments of the Enlightenment, people fought in revolutions for their right of liberty, including equality, which many considered as their first natural or innate right. Since those revolutions liberalism is considered as the right of the people combined with a collective right of revolution. Thus liberalism comes with an inherent pathos completely absent in anarchism and dictatorship. But there is another reason for this "pathos", and this is bound up with rationality and epistemology and breaks down to the rule of law.[15] In a political system where power is not linked with the will of the people, decision-makers need a different source of legitimacy. This can be closeness to a god, a higher wisdom by birth, or just tradition. The enlightenment disavowed those reasons by showing that man's perceptive ability is finite. Rulers who appeal to those reasons appeal to irrational sources of legitimacy and, in consequence, act arbitrarily.[16] As a new modus of legitimacy the idea arose that it was necessary to consider the interests of those who will be affected by a particular decision. To refuse those interests would be possible only for good reasons that convince those affected in a rational manner.[17] Due to this, liberalism and democracy so far go hand in hand.[18] And, further, everyone who wants to politically distance oneself from liberalism carries the burden of proof not to argue arbitrarily. This is the main function of liberalism in the discourse up to the present day, and explains why its use as a pattern of argument garners greater weight than anarchism or authoritarianism.

But there is also a negative aspect of liberalism highlighted in the discourse: its standing for economistic rationality which became suspicious of serving the greed of individuals instead of enhancing the well-being of social communities and ecosystems.[19] This economic side of liberalism is considered by some to be its very heart, since liberalism aims primarily at self-realization through individual property: Freedom becomes a genuine materialistic idea.[20] In consequence, liberalism mutates into the opposite of its original purpose: Instead of creating free, just and peaceful societies, it ends with unjust distributions where only an elite gathered wealth and power while not caring about the rest of society and nature.[21] And in this manner it threatens democracy itself, although “liberal democracy” is seen by many as the most successful combination in history so far.[22]

In sum, liberalism shows two functions in discourse: a positive one which is accompanied by a special “pathos” that shifts the burden of proof to everyone criticizing it, and a negative one which confronts liberalism with its economic dimension and its spin-offs which could lead to just the opposite of what liberalism seeks to accomplish. For the purpose of this paper I will distinguish between those two functions as the original political liberalism and the economistic liberalism.

1.2 Sustainability

The discussion surrounding the term sustainability is surprising for two reasons: (1) that it did not develop earlier than it did nearly 40 years ago[23] and (2) that it took a further 20 years before it was intensified and increasingly attracted public attention.[24] Looking at the basic statement of sustainability as it was established in forestry—not to use more resources than will grow again[25]—it is hard to understand why a general debate about sustainability came up so recently and is quite controversial right now. Sustainability appears to be such a simple but striking insight and is completely in line with what rational choice theory calls a preference of first order, to which every political and economical approach should adhere as the first and preferred pattern of rationality. The answer can only be of a speculative and psychological nature: At the end of the sixties and the beginning of the seventies the concept of national welfare states appeared to be at its zenith; some political scientists call this the "golden age" of the national welfare state.[26] Although confronted with the iron curtain and the Vietnam War, for many people, at least in the Western world, a global peace and, with it, economic wealth no longer seemed naïve illusions. But this prospect was suddenly confronted with warnings questioning nothing less than the dominant lifestyle and its political and economical consequences: the end of natural resources. Looking at the public environmental debates in the 1970s, they were mostly bifurcated along a "bad" and a "not too bad" message: natural resources and our current way of living will definitely come to an end ("bad"); but our generation (of the 70s) will, quite likely, not experience it, although our children, or theirs, may ("not too bad").[27] The last part of the message has changed in the last 40 years and with it the intensity of the discussions surrounding sustainability.

Two understandings of sustainability can be differentiated in the contemporary debate: a strong, or ecological, and a weak, or environmental, meaning.[28] The strong meaning is critical of growth and favors ecological sustainability; the weak assumes the validity of growth and places equal importance on environmental sustainability, social justice and economic prosperity.[29] The environmentalist approach refers to the international documents in which the notion of sustainable development was included. Significant in this regard is the Brundtland report.[30] While opening the justice perspective spatially and temporally (intra- and intergenerational), it kept the vision of an ongoing economical growth; it even promotes such growth as a necessary means to reach the goals of sustainable development.[31] Sustainable development as understood in this way asks the growing economy to respect environmental constraints.[32] Sustainable development therefore knows three equivalent dimensions: environmental, social and economical. The ecologist approach concedes the three dimensions but objects to their assumed equality. The environmentalist approach would overlook the ""simple truth, namely the interrelatedness of all life across boundaries between humans and non-humans.""[33] Sustainable development in the ecologist understanding is the practical outflow of the sustainability principle.[34] This principle gives the conception of sustainable development its basic meaning. This principle defines ""as the duty to protect and restore the integrity of the Earth's ecological systems.''''[35] As such, the principle of sustainability is the most fundamental of all environmental principles. It combines moral and legal dimensions and appears either as the elementary core of ecological justice or as an equally fundamental principle of justice.[36]

The dispute with regard to the proper understanding of sustainability reminds one of the debates between deep and shallow ecology. But while the latter showed traits of what could be called a mostly academic discussion, the results of the discourse surrounding sustainability are of direct relevance for social reality. In most states and supranational organizations (e.g. the EU) the notion of sustainability has entered high ranking legal documents. Public policy makers are thus forced to recognize it, at least, as one of several social aims. This means that society will change in consequence of encountering sustainability. Changes do promote fears but also hopes. As a discursive anchor for those fears and hopes, the question of the relation between the three dimensions—social, economical, ecological—is more than appropriate. Otherwise, it seems to elude explanation why some authors argue for the equality of the three dimensions up to the present day; this understanding obviously ignores the essential statement of sustainability: to design societies in such a way that they can durably coexist with nature.[37] Andrew Dobson's differentiation in technocentric and ecocentric approaches to sustainability probably reflects this best.[38] To this end, every approach to sustainability is a normative approach: Whether implicitly or explicitly, such an approach uses valuations. But with this constraint, the discourse surrounding sustainability unavoidably becomes a political one: Defining the notion of sustainability always means giving an answer regarding how we will live in the future. It is nearly impossible to disentangle this from political concepts. In consequence, critics can blame progressive theories of sustainability as attacking the predominant political and economical system, even if this was not the purpose of the theory at all. The function of sustainability in the discourse therefore becomes ambiguous: On the one hand to be the intellectual focal point for ecological thinking from which good reasons for social actions tackling the ecological crisis can be formulated; and on the other hand to create a new platform for the contest of political ideologies.

These two functions may be viewed as ecological and political. Sustainability in the ecological function serves as the notion wherein durable means to overcome the ecological crisis can be developed. The political function challenges the political ideologies with regard to whether or not they are capable of realizing the necessary changes. But this differentiation is, to say it frankly, an idealistic one. As mentioned above, the two functions are regularly mixed in discussion and it is practically impossible to disentangle one from the other. Since liberalism is, at least in the Western world, the predominant political doctrine which places burden of proof on any critics, the relation between sustainability and liberalism is especially at stake.

2. The Relation of Liberalism and Sustainability

As I have shown, sustainability challenges political doctrines. As the predominant doctrine, liberalism is particularly at stake. But liberalism places the burden of proof on the critic. In my attempt to clarify the relation between sustainability and liberalism I will take up what I have called original political liberalism and briefly consider the basic ecological critique of it. The thesis is that the main critique aims at economic liberalism. If this can be argued, I will eventually discuss the option of saving the original political liberalism and taking the critique of economic liberalism seriously.

2.1 Liberalism and the Ecological Critique

Original political liberalism (see above) embraces two entangled positions: every human being is born free; this includes legal equality. To avoid arbitrariness, state power must be bound to the rule of law. Some philosophers argue that this logically implies democracy and *vice versa*.[39] On the other hand, more and more ecologists blame liberal democracies for being unable to find sustainable solutions to the ecological crisis.[40] In a recent interview, James Lovelock pleaded for a more authoritarian world: ""We've become a sort cheeky, egalitarian world where everyone can have their say.""[41] For him, climate change is a situation comparable to war and in such situations democracy is put on hold.[42]

This is reminiscent of Carl Schmitt's argument which was very popular in Germany of the 1920s: In moments of crisis, democracies need a strong and sovereign authoritarian institution able to make the necessary political decisions. Schmitt favored the president of the state for this task as "guardian of the constitution"[43], although he argued for a legally prescribed formal realm of power.[44] Lovelock and Schmitt seem to be unified in a basic skepticism towards liberal democracies where—as a result of everyone's equality—sense and measure for a good or bad political option are threatened to be lost in the plurality of voices and thereby endanger society in situations of urgency. Unlike Schmitt,

Lovelock sees no alternative to liberal democracies, although he argues that they should be put on hold for a while.

But which aspect of liberal democracy does this critique target? Schmitt concedes that the execution of power in democracies can even be stronger and more efficient than in dictatorships as long as the government enjoys the consent and the confidence of the people.[45] And the essential core of Lovelock's plea blames an almighty individualism where ""everyone can have their say"". This focuses on the liberalistic part of democracy and with it one essential realm of modern legal philosophy while touching the ground of what Carl Schmitt has called political theology.[46] Indirectly the question is raised as to what can claim the higher priority: politics or law?[47] At least German history shows the real danger of a constitutional or provisional dictatorship which slides into a non-constitutional or sovereign dictatorship. Some constitutions, like the German *Grundgesetz* in Art. 20 IV, have established a right of resistance to prevent those scenarios. So why are liberalism and its individualism so fundamental to legal thinking that any critic soon encounters these very basic problems I have just named?

After all, individualism assumes everyone being able to reflect on their own actions. This assumption is fundamental to ethics and law. [48] It is justified, among others, by the epistemological insight that to be human means being fallible. Or to put it more provocatively: no human is God. This also means that no human is any closer to or further away from God than anyone else. As a consequence, a convincing ecological critique of original political liberalism needs to demonstrate two premises: (1) knowledge about the ecological crisis and the means to successfully overcome it are so strong as to justify authoritarian structures; and (2) contemporary individuals are in sum not reflective enough to behave as considered necessary. Considering the approximate status of most ecological knowledge, this appears to be a tall order. It thrusts those ecologists who insist on the necessity of authoritarian structures into the middle of very complicated and delicate discussions. Further, they have to take responsibility for the risk that authoritarian structures may slip into sovereign dictatorships. Is original political liberalism therefore really the correct aim of the ecological critique? Or does it rather question what I have called economic liberalism? And if this assumption is correct, should sustainability not even align itself with original political liberalism?

2.2. The Mutuality of Liberalism and Sustainability

Bertrand Russell advises us that: ""The problem, like all those with which we are concerned, is one of balance; too little liberty brings stagnation, and too much brings chaos.""[49] This alludes to the aspect of motivation I have not mentioned so far, but which is essential for political changes. In liberal democracies one mechanism for motivation is public debate guaranteed by the right to free speech. But those debates take time and the adverse power of lobbying groups makes it more and more difficult to speak of a culture of fair and rational discourse where the better argument will eventually succeed (as the ethics of discourse demand). In this regard, authoritarian structures are *prima facie* more promising as they can achieve political objectives without too much intervention from affected stakeholders. But is this sustainable? The thesis for which I would like to argue answers this question in the negative. Instead, in order to be successful, sustainability and liberalism are mutually dependent. Therefore I will present an empirical argument taking up the example of the yew tree and the anticancer medicine Taxol; and a theoretical argument claiming that cultural progress without armed conflicts is possible only in liberal democracies.

The comparison between liberal democracies and more authoritarian states on the path to being sustainable, although a quite negative contest so far, allows liberal democracies to take the lead by, say, half a nose. The economic exploitation of the yew exemplifies this.[50] In the 1980s it was discovered that the bark of pacific yews contains an active substance able to produce a very promising cancer medicine called Taxol.[51] Unfortunately, it takes 3000 trees to obtain one liter of Taxol; one literally must destroy an entire forest to safe human life.[52] But this is exactly what started to happen e.g. in the US. By now it is measured that over 80% of pacific yew forests in the US have been chopped down.[53] But resistance from environmental groups quickly started and, as a result, the yew was listed by CITES as an endangered species.[54] This put pressure on politicians and scientists. Eventually, a half-synthetic method was found which enables chemists to extract the needed ingredients just from the leaves of the European yew.[55] But this method requires highly specialized knowledge which is, of course, more cost-intensive than using the whole tree. The most interesting approach to the problem started in Canada, where the yew grows mostly as a bush. The tops of these bushes are chopped off every three years, giving them enough time to recover. And research is intensified to get plants yielding more active ingredient—not by genetically modifying but by traditional breeding. But so far, this happens only in Western states with active NGOs. In contrast, many other states see the yew tree as a highly economical resource and exploit it radically, like e.g. in China.[56] Just recently China finally announced its intention to respect CITES with regard to the yew. But environmental groups are skeptical as to whether it will eventually comply. After all, the Chinese government still suppresses critical groups that could pressure it from inside. This leads to the theoretical argument which focuses on the mechanism of conflict resolution.

A state where no problems arise and everybody appears happy is an illusion, because it would be a state of gods and not of humans.[57] Therefore every society needs a mechanism of conflict resolution which does not threaten the stability of the society and the life of its individuals—this means it must be peaceful.[58] Such a mechanism should serve, further, as a motor of progress by institutionalizing. In this manner societies can be understood as evolutionary.[59]

This can only be assured by communication and a culture of discussion. To stay with the yew example: the scientists followed the claim for "green medicine" and discovered the yew tree as a resource for a promising cancer treatment. They then took up the environmental critique and found a synthetic method while the yew was listed by CITES. This is, after all, the consequence of a culture where, ideally, everybody responds to critique and need not be stopped by violence. Only liberal democracies apparently have the power to face the "danger" while the "rescue" is already growing inside them.[60] And this is exactly the kind of culture sustainability needs, since it asks societies for a number of quite fundamental changes. And sustainability is reliant on the motivation of the people to get to those changes. But this can only be reached by discursively convincing the populace which is the basis for the desired institutions to render sustainable societies durable. Liberal democracies feature such mechanisms, although they take time. Authoritarian structures are at a major disadvantage in this perspective because the transaction costs are regularly imposed. And this is mostly due to a lack of communication which is responsible for a widening gap between government and people.

The ecological critique of liberalism should therefore not aim at the original political liberalism but at the economic one with its egoistical and greedy excesses. And this is why liberalism needs sustainability. Otherwise liberal entities have indeed the capacity to end at just the opposite they aimed for: at fragmented societies exploiting nature to serve the wishes of the few until everything is exhausted and the only remaining answer is violence of the underprivileged.[61] This quite drastic prospect aside, there is also a philosophical argument entangling liberalism with sustainability: Above I have argued for liberalism with its individualism using the epistemological insight that no human is God. But this argument contains a further aspect of liberalism. While liberal societies must deny that any one individual can be closer to God than everybody else, such societies put themselves as a whole in the role of God.

Only society as a whole is now able to tell the individual, using the means of law, how to behave appropriately. Societies become a "mortal god" as Hobbes called it. But with this they are not only responsible for individuals, as members of society, but for the nature around it and for the process of evolution taking place there. These mortal gods we created already enabled us to shape a world geologists consider as the beginning of an Anthropocene era.[62]

Sustainability in this perspective is nothing other than a discursive platform for liberal societies to become aware of their responsibility and to find institutions complying with it. Some authors even argue that the enlightenment philosophers were always aware of this responsibility but considered it as self-evident—no industrial revolution was imaginable at their time—and did not expressly focus on it.[63] But they did give us hints like Kant with his category of the sublime and the hidden purposes of nature in *Critique of Judgment* or Locke with his famous sentences: ""To understand political power right and derive it from its original, we must consider what state all men are naturally in, and this is a state of perfect freedom to order their actions and dispose of their possessions and persons as they think fit, *within the bounds of the law of nature*, without asking leave or depending upon the will of any other man.""[64]

Politically, this calls for a new understanding or at least further development of governance structures.[65] But the liberal modus to institutionalize political findings is the law. I will therefore finally discuss possible legal answers facing the collective dimension of sustainable societies. In the language of law this points to collective duties and rights.

3. Collective Duties and Rights for a Sustainable Society

In most societies law is organized around the notions of justice and solidarity. This is due to the fact that you cannot have the one without the other—you can even say they are two sides of one coin.[66] This exemplifies the supranational structure of the EC and EU where, for the purpose of establishing a common climate of justice, the principle of solidarity is placed centrally.[67] Liberal democracies are vividly dependent on solidarity.

The predominance of justice is only possible in the consciousness of solidarity.[68] At least every crisis reminds us of this fact. Sustainability has a similar relation to justice and is therefore unavoidable for liberal societies, as I have tried to show above. The ethical dimension of sustainability is referred to by some as "retinity", which basically means that everything is connected with everything.[69] This obviously alludes to the notion of solidarity but widens its perspective above the genuine social limits of society.[70] And it entangles solidarity, justice and sustainability in a way demonstrating their mutuality. What does this mean for a law serving the ideal of justice while ensuring the condition for solid and sustainable societies organized as liberal democracies?

3.1 Collective Rights to Nature

Law is often described as a corresponding relation of duties and rights. The duty of the state to protect nature is widely acknowledged. Most constitutions contain paragraphs obliging the state to protect nature—in Germany e.g. Art. 20a. Apparently, they all focus on the human being while opening the perspective of the law to include the interests of future generations.[71] To this end, most legal systems appreciate our collective responsibility to protect and restore nature or even Earth.[72] However, the legal quality of this duty is usually weak. In Germany, it is designed as a "state purpose" (*Staatsziel*) obliging governmental branches (executive, legislative and judiciary) to always consider the environment. But the state enjoys huge discretion in its attempt to comply with the state purpose and, of course, individual rights are not granted by it.[73] Therefore, most state duties of protection in Germany are not based on Art. 20a, but on Art. 2, the right of life and health.[74] The constitution of Venezuela takes a different approach in Art. 127[75]:

> "It is the right and duty of each generation to protect and maintain the environment for its own benefit and that of the world of the future. Everyone has the right, individually and collectively, to enjoy a safe, healthful and ecologically balanced life and environment. The State shall protect the environment, biological and genetic diversity, ecological processes, national parks and natural monuments, and other areas of particular ecological importance. The genome of a living being shall not be patentable, and the field shall be regulated by the law relating to the principles of bioethics. It is a fundamental duty of the State, with the active participation of society, to ensure that the populace develops in a pollution-free environment in which air, water, soil, coasts, climate, the

ozone layer and living species receive special protection, in accordance with law."

Art. 127 systematically codifies a contextual relation between individual and collective rights and duties. While everyone individually and collectively enjoys rights, the state has the duty to protect those rights. This conversely implies the obligation for everyone, both individually and collectively, to comply with the individual and collective rights of others, and of course with the environmental laws of the state.

Taking up the category of collective rights, Venezuela obviously tries to infer the result of the discussion around a "third generation" of human rights. This discussion articulated the social and environmental consequences caused by individual human rights and the culture they helped to create. As a possible answer, a collective dimension of human rights is the subject of many theoretical approaches.[76] The Canadian philosopher Will Kymlicka, for example, has deeply influenced this discussion with his attempt to establish an ethno-cultural concept of justice through minority rights.[77] On the other hand these discussions, especially with regard to minority rights, have led to confusion about the categories of individual and collective rights.[78] Roughly, collective rights serve as a main category for different cultural and/or group rights. This includes collective rights to the environment. To understand this development and the categories of collective rights discussed in this discourse we have to remember the universal character of liberalism with its innate right of freedom followed by legal equality and including autonomous choice towards personal and public well-being.[79] Since liberalism started its political triumph it has been confronted with different objections; one of the strongest is probably that it is too formal, abstracting too strongly from the individual and cultural context in which human beings exist, and from the social and environmental consequences the liberal way of life produces. The third generation of human rights were therefore considered to make individual human rights, and with them political liberalism, context-sensitive. But does this fit our message of durable sustainable societies?

Like most legal answers, the answer is "yes" and "no". To speak of a collective dimension of law needed for sustainable societies is basically the correct category. But it does not aim at greater context sensitivity. It aims at nothing less than a new or, more appropriately, morally correct understanding of human rights. Speaking of a collective dimension in our context is to focus on the universal character of human rights. As we have mentioned above, justice is only possible in socially stable and sustainable societies. This calls for a notion of freedom which takes both dimensions into account because they do not set limits to freedom but define it.[80] Liberal freedom never meant being free to do everything you want. Liberal freedom always was and is defined by moral and ethical presuppositions, mostly constructed by reason. Solidarity is a dimension without which the culture of justice so central to liberal societies becomes unthinkable. The same is true for sustainability.

As a universal dimension of human rights, sustainability changes the legal understanding of nature as a cheap and perpetually available resource. Regardless of the environmentalist's or ecologist's perspective on sustainability (see above), nature is seen as vulnerable and as a process that can come to an end due to mere human impact. This fundamentally questions the economic rationality underpinning the view of human rights so far.[81] And this means, for example, that property rights are not only defined through social responsibility[82], but also by considering sustainability.[83]

Humankind as a whole or as a global society therefore enjoys a collective right to nature, entailing the sustainability of the society they live in. This collective right corresponds to a collective duty to comply with the collective right of the others to nature. The state has the task of designing its infrastructure appropriately and to help establish the institutions needed. This might for example include further structures of governance; the establishment of a new supranational organization which is able to promote social and ecological justice (something which the WTO, so far, is obviously not); the granting of legal standing to more than human individuals; and so on.

Unfortunately I do not have the space to discuss this further. Finally I will have a brief look at possible objections to the concept of collective rights.

3.2 Problems of Collective Rights Concept

The discussion around collective rights has provoked several objections, all of which I cannot address here. I will therefore concentrate on probably the most fundamental one: paternalism. In my understanding this objection overlooks the negative side of moral objectivism, although this negative side is quite obvious for the realm of environmental ethics.

Supplementing an individual rights approach with a collective dimension can be seen as mutually reducing the prevalence of both. Since this happens in a quite common modus, it philosophically provokes the objection of paternalism.[84] The individual is no longer completely free in the execution of their right, but is forced to take into account expectations as to how he/she should execute his/her will. Regarding environmental laws, of course, this happens every day and on all levels. Examples: one is not free, e.g. in Germany, to fell an old tree in one' s garden, even though one may no longer like it, nor is one allowed to make international deals with CFCs. But national laws are justified through the sovereign parliament. We are arguing for a collective dimension globally. Can this be justified without provoking the objection of paternalism?

This would be possible if there were something like a moral objectivism or realism. This means there are ethical propositions which are true regardless of the human context. The thesis with which I would like to close is: in environmental ethics we do endeavor to achieve such moral objectivism, although in a negative sense.[85] The ecological crisis presents the objective limits of special kinds of actions. Because such actions endanger humanity, nature and earth as such they can be called morally wrong. In the manner of the ethics of discourse this places the burden of proof on those who want to stick with these kinds of actions. The language of legal ethics entrusts humankind with a shared responsibility to get away from these actions and to find new ways to durably comply with the earth. This also alludes to a new understanding of rationality which has to become ecological.[86] Sustainability serves in liberal democratic societies as the discursive platform to become aware of this negative objectivism and to discuss the solutions needed.

Notes

* I would like to express my gratitude to Steven Zeman and Vernon Tava for their assistance in checking this article.

[1] In Germany it was probably first brought up by: A, Gorz, Ökologie und Politik, Reinbek bei Hamburg 1977.

[2] See e.g. H. Zepp-LaRouche, Climate Campaign Is a Diversion; Real Danger Is Financial Meltdown, Executive Intelligence Review (EIR) Vol. 34 Nr. 12 (2007), who suspects the climate-activists of arguing for a state of emergency in which authors like Carl Schmitt asks for the authoritarian state. Further, many lobby groups, like timber associations etc., use the term "eco-dictatorship" to blame public policy. For a special and recent German case see Frank Drieschner, Ein bisschen Diktatur, in Die ZEIT of 3rd December 2009, p. 50; see also F. Ekardt, Zukunft in Freiheit, Leipzig 2004, p. 101 pp who, surprisingly, argues that Hans Jonas' "Principle of Responsibility" prepares the ground for concepts of eco-dictatorship.

[3] In Germany the warning of an upcoming threat of terrorism was made prominent by R. Bahro, Die Logik der Rettung, Berlin 1990, p.306.

---

[4] See e. g. D. Murswiek, Umweltschutz als Staatszweck, Bonn 1995, p.63 pp.
[5] For a comprehensive overview of this discourse see also K. Bosselmann, When Two Worlds Collide, Auckland 1995.
[6] There are even earlier documents. One of the most remarkable is "Iudicium Iovis" by Paulus Niavis from 1495. For more details see U. Grober, Die Entdeckung der Nachhaltigkeit, München 2010.
[7] F. Schiller, Über naive und sentimentalische Dichtung, Stuttgart 1978, p.4, my translation.
[8] I. Kant, Metaphysische Anfangsgründe der Rechtslehre, Hamburg 1986, p.47.
[9] Ibid, p.18 p.
[10] J. S. Mill, On Liberty, first published 1859.
[11] I. Kant, Über den Gemeinspruch: Das mag in der Theorie richtig sein, taugt aber nicht für die Praxis, in W. Weischedel [Ed.], Immanuel Kant Werkausgabe Vol. XI, Frankfurt a. M. 1977, p.145.
[12] In Germany, a human right of justification is discussed, see R. Forst, Das Recht auf Rechtfertigung, Frankfurt/M. 2007.
[13] Inspired by G. Radbruch, Vorschule der Rechtsphilosophie, Heidelberg 1947, p.26 pp.
[14] Michael Bakunin, who developed a kind of collectivistic anarchism also assumed, orientated by Kant, the individual freedom but considered that it can only be guaranteed in a collective polity.
[15] Compare for this also M. W. Schröter, European Legal Reasoning: A Coherence-based Approach, ARSP 2006, p.82 pp.
[16] See e.g. R. Dreier, Irrationalismus in der Rechtswissenschaft, in: R. Dreier, Recht – Staat – Vernunft, Frankfurt/M. 1991, p.120.
[17] To reach this requested rational modus several options are considerable. The probably two most famous are the "veil of ignorance" of J. Rawls, Theory of Justice, Cambridge 1971, and the "theory of communicative action" (*Theorie des kommunikativen Handelns)* by J. Habermas, two volumes: vol. 1: Reason and Rationalization of Society (*Handlungsrationalität und gesellschaftliche Rationalisierung*), vol. 2: Lifeworld and System: A Critique of Functionalist Reason (*Zur Kritik der funktionalistischen Vernunft*), Cambridge 1984 (Frankfurt/M. 1981).
[18] In the German Weimar Republic (1918-33) attempts to part the relation of the two terms and to think of democracy without liberalism were quite common, e. g. by Carl Schmitt (see also below). But they just made it easier for the national-socialist party to get into power.
[19] See e. g. C. Crouch, Postdemocrazia, Roma/Bari 2003 (engl.: Post-Democracy. Oxford 2004).
[20] See already C. B. MacPherson, The Political Theory of Possessive Individualism, Oxford 1962.
[21] This pattern of argument—with variations, of course—is used by many authors and political groups right now and has gathered new intuitive plausibility by the recent financial crisis. In the realm of political philosophy the worry of "atomistic

individualism" culminated in the theoretical approach of communitarism, see e.g. C. Taylor, Atomism, in Philosophy and the Human Sciences. Philosophical Papers 2, Cambridge 1985.

[22] This conviction is e.g. at the core of J. Rawls, The Law of the Peoples, Harvard 1997; compare further R. Alexy, Diskurstheorie und Menschenrechte, in: R. Alexy, Recht, Vernunft, Diskurs, Frankfurt/M. 1995, p.127.

[23] In my understanding the temporary discussion started with D. L. Meadows et al., The Limits to Growth, New York 1972.

[24] This intensification was not a consequence of the Brundtland-report but of the 1992 Rio-conference and its Agenda 21 which focussed on more regional concepts of sustainability and sustainable development. In consequence, more and more social groups (e.g. NGOs) and local administrations got involved in the discussion.

[25] The notion of sustainability is, for sure, much older; compare for a cultural history of sustainability Grober (2010) above.

[26] S. Leibfried/ M. Zürn, Von der nationalen zur postnationalen Konstellation, in: S. Leibfried/ M. Zürn, Transformation des Staates?, Frankfurt/M. 2006, p. 23 pp.; A. Hurrelmann/ S. Leibfried/ K. Martens/ P. Mayer, Die Zerfaserung des Nationalstaates: Ein analytischer Rahmen, in: A. Hurrelmann/ S. Leibfried/ K. Martens/ P. Mayer, Zerfasert der Nationalstaat?, Frankfurt/M./ New York 2008, p.25 pp.

[27] One prominent slogan of the young German green party in the 1980s was: We have merely borrowed the Earth from our children (*Wir haben die Erde nur von unseren Kindern geliehen*).

[28] A. Dobson offers an even more general categorization in, Drei Konzepte ökologischer Nachhaltigkeit, in: Natur und Kultur 2000 (1/2), p. 62 pp with the differentiation between technocentric and ecocentric approaches to sustainability.

[29] See e. g. K. Bosselmann, The Principle of Sustainability, Cornwall 2008, p.27.

[30] World Commission on Environment and Development, Our Common Future, 1987.

[31] See section part I. 3. IV, paras. 72 pp.

[32] Ibid para. 74.

[33] Bosselmann (2008) above, p.32.

[34] Ibid, p.62.

[35] Ibid, p.53.

[36] Different from Ekardt (2004) above, who understands sustainability as a general task which justice must take into account. This understanding retains traditional notions of justice and hinders changes in thinking; therefore this understanding must explain whether it sufficiently appreciates the challenges growing with and out of the ecological crisis.

[37] Similar Ekardt (2004) above, p.111.

[38] Dobson (2000) above.

[39] See e. g. J. Habermas, Faktizität und Geltung, Frankfurt/M. 1992, p.151 pp.

[40] See already M. Schröter, Rationality, Democracy, Ecology, Typescript Berlin 2009.

[41] James Lovelock on the value of sceptics and why Copenhagen was doomed, at: www.guardian.co.uk/environment/blog/2010/mar/29/james-lovelock (05/10).
[42] Ibid.
[43] *Hüter der Verfassung*, my translation.
[44] C. Schmitt, Die Diktatur, Berlin 1921; Politische Theologie, Berlin 1922; Verfassungslehre, Berlin 1928. See e.g R. Mehring, Carl Schmitt. Aufstieg und Fall, München 2009, p.157 pp. Compare for the theological background of Schmitt's theory e.g. D. v. d. Pfordten, Rechtsethik, München 2001, p. 160 pp.; H. Meier, Die Lehre Carl Schmitts, Stuttgart Weimar 2009[3].
[45] Schmitt (1928) above, p. 236. In Nazi-Germany this was perverted to the saying: The will of the *Führer* is the will of the people. One of the "legal" measures of Hitler to reach his power was the unification of the functions of president and chancellor, compare e.g. R. Dreier, Bemerkungen zum "Dritten Reich", in W. Baumann/ H.-J. V. Dickhuth-Harrach/ W. Marotzke (Ed.), Gesetz, Recht, Rechtsgeschichte, 2005, p.495.
[46] See e.g. R. Alexy, Die Natur der Rechtsphilosophie, in W. Brugger/ U. Neumann/ S. Kirste, Rechtsphilosophie im 21. Jahrhundert, Frankfurt/M. 2008, p.11 pp.; Schmitt (1922), above; H. Blumenberg, Legitimität der Neuzeit, Frankfurt/M. 1996 (first published 1966), p.102 pp.
[47] See also G. Agamben, Stato di eccezione, Torino 2003 (engl.: State of Exception, 2005).
[48] Similar v. d. Pfordten (2001), above, p.281 pp.
[49] B. Russell, The Authority and the Individual, London 1949, p.47.
[50] The most information given in this paragraph goes back to a French TV-documentary of the ARTE-series "Secrets de plantes", available under: http://videos.arte.tv/fr/videos/secrets_de_plantes-3230984.html (05/10).
[51] The history of this research goes back to the 1950s.
[52] To have the chance of saving one single human life you need around six fully-grown trees—each requiring at least a hundred years to grow.
[53] Since official statistics never existed before the chopping started, this is a pure guess. It is assumed though that the actual destruction was much higher.
[54] The yew is also on the red list of IUCN.
[55] By now synthetically methods exist, as well.
[56] Compare e.g. www.traffic.org/home/2007/12/11/has-the-yew-tide-turned.html (05/10).
[57] And would thereby represent just the opposite from a state of reasonable devils Kant was aiming at: I. Kant, Zum ewigen Frieden, Werkausgabe Vol. XI (ed. by W. Weischedel), Frankfurt/M. 1977; compare also P. Laberge, Von der Garantie des ewigen Friedens, in: O. Höffe (Ed.), Immanuel Kant. Zum ewigen Frieden, Berlin 2004, p.161 pp.
[58] See already Schröter (2009), above.
[59] Compare J. Habermas, Wahrheit und Rechtfertigung, Frankfurt/M. 1999.

[60] Free translation of a quote by Hölderlin (""*Wo aber Gefahr ist, wächst das Rettende auch!*"") which deeply inspired Martin Heidegger, see e.g. Die Technik und die Kehre, Pfullingen 1991[8], p.31.
[61] In Germany this was clearly articulated by Bahro (1990), above.
[62] This notion, going back to Paul Crutzen, is now acknowledged e.g. by the Geological Society of London (est. 1808), compare M. Davis, Wer baut uns jetzt die Arche?, in: M. Vogt/ F. Uekötter/ M. Davis, Prinzip Nachhaltigkeit: Ethische Fragen im interdisziplinären Diskurs, München 2009., p.47.
[63] E.g. O. Höffe, Moral als Preis der Moderne, Frankfurt/M. 1993.
[64] J. Locke, The Second Treatise of Government, Ch. II, para. 4, first sentence.
[65] See e.g. K. Bosselmann/ R. Engel/ P. Taylor, Governance for Sustainability, Gland 2008, IUCN Paper No. 70.
[66] The right relation of the two notions is subject of many debates, though. The most well-known example is probably the dispute between liberalistic and communitarian thinkers. But the focus is not on whether you can have the one without the other but what enjoys morally higher priority: the just or the good, see e. g. E. Özmen, Moral, Rationalität und gelungenes Leben, Paderborn 2005.
[67] Compare formerly Art. 2 EC and formerly Art. 1 EU; now (since contract of Lisboan) Art.2 EUC.
[68] Compare e. g. A. Gewirth, The Community of Rights, Chicago 1996, p. 6; J. Habermas, Zwischen Naturalismus und Religion, Frankfurt/M. 2005, p.106 pp.
[69] Sachverständigenrat für Umweltfragen (SRU), Umweltgutachten 1994. Für eine dauerhaft-umweltgerechte Entwicklung, Stuttgart 1994, p.54.
[70] Compare already Schröter (2009), above; for a theologican view compare e. g. M. Vogt, Nachhaltigkeit theologisch-ethisch, in: M. Vogt/ F. Uekötter/ M. Davis, Prinzip Nachhaltigkeit: Ethische Fragen im interdisziplinären Diskurs, München 2009, p.25.
[71] Therefore I neglect the discussion around rights of nature although I have sympathy for this concept; see C. Stone, Should Trees Have Standing?, Southern Californian Law Review 1972 (45), p.450 (3rd reworked edition Oxford 2010); M. W. Schröter, Mensch, Erde, Recht, Baden-Baden 1999; for further discussion compare Bosselmann (2008), above.
[72] Compare P. Taylor, An Ecological Approach to International Law, London New York 1998.
[73] Compare e.g. H. D. Jarass/ B. Pieroth, Grundgesetz Kommentar, München 2007[9], Art.20a.
[74] Ibid, Art. 2 para 14, 91.
[75] Similar Art. 33 and 34 of the 2009 Bolivian Constitution. A comprehensive study was prepared by V. Tava, Resisting Enclosure: The Emergence of Ethno-Ecological Governance in a Comparative study of Venezuela, Ecuador and Bolivia (Master Thesis, Auckland 2010, will be published soon).
[76] See also L. Westra, Human Rights, The 'Commons' and the Collective, (will be published soon).
[77] Starting with W. Kymlicka, Liberalism, Community and Culture, Oxford 1989.

---

[78] Compare e.g. J. T. Levy, Classifying Cultural Rights, in: I. Shapiro/ W. Kymlicka, Ethnicity and Group Rights, New York London (Nomos XXXIX) 1997, p.22.
[79] See also Alexy (1995), above, p.127 p.
[80] Compare already K. Bosselman/ M. W. Schröter, Umwelt und Gerechtigkeit, Baden-Baden 2001.
[81] See Bosselmann (2008), above, p.111 pp.
[82] Acknowledged e.g. by Art. 14 German constitution.
[83] For further discussion see Bosselmann ibid.
[84] Compare for a similar problem regarding the construct of minority rights A. Addis, On Human Diversity and the Limits of Toleration, in: I. Shapiro/ W. Kymlicka, Ethnicity and Group Rights, New York London (Nomos XXXIX) 1997, p.120 pp.
[85] Compare already Schröter (1999), above, p.186 pp.; Schröter (2009), above; see also A. Kaufman, Rechtsphilosophie, München 1997^2, p.176 pp.
[86] See e.g. P. G. Brown/ G. Garver, Right Relationship. Building a Whole Earth Economy, San Francisco 2009.

CHAPTER SEVENTEEN

HOW CAN AFRICAN COUNTRIES BENEFIT FROM THEIR NATURAL RESOURCES IN ORDER TO DEVELOP AND MODERNIZE THEMSELVES IN A CONTEXT OF GLOBALIZATION AND FINANCIAL CRISIS?

PHILIPPE CRABBÉ

The abundance of natural capital, especially of non-renewable resources, is an opportunity for Africa's development. However, recent history seems to indicate that the opportunity has been missed up to now: African resource-poor countries have outperformed resource rich countries in terms of growth since the 1960's (Auty 2001). This is certainly due to difficulties in managing resource rents. Some of these difficulties point to the volatility of primary commodity prices and the large share of these resources in African exports. But difficulties also arise from misguided governance and policy frameworks promoted by both international and African national policies. To turn natural capital, and especially mineral resources-the main focus of this chapter-into an opportunity requires the fine-tuning of various complementary policies which can easily go astray (Collier 2010). The first section of this chapter is devoted to the African socio-economic context using various forms of capital as an organizing principle. The second section conducts an inventory of African non-renewable resources and identifies the source of their price volatility. The third one evaluates the institutional framework for the extraction of mineral resources which was in place until very recently. The fourth one looks at the macro-economic implications of revenue volatility. The fifth one looks at the Natural Resources Charter proposed by P. Collier and M. Spence for adoption both by African countries and the foreign corporations which operate in African countries. The sixth one examines briefly the sources of the principles of corporate social responsibility developed by international institutions and some developed countries hosting the headquarters of these foreign corporations. A seventh one examines in detail the "African Vision 2050" which is a very encouraging document focusing exclusively on the responsibilities of African countries. Finally, an eighth section documents the important role Canadian companies may play in the implementation of the Natural Resources Charter. A concluding section emphasizes hopeful signs in favor of the implementation of the "African Vision 2050" and the Charter.

Introduction: The African Structural Context[1]

The African structural context will be examined from the capital endowment perspective: first, the social capital which encompasses institutional and governance issues; second, the demographic/human capital perspective, which encompasses health, education and, generally, human development issues; third, manufactured and financial capital and, finally, the natural capital perspective which encompasses extractive, renewable natural resources and environmental issues.

Social Capital

Social capital refers to society's social and political institutions. It revolves around trust, backed by sanctions, that provides information and security to facilitate economic transactions by reducing risk. It includes the building of trust in order to lower transaction costs, to improve knowledge externalities from social networks and to enhance the capacity for collective action. One subcomponent is civic social capital, which functions at the level of the local community. The second subcomponent is public social capital, which includes the maintenance of property rights, government accountability, and the legal framework. Their effective provision is governed by the political economy.[2]

77% of African population lives under the poverty line (US $ 2.15/d) and 45 % live in absolute poverty (US $ 1.08/d).[3] 90 % of the population is not covered by any form of social protection (Delcourt 2009).

This situation enhances, of course, the vulnerability of the population to any form of crisis whether fiscal, institutional, legitimacy, health, natural or food price. A fiscal crisis may be due to the external debt burden (risk considerably decreased to-day in Africa by more prudent management), fiscal revenue shortfall due to exchange rate variations, lighter taxation of financial capital and fraud due to deregulation (within a globalization context). An institutional crisis may result from the regression of public utilities, from lack of equity due to politically favored constituencies, and from corruption. A legitimacy crisis occurs in time of civil wars creating refugee problems within a country or for neighboring countries. Finally, AIDS (health crisis), floods and droughts (natural crisis) and, prominently, food price increases (food price crisis as recently occurred) are familiar causes of vulnerability for the poor.

The growth, development and formal sector employment actually regressed in the 1990's under globalization while income inequality, the size of the informal sector, poverty, hunger and, generally, vulnerability of the South actually increased. North-African welfare state systems have been decimated and replaced by public/private charity (Delcourt 2009).

These negative consequences have led to the end of the promotion of the unfettered self-regulating market economy and globalization (Neo-Liberalism) by international institutions. Keynesian policies and protectionism are being reintroduced and privatized anti-poverty measures (previously International Labor Organization (ILO)'s exclusive concern) replace Structural Adjustment Programs. In other words, the roles of the state and of social capital are being reinvented. This marks the end of the Washington Consensus. The "Post-Washington Consensus" emphasizes capacity-building of national governments, empowerment of the most vulnerable populations and social risk management to remedy market deficiencies (Delcourt 2009).

At the national level, social policies (already existing in Latin-America after the 2nd world war) and their impact on development are re-emphasized. This is called the New Development Paradigm, following the 1995 Copenhagen Social Development Summit. One finds these social policies in Southern Africa and, to some extent, Eastern Africa, Nigeria and Ghana, in the African Union initiative (2006) and the 2008 ILO initiatives. International Emergency Aid (e.g. food) is being managed by international institutions but also by Non-Governmental Organizations (NGO) supplemented by civic social capital, i.e. community and family institutions (Delcourt 2009).

While, in the not too distant past (e.g. 1973), African governments would in case of crisis increase their external debt burden or run down their foreign exchange reserves, this time around, at least twenty African governments have decided to cut their budget instead. While budget cuts have clearly implications for social protection, governments have decided to sit with civil society in order to determine the areas where cuts should be made and what constituency should be protected. Governors of central banks are now more independent from their governments and have suggestions about how to cope with the crisis. (Brookings 2009)

Demographic/Human Capital Perspective

The population of Africa is expected to roughly double by 2050 given a scenario which is AIDS-dependent. (Udjo 2008; UN 2006) In 2005, African population was 905 millions or 14 % of world population. By 2050, it is projected to be 1.7 billion or 22 % of world population. The African population growth-rate is currently high but steadily decreasing. The proportion of the elderly population is expected to increase.

This demographic projection has clear implications for food, water, clothing, housing, health and sanitation, school and employment, and pension needs. It also affects the implementation of the Millennium Goals.

Regional Distribution

About 29% of the population lives in West Africa (WA), 27% in East Africa (EA), 18% in North Africa (NA), 17% in South Africa (SA), and 10% in Central Africa (CA) (see Figure 17-1). NA is the most urbanized region of Africa while EA is the least urbanized. Vast desert and densely forested regions in Africa are nearly un-habited. Population density is very high in places like Nigeria, the Nile River valley, and the Great Lakes region. Cairo and Lagos are two of the largest cities in the world. Estimates of the number of different African languages range from 700 to 3,000.

Economic Output and Employment

Africa, as a whole, accounts for about 2% of world economic output, i.e. the equivalent of Belgium's. Overall, the agricultural and mining sectors employ the largest numbers of Africans. Around two-thirds of Southern Africa's population is dependent on agriculture for employment.

Human Capital Endowment

In NA, primary education enrollment is 100 % of relevant age group while secondary education enrollment is 77 % and tertiary, 22 %. In Sub-Saharan Africa (SSA), the figures are respectively 93 % (with a 60 % completion rate), 29 % and 6 %.

In NA, life expectancy at birth is 71 years and the infant mortality rate is 3 %. In SSA, the comparable figures are: 50 years and 9.4 % respectively. In both regions, the age-dependency ratio, i.e. the proportion of dependents to working age population, is 1 (Worldbank 2009).

Figure 17-1. Regions of Africa

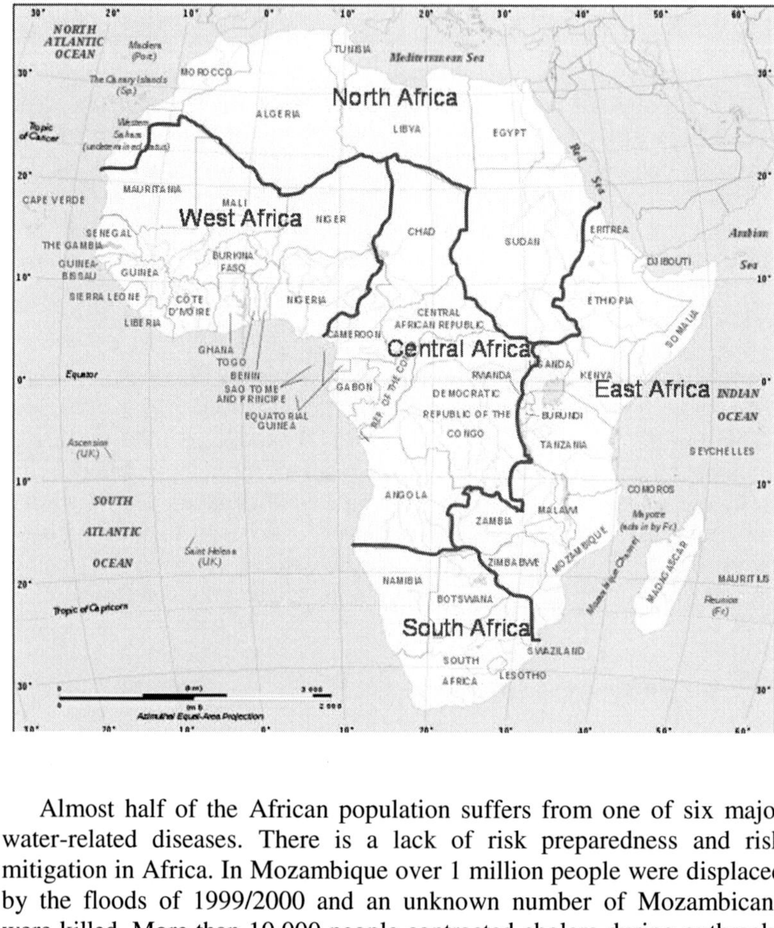

Almost half of the African population suffers from one of six major water-related diseases. There is a lack of risk preparedness and risk mitigation in Africa. In Mozambique over 1 million people were displaced by the floods of 1999/2000 and an unknown number of Mozambicans were killed. More than 10,000 people contracted cholera during outbreaks in South Africa in 2001. Every day, 650 people die from diarrhea, mainly children under five years of age.

During the last ten years, provision of sanitation in rural Africa has decreased by 2 per cent, and the low levels of urban water supply and sanitation have hardly improved.

Women and children suffer disproportionately from negative health effect due to the smoke generated by the use of fuel-wood for cooking (smoke is a carcinogen and causes respiratory problems). About 75% of wood harvested in sub-Saharan Africa is used for household cooking.

For Cameroon, Ghana, Kenya, Zimbabwe, it has been argued that the underlying cause of the lack of manufacturing exports from African economies is the relative scarcity of skilled labor that ensures that Africa has a comparative advantage in natural resource exports. This argument has recently been extended from a narrow definition of manufacturing to one that includes the processing of primary products. If skilled labor is scarce, then the implication would be that the returns to skilled labor in Africa should be relatively high. Inferences of this form are the basis for the view that expanding educational provision is a requirement for a successful program that will accelerate growth in sub-Saharan African economies. But returns to manufactured capital (23 %) are higher than returns to human capital (8 %). Why, if skilled labor is relatively scarce, is not the return to education high? The high rate of return in manufacturing may be due to the high cost of capital (Bigsten et al. 2000).

Human Development Index (HDI)

The HDI is a weighted average of life expectation years, education years and national income per capita for a given country.[4] The index is normalized to fall within the interval 0 - 1 and countries are ranked according to its value. No African country is located in the high development index trends category (.8 - .968) except Libyan Arab Jamahiriya (ranked 52nd out of 75) and Mauritius (74th).

In the medium development index trends category, Tunisia (NA) is ranked 95th followed by Algeria (NA), one hundredth (among 179 countries), and is the highest-ranked in the medium development index trends category (.5 - .8).[5]

In the low human development index trends category, Sub-Saharan Africa scores .472 in 2006 and 81 % in the category are SSA countries.[6]

It is difficult to identify regional clusters of performance except for South-Africa. Positive long-run trends (1980-2006) are recorded for Guinea-Bissau, Burundi, Burkina-Faso, Benin, Mozambique, Malawi, Rwanda, Central African Republic, and Liberia. Negative trends are recorded from 1990 to 2006 for Zambia, Central African Republic, and Côte d'Ivoire (Figure 17-1).

Manufactured and Financial Capital

In NA, industry contributes 41.9% of GDP, Agriculture 10.3 %, services, 40.9 % while, for SSA, the figures are respectively 34 %, 14.5 % and 42.6 %. Gross Fixed Capital Formation is 23.5 % in NA and 18.9 % in

SSA. Electric power outages are on average 4 per month in North-Africa and 7 for SSA (World Bank 2009).

African banks came out relatively unscathed from the world financial crisis. They retain loans they originate on their balance sheets. The interbank market is small, and the market for securitized or derivative instruments is either small or nonexistent. Even though some African countries' banking systems have significant foreign ownership, the parent banks are typically not in the U.S. Furthermore, the foreign ownership share in the largest economies, Nigeria and South Africa, is less than five percent (compared with a developing-country average of 40 percent).

Of greater concern in Africa is the resurgence of inflation and macroeconomic imbalances in some countries. For example, Ethiopia's inflation rate is 61 percent and its trade deficit is 30 percent of GDP. Though these are closely related to the food and fuel price increases of earlier 2008, these developments will require early and decisive actions to avoid the situation getting worse.[7]

Foreign direct investments (led by oil investments) trebled in Africa at the beginning of the 21rst century, more than in any other region of the world. These were motivated by natural resources development, the size of some African economies (Nigeria, Egypt, South-Africa), and by governance improvements, both public and private. Foreign direct investments in manufacturing are also not negligible in Zambia, Ghana, Angola, Nigeria, Tunisia, Egypt, South-Africa and are directly related to governance improvements.

Natural Capital

The Environmental Performance Index (EPI) rankings computed by Yale and Columbia University show a remarkable correlation with the HDI rankings.[8] Mauritius (58) is the highest ranked African country according to EPI followed by Tunisia (59). 8 African countries are at bottom with Niger last (score of 39.1).[9]

Climate Change

Five African countries have signed, but not ratified, the Kyoto Protocol: Egypt (NA), Mali (WA), Niger (WA), Seychelles (EA), and Zambia (EA). Africa contributes only about 3.7% of total world energy-related carbon dioxide emissions. Gas "flaring" is still practiced. Because the gas infrastructure in Africa is extremely limited, the gas is often burned off rather than captured for use. The World Bank estimates that every day

Africa flares gas equivalent to twelve times the energy used by the continent. This practice is being phased out, especially in Nigeria, through dramatically increased fines for flaring. SSA is home to the world's second largest rain forest in West Africa. Climate change could expand the home range of malaria-bearing mosquitoes (African Partnership Forum 2007).

Renewable Resources

African *biodiversity* is a unique natural capital asset for tourism, pharmaceuticals, and ecosystems and thus needs to be protected. It is unique in that it has great potential for rural and local people's economies because it is labor-intensive and requires local input including indigenous knowledge and management.

African *forests* have great potential as carbon sinks as they are one of the world's most important. There is great potential for *solar* energy in Africa at the local level. African *hydroelectricity* potential remains large and essential for mining development.

Africa's Non-Renewable Resources

If one makes the reasonable assumption that the amount of mineral remaining to be discovered is proportional to the land area, most minerals are yet to be discovered in Africa. Indeed, the average square kilometer of the African landmass has beneath it only around $25,000 of known sub-soil assets, whereas the corresponding figure for the OECD landmass is $125,000 (Collier 2010).

Non-renewable resources will thus remain a large component of African asset portfolio (Tamufor 2009). Per capita African endowment in human and manufactured capital is smaller than in the rest of the world.

Exploration for non-renewable resources is a costly and risky investment and so known reserves are determined by the institutional and economic environment rather than simply being a geological given (Collier 2010).

Africa holds (Muehlberger 2007):

- 88 % of world *platinum* (mainly in South-Africa), a precious metal without industrial use (US $(2010) 1600/oz, double of 2005).

- 60.6 % of the world *diamonds* which, besides their precious gemstone character, are used for cutting, drilling, grinding, and polishing (Congo DR, Botswana, South Africa).
- 48.5 % of the world *cobalt*, a strategic material used as in alloys, magnets and magnetic tapes, some batteries, as drying agent for paints, as catalyst in industrial chemical reactions, and in radiotherapy (Congo, DR and Zambia).
- 40 % of the world *phosphate*, a fertilizer (Morocco, South-Africa).
- 30 % of *bauxite* (Guinea, Ghana), a fertilizer, the feedstock for aluminum.
- 19 % of *gold* (South-Africa), a precious metal with a few industrial uses such as electrical contacts, (US $(2010) 1300/oz, ten times what it was 10 years ago).
- 9.5 % *uranium* (Namibia, Niger), the feedstock for nuclear plants and some military uses.
- 8.4 % of *manganese* (South-Africa and Gabon), used as alloy and fertilizer.
- 6 % of *nickel* (South-Africa), used mainly as alloy.
- 5.4 % of *copper* (Zambia, Congo DR), used as alloy and for its electrical and thermal conductivity.
- 3.2 % of *zinc*, used as anticorrosion agent through galvanization and as fertilizer (Namibia, Congo DR, and Morocco).

Moreover, Africa holds significant non-renewable energy assets:

- 10% of the world oil proven reserves (2/3 in Libya, Nigeria, and Algeria in the order) and
- 8% of proven natural gas reserves (80 % in Nigeria, Algeria and Egypt).

Revenues from the extraction of non-renewable natural resources are highly volatile. The reason for the volatility is illustrated by Figure 17-2. The quantity supplied of mineral resources cannot be changed quickly in times of prosperity. Accordingly, all market adjustments must be made through prices. In recession periods, the quantity of mineral supplied cannot be adjusted downward. Therefore, a large portion of what is extracted goes into inventory. Once demand picks up again, the inventory of extracted mineral resource is first depleted.

Figure 17-2. Volatility of mineral prices

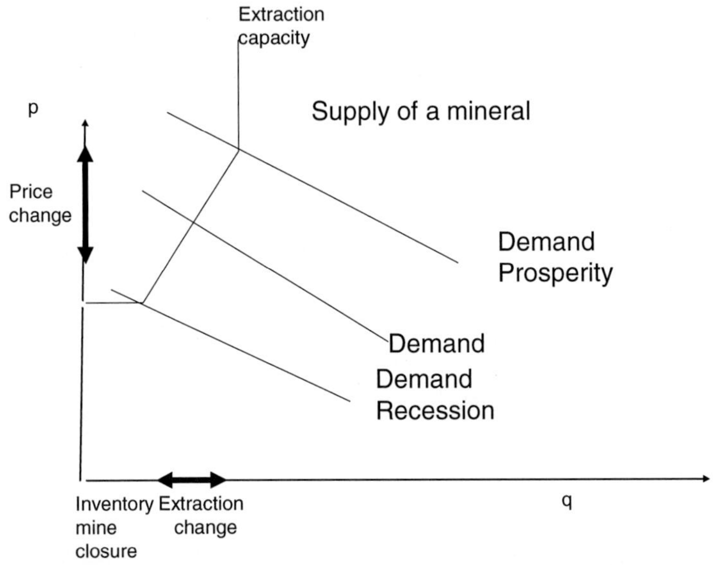

Currently crude oil and unprocessed minerals contribute 70 % of African exports vs. 30 % for manufactured goods. Because of this lack of diversification, Africa is highly vulnerable to adverse terms of trade (price of exports/price of imports) as it is in the current recession. According to the Commission for Africa 2010 Report, the natural resources sector grew 24 % between 2003 and 2008, i.e. four times Africa's annual growth rate (Commission for Africa Report 2010).

Institutional Environment for Mineral Resources

In the 1980's, Africa attracted only 5 % of the world exploration and mining capital expenditures. Ghana ushered in a wave of reforms across Africa's mining industry which turned Ghana, Africa's second largest producer of gold, into the pin-up model of purveyors of neo-liberalism. The neo-liberal reforms package of liberalization, privatization, deregulation and redefinition of a minimalist role for the state was essentially predicated on the assumption that the future expansion of the extractive sector in Africa would be led by foreign direct investments (FDIs). It is true that the huge expansion in Africa's production and export of minerals and metals in the 1990s and this decade were largely fuelled by FDI inflows. According to the World Bank, the objectives of mining policy were threefold. First, generate tax revenues and foreign exchange under the assumption that benefits would "trickle down" to benefit society more broadly (World Bank 1992). Second, deepen and sustain reform to enable a private sector led industry and implement private sector strategies and competitive investment (World Bank 1997). Third, the terms and conditions of the reform should be modeled on countries that have been successful in attracting investment (including Chile) (World Bank 1992). The main objective of the reform was to balance risks and rewards for the industry. The reform was thus top down and prescriptive.

Mining agreements went from being negotiated and discretionary to standardized legal regimes in which governments had minimal discretion. Mineral Investment Agreements are formulated to fill gaps in general mining law, set limits on the manner in which governments can exercise discretion in the absence of more specific mining legislation, and aim at stabilizing fiscal and (in a few cases) regulatory terms applicable to investment. These agreements are subject to parliamentary ratification to gain the status of law.

African efforts to attract FDI led to the formulation of overly generous investment laws and regulations, according to the Economic Commission for Africa/African Development Bank 2007 report.[10] They led to reductions in corporate income taxes and royalties, to increases in capital allowances, to exemption of import duties for plant and equipment, to tax exempt personal remittance for foreign personnel, and to large allowance for foreign exchange earnings in external accounts (Campbell et al. 2004).

This regime largely failed African countries. It was too sectoral (in attracting mining investment and reducing risks) and gave no consideration for forward and backward linkages in the economy.

Macro-Economic Implications of the Extraction of Mineral Resources for Developing Countries

In recent decades, resource rich economies have underperformed with respect to resource-poor developing countries. Yet the mineral economies have the potential for rapid growth because most of them also have ample cropland so that mineral exports further enhance their capacity both to invest and to import compared with the non-mineral economies.

The deterioration among the resource-abundant countries is more severe where the natural resource rents emanate from 'point' resources, such as mining, rather than from 'diffuse' source resources like land under peasant farms. Point rents are associated with staples that are relatively capital-intensive and thereby concentrate ownership.

Small resource-abundant countries are more vulnerable to growth collapses than large ones are. Large resource-abundant countries have two main advantages over small ones in avoiding a growth collapse. First, the probability of depending on one or two primary exports is lower for large countries compared with small ones. This is because, all else being equal, the large country has a more diversified natural resource endowment than a small one while it can also diversify earlier into manufacturing because its large market affords scale economies. Second, trade accounts for a smaller share of Gross Domestic Product in large economies so that they are more self-contained and less vulnerable to external shocks.

The boom in commodity prices since 2001 along with huge profits accruing to mining conglomerates all set against insignificant revenues to government coffers, high environmental destruction and displacement of local communities living in mining areas, may have coalesced to play an important role in a new found convergence of opinion.

Civil society organizations (CSOs), NGOs and communities affected by mining, in the immediate past, were easily passed off as detractors with narrow interests, anti-government and inter-governmental establishments and anti-mining companies.

The "lesson of Africa's contemporary mining industry experience-FDI attracting policies- [should] not be confused or conflated with strategies for integrated development which provide a role for FDI that potential investors would find attractive" (Tamufor 2009).

Dr. Graham from the Third World Network Africa outlined five key issues that must engage the attention of African countries if the continent was to maximize the benefits of mining to the region. These are:

- equity in mining codes and contracts;
- respect for the rights of local communities living in mining areas and other human rights enshrined in national constitutions as well as other international human right conventions and
- re-defining the role of the developmental state in mining.
- breaking out of the export enclave so characteristic of the mining industry to link the extractive sector directly to the larger economy to spread the benefits of mining;
- preserving Africa's policy space from encroachment by other regions or countries through international agreements such as the Free Trade Agreement the European Union is pushing to

African countries and other regions of the African, Caribbean and Pacific (ACP) group of nations (Tamufor 2009).

The Natural Resource Charter

In 2009, Michael Spence, Nobel Prize in Economics and Paul Collier, a well-known development economist, came up with the idea of "… an international convention in the making, but one that will be built by a participatory process guided by academic research" (The Natural Resources Charter 2009).

Its two first principles are that natural resources are public assets (not owned by private companies but by the state). The exploitation of natural resources should be carried out to help a country meet its broader social and economic goals, not as an end in itself (e.g. not to maximize enterprise profit). This means having a vision of how the resource sector fits in a country's economic future. For some countries, it may be to extract rapidly to generate revenues to sustain the investment necessary for growth and to meet urgent human needs. For others, the best use of resource endowments may be to leave it in the ground for future use (investment by waiting).

Principle 7 says that the first priority for use of resource revenues is to promote sustained growth in incomes, employment and human development throughout society. Achieving such growth requires continuing high levels of domestic investment in manufactured, social and human assets. This can be accompanied by some increase in current household consumption (e.g. food, health, education, communication media, time-saving devices such as electronic banking, etc.).

Its principle 8 says that effective utilization of resource revenues requires that domestic expenditure be smoothed over time to take account first of revenue volatility and second of a country's growing ability to spend efficiently. This requires management of liquid foreign assets, possibly in the form of a Sovereign Liquidity Fund.[11]

Its principle 9 says that Government should use the ability to increase public spending as an opportunity to adopt systems that raise the efficiency of public spending and enable the private sector to respond to structural change in the economy.

Principle 10 says that the home governments of extractive companies and international capital centers should adopt policies that are supportive of the Charter, requiring extraction companies to report full and transparent accounts, including payments to governments, and, internationally active banks to report transactions that may involve looted assets.

Complementary principles require resource projects which may have serious environmental and social effects, to account for and mitigate these effects at all stages of the project cycle–from the decision to explore and extract, through development and to decommissioning. Principle 10 and these complementary principles lead to social corporate responsibility.

Corporate Social Responsibility

Mining companies must respect human rights and environmental responsibility. These principles of Social Corporate Responsibility are formulated in various international and Canadian documents:

- Extractive Industries Transparency Initiative (2003)[12]
- International Finance Corporation (IFC), Performance Standards on Social and Environmental Sustainability (2006)[13]
- Equator Principles (2006)[14]
- UN Principles for Responsible Investment (2006)[15]
- National Roundtables on Corporate Social Responsibility (CSR) and the Canadian Extractive Industry in Developing Countries (Advisory Group Report 2007)[16]
- Organization for Economic Cooperation and Development (OECD, 2008)[17],
- "Building the Canadian Advantage: A Corporate Social Responsibility (CSR) Strategy for the Canadian International Extractive Sector" (Government of Canada 2009)[18]
- ISO 26000 (voluntary principles under development)[19]

African Vision 2050

African ministers for mineral resources development in October 2008 endorsed a continental mining framework, the African Mining Vision 2050, at their meeting of the African Union Commission in Addis Abeba, Ethiopia. The meeting was the first ever of African ministers of mineral resources development. It also marked the most visible sign yet of a growing convergence of opinion on the extractive sector in Africa (Inngs 2009;see also The Big Table 2007).

The African Mining Partnership (AMP), a voluntary organization of mineral ministers under the auspices of the 2002 New Partnership for African Development (NEPAD), which originated at the Kananaskis G-8 meeting, was merged with the African Union (AU) Conference of African Ministers Responsible for Mineral Resources Development (Durban,

February 2009). This Conference will be the ONLY African Union Ministerial organ responsible for mineral resources development and management (AMP, February 2010). This decision will give resources ministers access to the African Union's financial resources. The Durban meeting also recommended that the 2004 African Mining Policy Framework should be reconsidered and aligned with the African Mining Vision and that the Sustainable Development Charter for the Minerals and Mining Sector in Africa (2002) should be implemented.

The Assembly welcomed the African Mining Vision and asked African Union Ministers responsible for mineral resources development to develop a concrete action plan for its realization (AMP, February 2010).

The Vision aims at improving African governance of the mining industry and increasing the government share of the rent of mineral resources.

This requires first to remedy the lack of resource knowledge infrastructure, allowing distinction between high risk (low geological knowledge) and low risk areas.

It requires second to improve contract negotiation capacity.

Third, competitive auctioning of prospective resource "blocks" for high risk areas (i.e. blocks large enough to avoid information externalities, which provide incentives to delay exploration until one can observe the exploration results of one's close neighbor) is needed.

Fourth, self-adjusting resources tax regimes should be set up, increasing taxes in periods of prosperity and decreasing them in recession.

Finally, extraction contracts should also be comprehensive to decrease investor's uncertainty.

Investment in maintaining mining infrastructure is also needed as well as allowance for third party access.

One should remedy failure of downstream value addition due to lack of minimum resource beneficiation agreement and of effective competition regulation. One should remedy lack of upstream value addition due to knowledge-intensity of these industries. Developing the resource sector upstream linkages into the domestic economy could be achieved through skill transfer provisions in consultant contracts.

Investment in technology/product development is also needed.

African capacity for ongoing auditing, monitoring, regulating and improving resource exploitation regimes should be created.

Capacity to manage resource rents has to be improved, e.g. through state mining companies, and portions of the rent should be apportioned to local mining communities.

State owned natural resource companies should have clear objectives, responsibilities, professional management, transparency in operation and be subject to competition. State owned natural resource companies should not be charged with regulatory or other functions which properly belong with government (see also Natural Resource Charter).

Governance and macroeconomic policy should be improved, to address problems such as the Dutch Disease (inflation due to increased export revenues crowding out domestic production in favor of imports), rent seeking (ways to acquire ownership by any mean), corruption, and the impact of natural resources exploitation on conflict and externalities such as unstable commodity prices.

Transport and energy that are development bottlenecks require attention, e.g. through integrated multi-state Development Corridors.

Mining in Africa over the last century and particularly since the 1980s has not delivered broad-based development, except for a few cases such as Botswana (relatively exceptional governance).

But there is hardly any other sector that compares with the mining industry in terms of ability to bring about rents. Mineral rents can catalyze growth of other less competitive sectors such as agriculture, for now the backbone of many African economies.

The Vision acknowledges the failure of the Washington Consensus to deliver sustainable growth and development to Africa, the increased role of the state as exemplified by the intervention of United-States and European Union governments in their markets in the wake of the financial meltdown, and the rising strength of Africa's bargaining power as a result of the Chinese and Indian competition with the West for Africa's mineral resources.

All this means that the continent stands a better chance of increasing the proceeds from mineral resources.

Canada's role in developing Africa's mineral resources[20]

Canada had the largest share of the world exploration effort in 2008 (43 %).

The Toronto Stock Exchange (TSX) is the largest financial market in the world for mining finance: 81 % of market *transactions* for equity financing (followed by Australian Securities Exchange, 10%) and 40 % of *value* of equity financing (followed by London Stock Exchange, 26 %). $369 billion in mining company stocks were traded in 2006. The Toronto market is mature and sophisticated. The TSX Venture (junior exchange) is good for early stage projects while the TSX is–good for later stage, or producing mine or greater capital requirements.

There is also an excellent service sector in Toronto for exploration, technical, legal, financial, marketing, analysts, and media.

Canada's share of future mining investments in Africa is expected to be 29 % in 2010.

1300 large and small Canadian mining companies are listed on the Toronto Stock Exchange. In 2008, the Canadian larger-company mineral exploration market in Africa was valued at US $1.74 billion. From 2007 to 2008, the Canadian larger-company market in Africa grew by US $421 million, or by 32%.

In 2008, Canada was a relatively small player in exploration expenditures in Africa (Canada's share was 1/4). Canada large mining companies (over 3 million in exploration expenditures) are responsible for 20 % of world exploration expenditures in Africa. Smaller mining companies are responsible for only 7 % of the world African and Middle-East exploration budget.

Large Canadian companies' African exploration expenditures are distributed on base-metal (49 %) and diamond (27 %), gold and platinum Group Metals (24 %).

Countries counting for 90 % of Canadian firms' African exploration budget are DR Congo, Tanzania, Burkina Faso, Ghana, Zambia, Guyana, and Mali.

150 TSX listed mining companies are active in 37 African countries.[21]

A world total of $45 billion of new mining investments is planned for the next five years, Canadians accounting for $13 billion of this total (29 %). By 2010 Canadian mining interests in Africa will double with capital assets of $10 billion.[22]

Canadian industry, being a major mining player in Africa, needs to ensure that corporate social responsibility is respected according to Principle 10 of the Natural Resource Charter.

Conclusion

The African Mining Vision addressed specific issues in the general perspective of the Natural Resources Charter. It is more than rhetoric as the Vision has the unanimous endorsement of African mineral resources ministers that meet as one single responsible body. Moreover, the Vision merges all previous policy documents. It sets the objectives and constraints on the policy space. It encourages the continuation of the FDI role. Its drawback is that it did not address corporate social responsibility.

References

African Economic Outlook 2009, OECD, www.oecd.org/document/59/0,3343, en_2649_15162846_42689915_1_1_1_37413,00.html.

African Partnership Forum, *Climate Change and Africa, 8th forum,* Berlin 2007, www.africapartnershipforum.org/dataoecd/57/7/38897900.pdf.

Arye, B. (no date), *Sustainable Development Charter for the Minerals and Mining Sectors in Africa*, African Mining Partnership.

Auty, R.M.(2001), *Resource Abundance and Economic Development*, Oxford University Press e-book, UNU wider studies in Development Economics.

Bigsten, A., et al, (2000). "Rates of Return on Physical and Human Capital in Africa's Manufacturing Sector," *Economic Development and Cultural Change*, vol. 48(4), pages 801-27.

Boocock, C.N. (2002), *Environmental Impacts of Foreign Direct Investment in the Mining Sector in Sub-Saharan Africa*, OECD, www.oecd.org/dataoecd/44/ 40/1819582.pdf.

Campbell, B., P. Hatcher, A. Lafortune and B. Sarrasin,(2004), The Challenges of Development, Mining Codes in Africa and Corporate Responsibility, chapter in E. Bastida, ed., *International and Comparative Law and Policy*, Kluwer , p. 801-822.

Canadian Intergovernmental Working Group on the Mineral Industry (2008), *Overview of Trends in Canadian Mineral Exploration*, Government of Canada, www.nrcan-rncan.gc.ca/mms-smm/busi-indu/pdf/explor/2008/explor-2008-eng.pdf.

Collier, P. (2010), *The Plundered Planet,* Oxford University Press.

Collier, P. and A. J. Venables (2008), *Managing the Exploitation of Natural Assets: Lessons from Low Income Countries,* paper presented at the African Economic Research Consortium 2008 Annual Conference, Nairobi.

Collier, P. and B. Goderis (2008) *Commodity Prices, Growth, and the Natural Resource Curse: reconciling a conundrum,* Oxford University research paper.

Commission for Africa, *2010 Report: Still our Common Interest* www.commissionforafrica.info/2010-report. (accessed october 17, 2010).

Darimani, A., *Mineral resource extraction in Africa: Continuity and shifts, lessons*, Third World Network Africa, 2008, www.africafiles.org/article.asp?ID=19213.

Delcourt, L. (2009), *Retour de l'État. Pour quelles politiques sociales?*, Centre Tricontinental, Louvain-la-Neuve.

Economic Commission for Africa/African Development Bank, *2007 Report.*

Government of Canada (2009), *Building the Canadian Advantage: A Corporate Social Responsibility (CSR) Strategy for the Canadian International Extractive Sector*, www.international.gc.ca/ trade-agreements-accords-commerciaux/ds/csr-strategy-rse-stategie.aspx (accessed September 28 2010).

Inngs, M. (2009), African Mining Partnership merges with AU body to improve access to funds, Mining Weekly, February 13, www.miningweekly.com/article/african-mining-partnership-merges-with-au-body-to-improve-access-to-funds-2009-02-13.

Leipziger, D. (2010), *The Corporate Responsibility Code Book.* Revised Second Edition, Greenleaf Publishing Limited. Database of Press Releases Related to Africa (2008), *Ministers Endorse ECA Inspired Mining Vision*, October 21, http://appablog.wordpress.com/2008/10/21/ethiopia-africa-s-mineral-resources-ministers-endorse-eca-inspired-mining-vision/.

Muehlberger, M. (2007), *Africa's Natural Resources in the Spotlight Again*, Deutsch Bank Research, www.dbresearch.com/PROD/DBR_INTERNET_DE-PROD/PROD0000000000207813.pdf.

Niyiragira, Y (2010), Convergence of Opinion on Africa's Mining Industry, *Au |Monitor Daily*, January 12, www.pambazuka.org/aumonitor/comments/2050/.

Obeng, K. (2008), Convergence of Opinion on Africa's Mining Industry,Third World Network Africa, December 23, www.twnafrica.org/index.php?option=com_content&view=article&id=85:convergence-of-opinion-on-africas-mining-industry-&catid=54:mining&Itemid=60.

OCDE (2009), *African Economic Outlook.*

Sachs, J. and A.M. Warner (2001), The Curse of Natural Resources, *European Economic Review*, 827-838.

Shapiro, B., *Canada in Africa- Benchmarking Africa's Progress and Development Against Best Practice*, ppt presentation to African junior Mining Congress, www.mineafrica.com (accessed 12/8/2009).

Sinkala, T. (2009), *Mining and Environment in Africa*, UNEP, www.scribd.com/ doc/23618617/Mining-and-Environment-in-Africa.

Stijns J.P.C.(2005), Natural Resource Abundance and Economic Growth Revisited, *Resources Policy*,107-130.

Tamufor, L. (2009), *The African Mining Vision and the Continent's Development, Extractive Sector - Mining* , Third World Network Africa, 10 July, www.twnafrica.org/index.php?view=article&catid=54

%3Amining&id=173%3Athe-african-mining-vision-and-the-continent s-development&tmpl=component&print=1&page=&option=com_cont ent&Itemid=60.

The Brookings Institution, Africa Growth Initiative (2009), *The Impact of the Financial Crisis on Africa*, Washington, February 13 (Anderson Court Reporting).

The Natural Resources Charter, www.naturalresourcecharter.org/ (accessed September 28,2010).

Udjo, E.O.(2008), *Demographic projections of Africa's population for the period 2000-2050 taking account of HIV/AIDS and its implications for development*, www.unisa.ac.za/contents/faculties/service_dept/docs/SABVI12_3Chap4.pdf.

United-Nations (2006), *World Population Prospects: The 2006 Revision Population Database*. New York: Department of Economic and Social Affairs, UN Low Variant.

United-Nations Economic Commission for Africa(2007), The Big Table, www.uneca.org/thebigtable/summary-report.htm.

The *World Bank Group's Africa Action Plan Progress in Implementation*, Africa Region, The World Bank, 2007.

World Bank, Africa (Devarajan) *http://africacan.worldbank.org/financial-market-turmoil-and-africa* (accessed October 20,2010).

World Bank (1992), *Strategy for African Mining.*

World Bank (1997), *A Mining Strategy for Latin America and The Caribbean.*

—. (2009), *The Little Data Book on Africa 2008,* http://siteresources.worldbank.org/INTSTATINAFR/Resources/LDB-Africa-12-2-08.pdf (accessed October 20 2010).

—. (2010), Global Economic Prospects Report—Summer 2010, Regional Appendix: Sub-Saharan Africa, http://siteresources.worldbank.org/INTGEP2010/Resources/GEP2010Summer2010-SSAAnnex.pdf. (accessed October 18 2010).

Notes

---

[1] For a recent evaluation, see *Commission for Africa Report*, September 2010 and World Bank 2010.

[2] Knack, S. and P. Keefer (1997), Does social capital have an economic payoff? A cross-country investigation, *Quarterly Journal of Economics*, 112, 4, 1251-1289.

[3] The *World Bank Group's Africa Action Plan Progress in Implementation,* Africa Region, The World Bank, 2007.

[4] http://hdr.undp.org/en/statistics/ HDI figures are for 2006.
[5] followed by Gabon (CA, 107), Equatorial Guinea (CA, 115) and Egypt (NA, 116); South Africa (SA, 125) and Botswana (SA,126), Morocco (NA, 127), Namibia (SA, 129) and Congo (CA, 130); Swaziland (SA(excluding Lesotho), 141), Ghana (142), Madagascar (EA, 143), Kenya (EA, 144), and Sudan (EA, 146); Cameroon (WA, 150), Djibouti (EA, 151), Tanzania (EA, 152), and Senegal (WA, 153).
[6] One finds Nigeria (WA, 154), Lesotho (SA, 155), Uganda (EA, 156), Angola (CA, 157) followed by the remaining twenty-one African countries with Sierra Leone (WA) at the bottom (.329). .218 was recorded by Liberia (WA) in 1995.
[7] World Bank, Africa (Devarajan) *http://africacan.worldbank.org/financial-market-turmoil-and-africa*
[8] http://epi.yale.edu/. The figures are for 2008.
[9] Gabon (WA, 64), Algeria (NA, 66), Egypt (NA, 71), Morocco (NA, 82), Ghana (WA, 86), Namibia (88), Congo (CA, 93), Zimbabwe (EA, 95), Kenya (EA, 96), South-Africa (SA, 97), Botswana (SA, 98), Cote d'Ivoire (WA, 103), Cameroon (WA, 114), Senegal (WA, 115), Togo (116), Uganda (EA, 117), Swaziland (SA, 118), Malawi (EA, 121), Eritrea (EA, 122), Ethiopia (EA, 123), Nigeria (WA, 126), Benin (WA, 127), Central African Republic (CA, 128), Sudan (NA, 129), Zambia (EA, 130), Rwanda (CA, 131), Burundi (EA, 132), Madagascar (EA, 133), Mozambique (EA, 134), Guinea (WA, 138), Djibouti (EA, 139), Guinea-Bissau (WA, 140).
[10] Dr. Yao Graham, Co-ordinator of Third World Network-Africa (TWN-Af).
[11] A Sovereign Wealth Fund (SWF) is a state-owned investment fund composed of financial assets such as stocks, bonds, real estate, or other financial instruments funded by foreign exchange assets.

These assets can include balance of payments surpluses, official foreign currency operations, the proceeds of privatizations, fiscal surpluses, and/or receipts resulting from commodity exports.

The definition of sovereign wealth fund excludes, among other things, foreign currency reserve assets held by monetary authorities for the traditional balance of payments or monetary policy purposes, state-owned enterprises in the traditional sense, government-employee pension funds, or assets managed for the benefit of individuals.

Some funds also invest indirectly in domestic state-owned enterprises.

In addition, SWF tend to prefer returns over liquidity, thus have a higher risk tolerance than traditional foreign exchange reserves.
[12] http://eiti.org/eiti/principles (accessed October 14, 2010)
[13] www.ifc.org/ifcext/sustainability.nsf/AttachmentsByTitle/pol_PerformanceStandards2006_full/$FILE/IFC+Performance+Standards.pdf
(accessed October 14, 2010)
[14] www.equator-principles.com/documents/Equator_Principles.pdf
(accessed October 14, 2010)
[15] www.unpri.org/principles/ (accessed October 14, 2010)

---

[16] www.pdac.ca/pdac/misc/pdf/070329-advisory-group-report-eng.pdf (accessed October 14, 2010)

[17] OECD *Guidelines for multinational enterprises*, www.oecd.org/dataoecd/56/36/1922428.pdf (accessed October 14, 2010)

[18] www.international.gc.ca/trade-agreements-accords-commerciaux/ds/csr-strategy-rse-stategie.aspx (accessed October 14, 2010)

[19] www.iso.org/iso/iso_catalogue/management_and_leadership_standards/social_responsibility/sr_resources.htm (accessed October 14, 2010)

[20] This section is based on the Canadian Mining Asssociation Reports and on Shapiro,

[21] Anooraq, Anvil, Banro, Barrick, Cassidy Gold, Dynatec, Etruscan, First Uranium, Forsys Metals, Goldbelt Resources, Homeland Energy, High River Gold, Hunter Dickinson, MAG Industries, Nevsun Resources, Platinum Group Metals, Riverstone Resources, Rockwell Diamonds, Southern Era Diamonds, SXR Uranium One, Teal Exploration and Mining.

[22] Some key countries for Canadian mining interests are: Algeria, Angola, Burkina Faso, DRC, , Ghana, Guinea, Kenya, Madagascar, Mali, Nigeria, South Africa, Tanzania, Uganda, and Zambia.

CHAPTER EIGHTEEN

INTERBASIN WATER TRANSFERS AND SUSTAINABLE WATER USE: A RELATIONSHIP OF CONTRADICTION OR COMPATIBILITY?

VICKY KARAGEORGOU

Introduction

It is widely recognized that one of the overriding problems of the twenty first century is that of the protection and sustainable management of water resources.[1] Increasing water demand, which has resulted from the multiplication of human needs, has led to an increasing depletion of the scarce water resources and to ever increasing pollution. Climate change has also started to impact significantly the water availability and quality.[2] The water crisis is much more intense in areas where the water demand exceeds significantly the available water supplies, for example regions, which are strongly dependant on irrigated agriculture. In such circumstances, interbasin water transfers (IBTs)[3] become relevant, as they have been viewed for a long time as a solution to meeting increasing water demands in water-stressed regions.[4]

The central question that arises is whether IBTs can be an effective and sustainable solution to the problem of increasing water demand in certain water stress areas under the new circumstances of the emerging water crisis. The scope of the paper is to make steps towards answering that question by taking into consideration the basic principles and regulatory directions set in the legislative frameworks for water protection and management at regional and national level (Part II). Furthermore, the designed and partially implemented diversion of the Acheloos River in Greece is chosen as a case study, with a view to demonstrating not only its interesting legal perspectives but also the emerged governance issues between the recipient and the donor region (Part III). Finally, certain conclusions are drawn concerning to whether and if yes, to which extent IBTs can be acceptable as an option to satisfy water demand in arid regions (Part IV).

Regulatory Trends and Basic Principles in Water Law-Relevance with IBTs

The Paradigm Shift in Water Law

Water law constitutes, undoubtedly, the cornerstone for achieving effective and sustainable water management. The economic, societal and environmental changes that have taken place in the last decades, have posed new challenges to the water law regimes, at both international, regional and national level. Climate–induced changes to water resources require, for example, that effective tools and mechanisms be in place, in order to cope with increased uncertainty and extreme weather events. All the above-mentioned factors have substantially influenced the "paradigm shift" concerning the adopted regulative concepts of both national water laws and transboundary water agreements. In particular, there is a gradual shift from water laws, which are primarily centered on the satisfaction of human water needs, to legislative regimes that have adopted a more integrated approach in managing and protecting freshwater[5] and give emphasis to the introduction of integrated supply- and demand policies. Furthermore, the recognition of a human right to water has also gained importance as an emerging regulative concept[6]. The above-mentioned regulatory trends and concepts, although important in terms of guiding the legislators concerning the basic directions to be adopted, are not precise or suitable enough to provide concrete normative criteria for evaluating IBTs from the perspective of sustainable water management. It is thus important to search for any specific rules or general principles relevant to the examination IBTs as an option.

Rules and Principles Relevant to IBTs

In any analysis concerning the relevant rules and principles for assessing IBTs, the distinction between legal principles and rules should be at first taken into consideration, because of its great relevance for the legal effect that each category of legal norms has. In particular, in spite of the existence of divergent views within legal theory[7], it is at least acknowledged that legal principles incorporate certain fundamental ethical values so that their introduction into the legal system does not lead to a particular consequence but only provides justification for taking a decision. Legal rules, instead, set certain standards that lead to particular consequences, while their non-observance is associated, in the majority of cases, with the imposition of sanctions.

1. No direct applicable rules for IBTs

Taking the afore-described distinction into consideration, it should be first examined whether any legal rules that either set limitations for the development and implementation of IBTs or even concrete requirements for their acceptability from the perspective of sustainable water management, are in place. The systematic analysis of the provisions of the legal instruments that regulate the management and protection of transboundary waters at global and regional level leads to the conclusion that *they do not contain any specific rule, which regulates water transfers between transboundary waters.*[8] A characteristic example thereof is the Water Framework Directive (2000/60/EC), which introduced an integrated and innovative approach for water management and protection. It does not contain, though, any specific provision for water transfers both at interstate and transboundary level. Furthermore, the relevant national legislative frameworks do not set any specific norms or standards that relate directly to the acceptability of IBTs as an option for satisfying water demand. Subsequently, an examination should be contacted into whether any general principles of water and environmental law can provide guidance towards this direction.

2. Critical principles of water and environmental law

One of the most relevant principles for evaluating IBTs is that of the *sustainable use of water resources.* The fact that the above-mentioned principle is recognized as a guiding principle in several important legal instruments, such as the Water Framework Directive (WFD)[9] and the Revised SDAC Protocol on Shared Watercources[10] is of utmost importance. In an effort to define the precise regulative contours of the principle of sustainable water use, it seems necessary to take into consideration the context of both the principle of *sustainable use of natural resources*, as the former can be regarded as a specification of the latter in water law, and of the *sustainability principle* being viewed as an emerging fundamental principle of environmental law.

The principle of sustainable use of natural resources has gained significant recognition at the international level[11], since it is strongly reaffirmed in several International Conventions related to environmental protection and sustainable development, such as the Convention on Biological Diversity[12], the UN Convention to Combat Decertification[13] and the Preamble of the WTO Agreement. Furthermore, the introduction of the principle in relevant national laws[14] is mainly based on the distinction between renewable and non-renewable natural resources and on the adoption of certain standards and mechanisms for controlling rates of natural resource use. The most important elements of the principle, viewed as the result of the systematic interpretation of the provisions of the legal instruments, where it is established, can be described as follows: i) the recognition of the guiding function of the so-called management rules[15] in terms of providing clear-cut criteria for the use of the natural resources ii) the adoption of a holistic and implicit precautionary approach and iii)the long-term perspective, with a view to meeting the foreseeable needs of the future generations.[16]

Furthermore, the regulative standards of the principle of sustainable use should not be seen separately but in association with the sustainability principle, not only because the former constitutes a specific application of the latter[17] but also because of the emerging recognition of the fundamental character of the sustainability principle. At this stage an in-depth analysis of the regulative context of the sustainability principle cannot take place due to space limitations. In brief, it should be noted that the core of the principle is to protect and restore the integrity of the planetary eco-systems, although it is not irrelevant, both from ethical and legal perspective, which approach (strong or weak sustainability[18]), is followed. Subsequently, the principle, especially in its strong version, can provide important regulative standards not only concerning the rates of natural resource use but also the most important government decisions (e.g Infrastructure projects) as well as the model of development that has to be followed, namely a model that is within the limits of the ecological systems.[19]

In the light of the above analysis, a search for the regulative contours of the principle of *sustainable water use* is attempted with a view to finding any relevant criteria for assessing IBTs. In addition to the regulative core of both the sustainability and the sustainable use principle, the systematic interpretation of the provisions of the legal instruments, where the critical principle is established, is also of utmost importance. An important remark, in this context, is that the legal instruments with regional scope of application, such as the WFD (Art 1.lt.b), the UNECE

Convention on the Use of Transboundary Waters and Lakes (Art.2 par. 2 lit. b) and the revised SCAD Protocol on Shared Watercources (Art.2 lit.b) *are not enlightening enough in terms of setting the regulative contours of the principle of sustainable water use*. This is though, not the case for certain national water laws, especially when they have been recently subject to reform. A comparative overview of the existing water laws[20] leads to the conclusion that the newly revised German Federal Water Act (Gesetz zur Ordung des Wasserhaushalts, amtlich Fassung von 31.07.2009) constitutes the most characteristic example in terms of not only recognizing the principle of sustainable water use as the most important guiding principle (Art.1) but also specifying its context by setting several sub-principles. One such sub-principle is for example the safeguard of the performance and functioning of the water resources both as an integral part of the natural capital and as a place for other living organisms (Article 6). In view of the specification of the principle of sustainable water use in certain national laws as well as its inter-connectedness with the regulative concept of the sustainability principle, it can be argued that its most important elements, are the following: 1) *the adoption of certain rules and criteria for the acceptable rates of water use, with the aim to ensure that use for various purposes does not exceed the carrying capacity of the aquatic ecosystems* 2)*the adoption of the necessary protection measures, with the aim to preserve the aquatic eco-systems in favour of both their own intrinsic value and the satisfaction of the water needs of the future generations.*

3. Evaluating IBTs

The critical criterion for the evaluation of IBTs, which arises from the principle of sustainable water use, relates to the *limits* that have to be set for their implementation, when it is scientifically grounded that such a project will pose "significant" threats to the carrying capacity of the relevant water bodies and subsequently their ability to satisfy future water needs for both humans and the environment. Furthermore, the sustainability principle as a crucial principle for guiding decisions for key infrastructural projects also sets limits to their implementation. This is the case when conducted scientific assessments demonstrate that such projects are expected to impact significantly not only the aquatic but all the other eco-systems as well, especially those in the donor region.

In this context, the following important conclusions concerning the acceptability of IBTs can be extracted: 1) *the development and implementation of such projects* is *almost incompatible with both the principle of sustainable water use and the sustainability principle due to their ability to cause significant harm to the functioning and the carrying capacity of aquatic and all other eco-systems. Such an acknowledgement is also justified by the relevant international experience*[21]. 2) IBTs *can be only exceptionally allowed, when all other viable and small-scale alternatives for satisfying water demand at river basin level, which have to be examined before deciding for an IBT, are not possible.* The consideration of the effects of the climate-induced changes on the water cycle speaks also for choosing reversible and flexible options instead of large-scale water infrastructural projects.[22] 3) The context of the principle of sustainable water use especially in terms of safeguarding the carrying capacity of aquatic eco-systems in association with the directions of the IWRM can provide guidance concerning the priority, according to which alternative solutions to IBTs should be considered. In this context, a wide variety of measures aiming at reducing water demand in the water stress region, such as an effective system for water permits and pricing policies should be at first examined, while recycling waste water and supplementing water supplies locally should come as "second best solutions".[23] In this context, the case of Acheloos is presented, with the aim to examine to which extent the above remarks are also justified by the relevant practical experience.

The Case of Acheloos

The 220 km long Acheloos river is one of the biggest rivers in Greece. The upper reaches of the river are already developed for hydro-electricity. Since the eighties a major composite project which involves several dams and other interventions (pipelines) has been planned and started partially being implemented. The main aim of the project is to divert water from the Upper Acheloos River to the Pinios River and from there to the Plain of Thessaly, the biggest agricultural region of Greece, in order to boost agricultural production. The Supreme Administrative Court, the so-called Council of State[24], has, though, set limits to its implementation, also by employing dynamic methods of interpretation of the EU environmental and water legislation.[25] In this context, the Council of State twice annulled the relevant environmental licenses granted for specific components of the diversion project on the grounds of the violation of the Environmental Impact Assessment (EIA) legislation. In particular, in its Decisions Nos. 2759 and 2760, the Court held that the available assessments were not valid, because each of them referred only to each separate dam planned, without assessing the overall environmental impact of the composite project. Therefore, it ruled on the necessity of a composite and comprehensive EIA Study of the highest scientific standard. Following the same line of argumentation, the Council also annulled the environmental licenses that were based on a composite EIA Study, because the critical Study did not examine any alternatives for the construction of the dams that could be less harmful to the natural and cultural environment (Plenary Decision 3478/2000). It is worth noting, though, that the Council rejected the argument of the petitioners concerning the incompatibility of the whole project with the principle of sustainable development. The principle was perceived in its currently most acceptable version, namely as a principle consisting of three pillars (economic, social, environmental), which are recognized as equal and their context has to be balanced in conflicting situations.

The environmental permits granted in 2003 were also annulled but with different reasoning to that of the previous decisions. At that time the Court ruled that the current design of the diversion project is not in line with the provisions of the Water Framework Directive (WFD) and especially *the principle of sustainable water use* as its guiding principle, although the WFD was not yet transposed into the national legal order. The relevant provisions require that any project for the utilization of the water resources should be in line with the directions and the criteria set in the water management plans (Plenary Decision 1688/2005). As no regional water plan was in place resulting from the lack of implementation of the then existing law, the Court ruled that the granted environmental licenses for several components of the project were invalid on the grounds that they were not based on certain directions or criteria set in a water plan.

As a response to the latest decision of the Council of State, the then Ministry of the Environment and Public Works initiated the authorization of a river basin management plan (RBMP) for the basins of Achelloos and Pinios (River) and the relevant environmental licenses based on the water plan by enacting a formal law (Law 3481/2006) instead of following the standard administrative procedure that also provides for public consultation procedures. It is worth noting that the legislative practices that were regarded as unconstitutional by the Supreme Administrative Court, have also been used in other cases to grant authorization to certain projects that could be subject to strong public resistance.[26] NGOs and citizens of the affected region brought again before the Court the administrative acts that were issued on the basis of either the authorized RBMP or the environmental licenses. The Council of State requested that the Court of the European Union give a **preliminary ruling** concerning the interpretation and the application of the following European Directives relevant to the case :a) The Water Framework Directive[27] b) The Environmental Impact Assessment Directive[28] c) The Strategic Impact Assessment Directive[29] and d) The Habitats Directive.[30] The case is still pending on the Court. In conclusion, it is to emphasize that the design and the implementation of the diversion project up to now demonstrate that the lack of clear-cut criteria, the ineffective implementation of the water and environmental legislation and the absence of effective governance arrangements can lead to disputes that have not been resolved for more than two decades. Furthermore, the partial development of the project is characteristic to indicate that such a scheme can have serious impacts on rare riverine and forest habitats as well as on the downstream freshwater eco-systems due to reduced flow.

Concluding Remarks

As already indicated, the emerging water crisis has resulted mainly from the utilization of water to its maximum economic potential. The large dams and the complex diversion projects that have been implemented worldwide can be classified as some of the most characteristic examples of maximum water utilization without taking into consideration the environmental impacts.

The analysis of both the normative and the empirical aspects of Interbasin Water Transfer (IBTs) Schemes that aimed at shedding light on the question of their acceptability as a solution to meet increasing water demand, leads to certain conclusions that can provide guidance for answering the central question of this paper. These conclusions are the following: 1) .From a normative point of view, it is worth noting that the "greening" of water law both at transnational and national level has not reached the point of containing, as already indicated, clear-cut rules concerning the permissibility of IBTs or at least certain criteria for their acceptability. General principles of water and environmental law become, thus, relevant for assessing IBTs. 2) IBTs can be regarded as an option that can be only exceptionally acceptable in terms of its compatibility with the principle of sustainable water use, which is recognized as the most important guiding principle in water law. In particular, although the above-mentioned principle has not gained its precise regulative contours in the majority of the legal instruments, *its regulative core in terms of safeguarding the carrying capacity of the aquatic eco-systems from a long-term perspective can set significant limits to the implementation of IBTs*. One can also reach the same conclusion concerning the compatibility of

IBTs with the sustainability principle as an emerging fundamental principle that provides guidance to the development of large infrastructural projects. 3) The exhaustive examination of other small scale alternatives at river basin level before deciding in favor of an IBT is an imperative. When all other alternatives fail, IBT can be seen as a last resort. Climate-induced changes on water-cycle should also be taken into consideration, when several options of water infrastructure are examined. 4) The majority of the existing legal systems provide for certain legal instruments that are crucial mainly in terms of assessing the expected environmental cost of the variety of options. The most important one is the Environmental Impact Assessment (EIA), which provides a systematic methodology mainly for evaluating possible environmental consequences of a proposed project and including that assessment in the decision-making process, while the exclusion of other possible alternatives also has to be justified.[31] Additionally, Strategic Impact Assessment can also be useful, because it can substantially contribute to the specification of the context of the river basin management plans, so that they can provide clear directions concerning the acceptability of IBTs as a solution to satisfy water demand.[32] 5) In the case that an IBT is decided, all necessary measures should be taken with the aim to minimize the environmental cost associated with the project development and implementation. An important measure in this direction is to ensure a minimum volume of water (environmental flow) as a means of protecting the ecological integrity of the eco-systems in the donor region.[33] 6) Progressive jurisprudence, as the case of Acheloos also demonstrates, can play a crucial role in shaping the regulative context of relatively new principles of water law, such as the principle of sustainable water use. Furthermore, it can set limits to the implementation of projects that can have multiple environmental consequences for the natural environment. 7) Governance arrangements are of utmost importance not only in terms of minimizing and effectively managing any risks associated with the scheme but also of preventing any possible social conflict between recipient and donor region. As the case of Achelloos demonstrates, poor governance arrangements, such as the lack of constructive dialogue between the recipient and the donor region and the lack of any compensatory benefits can lead to social conflicts that cannot be resolved for a long time. 8) Legislators both at national and transnational level should adopt a more sophisticated and integrated approach when regulating water management and protection in terms of recognizing, among other things, the water needs of the environment alongside with the various human needs. Only when such an approach, which is based on the assumption that people's well-being is interrelated with the existence of healthy aquatic eco-systems, is adopted, clear-cut rules concerning complex projects aiming at water utilization can be set. Finally, it should be emphasized that drafting or reforming relevant water laws towards embracing more precisely the context of the principle of sustainable water use should not be seen as an solated endeavor but instead as a part of a general effort towards reforming laws regulating management and use of natural resources within the context of an ecologically sustainable development. The re-orientation of environmental law is, though, not so simple, as it pre-supposes the re-orientation of the values of the society both at global and national level and especially the human-nature relationship that has to be reflected in the reformed legal documents.

Notes

[1] See G, Eckstein, *Water Scarcity, Conflict and Security in a Climate Change World: Challenges and Opportunities for International Law and Policy*, Wisconsin International Law Journal, Vol. 27 2010, p. 409 ff.

[2] IPCC Report 2007, *Climate Change 2007 : Climate Change Impacts, Adaptation and Vulnerability.* Contribution of Working Group II to the Fourth Assessment Report of the IPCC.

[3] Under the term "interbasin water transfer" can be described the development of the necessary technical infrastructure, such as pipelines, dams and canals, in order to transfer water from areas, where water supplies exceed the demands (donor areas) to the arid areas (recipient areas). See J. Jupta/P.van der Zaag, *Interbasin water transfers and integrated water resources management: where engineering, science and politics interlock*, Physics and Chemistry of the Earth Part A/B/C, Vol. 33, 2008, p.28-40.

[4] More than 364 large-scale interbasin water transfer schemes have been established worldwide and transfer around 400 km³ per year. See WWF Document, Interbasin water transfers and water scarcity in a changing world –a solution or a pipedream?, available at: http://assets.panda.org/downloads/pipedreams18082009 .pdf (last accessed 4 October 2010).

[5] The concept is mainly characterized by the adoption of an integrated approach in terms of regulation, planning and governance. For the definition of IWRM see Global Water Partnership, Technical Advisory Committee (TAC), *Integrated Water Resource Management*, Background Paper No 4, 2000, available at: www.cepis.ops-oms.org/bvsarg/i/fulltext/tac4/tac4.pdf, (Last accessed4 October 2010).

[6] See Draft Resolution A/64/L.63/Rev.1 adopted by the General Assembly of the United Nations on 28 July 2010. See also J. Scanlon/A.Cassar/N.Nemes, *Water as a Human Right* ?, IUCN Environmental Policy and Law Paper No. 51, IUCN 2004.

[7] For the distinction between legal principles and rules see K.Larenz/Cl-W. Canaris, *Methodenlehren der Rechtswissensschaft*, 3 Aufl, Berlin 1995, p. 240; . R. Alexey, *Zum Begriff des Rechtsprinzips* in: R. Alexey (Hrsg), Recht, Vernuft, Diskurs, Frankurt am. Main 1995, p. 177 (182); R. Dworkin, *Taking Rights seriously,* Cambridge, MA, Harvard Univrsity Press, 1977, p.22ff.

[8] See P. van der Zaag, *Southern Africa: Evolving Regional Water Law and Policies* in : J. Dellapenna/J. Gupta (eds), The Evolution of the Law and Politics of Water , Springer Science and Business Media B.V, 2009, p. 245 (258-9).

[9] Article 1 lit. b of the WFD provides that the purpose of the Directive is to establish a framework for the protection of inland surface waters, which promotes *sustainable water use* based on a long term protection of available water resources (emphasis inserted).

[10] Article 2 lit . b of the SDAC Protocol Article 2 lit. b of the SCAD Protocol stipulates that one of the main objectives of the Protocol is to achieve *the sustainable, equitable and reasonable utilization of the shared resources* (emphasis inserted).

[11]The principle of sustainable use is one of the seven principles that have been adopted by the International Law Association (ILA), in order to conceptualize the legal contours of sustainable development. See ILA New Delhi Declaration of Principles of International Law Relating to Sustainable Development. In the context of international law, the principle refers to the obligations of the states to make prudent use of the natural resources that are in their jurisdiction but can have transboundary effects. See M. C Cordonier Segger/A. Khalfan, *Sustainable Development Law: Principles, Practises and Prospects*, Oxford University Press 2004, p. 117-122 qualifying the principle as a part of the customary law.

[12] Article 2 of the CBD gives the only treaty definition that refers to sustainable use. As sustainable use of natural resources is, thus, defined "the use of components of biological diversity in a way and at a rate that does not lead to the long term decline of biological diversity, thereby maintaining its potential to meet the needs and aspirations of present and future generations".

[13] Article 3 lit. b of the UN Convention to Combat Desertification states that Parties agree on a principle to work towards *sustainable use of scarce water and land resources* (emphasis inserted).

[14] A characteristic example of a national law, which gives a clear definition for "sustainable management" and sets clear-cut criteria concerning the sustainable use of natural resources, is the New Zealand Resource Management Act, which was enacted in 1991. See K. Bösselmann, *The principle of sustainability,* Ashgate Publications 2008, p. 64. Furthermore, the principle of sustainable use of natural resources is also established in several federal german laws related to the protection and use of natural resources, such as the Federal Act on Nature Conservation and Landscape Management of July 2009 (Art. 1 par. 1 lit. 2).

[15] The so-called management rules were at first introduced by the famous ecological economist H. Daly and were further developed by the ecological science. These rules can be described as follows: i) renewable resources such as soil and groundwater must be used no faster than the rate at which they regenerate

ii) Non-renewable resources must be used in such a way, so that they can be conserved for the future generations iii) Pollution and wastes must be emitted no faster than natural systems can absorb them, recycle them, or render them harmless. The management rules have gained significant acceptance in the German environmental law theory. See Sachverständigen Rat für Umweltfragen, 2002, p. 6ff; Beaukamp, *Das Konzept der zukunftsfähigen Entwicklung*, 2002; K.Lange (Hrgs), *Nachhaltigkeit im Recht*, 2003.

[16] For certain concrete regulative standards arising from the principle of sustainable use of natural ressources E.Rehbinder, *Nachhaltigkeit als Prinzip des Umweltrechts: Konzeptionelle Fragen i*n : Dolde (Hrsg), Umweltrecht im Wandel,2001, p. 721 (728).

[17] See K. K. Bösselmann*, supra* note 14, p. 62.

[18] The main idea behind the concept of the strong sustainability is that natural resources should be conserved for the future generations, without taking into account whether the technical capital could substitute, to some extent, any losses of the natural capital. The notion of weak sustainability places, though, specific emphasis on the technical evolution in terms of providing solutions for compensating any losses of the natural capital For the concept of weak sustainability see P.M Solow, *Sustainability: An Economist's perspective* in : R. Dörman/N. Dörman (Eds), Economics of the Environment-Selected Readings, New York (W.W Norton), 1997, p.179-187. For the concept of strong sustainability H.E Daly, *Beyond Growth*, Beacon Press 1996. For the influence of both concepts on the context of environmental law see R. Rehbinder, *supra* note 16, p. 730ff ; K. Bösselmann, *supra* note 14, p. 100ff who analyses also the ethical implications concerning the choice of each concept.

[19] Both in the international law as well as in the vast majority of the national legal systems the priority of environmental concerns over the economic and social considerations has not yet been recognized within the three pillar concept of sustainable development. Several arguments have though come into the fore in the legal discourse, which strongly demonstrate that the recognition of equality within the three pillar concept neither offers clarity in cases of balancing competing interests, nor reflects the specific nature of the risks posed to the environment. See I.v Appel, Staatliche Zukunfts-und Entwicklungsvorsorge, 2005, p.339ff; K. Bösselmann, *supra* note 14, p. 53f, who argues that only if we accept the fundamental character of sustainability, the concept of sustainable development becomes operable; F. Ekhardt, *Nachhaltigikeit und Recht*, Zeitschrift für Umweltpolitik, 2009, p. 223 (228).

[20] Quite innovative in terms of recognizing the importance of water for satisfying both human and environmental needs is the South African National Water Act of 1998, which defines as "Reserve" the quantity and quality of water required to satisfy "basic human needs" and to protect the "aquatic eco-systems". It does not provide, though, more concrete criteria concerning sustainable water use. See M. Kidd, *South Africa: The development of water law* in: J. Dellapenna/J. Gupta (Eds), The Evolution of Law and Politics of Water, Springer 2008, p. 87 (93-94).

[21] See UNEP, *The Greening of Water Law: Managing Freshwater Resources for People and the Environment,* 2010, p.83-84, where it is argued that the implementation of diversion projects has led, in the majority of cases, to serious environmental degradation of the donor region, with the most characteristic example being the Aral See as a result of the diversion of the Amu Darya and Syr Darya.

[22] See J. Gupta/P.van der Zaag,*supra* note 3, arguing that large-scale projects should be the last resort, when all other solutions fail. After applying certain criteria from different disciplines, they have proposed the following five criteria that have to be fulfilled in order to decide for an IBT : 1) real surplus in the donor region and real deficit in the recipient region 2) sustainability of the transfer scheme in economic, social and environmental terms 3)good governance as the key process for developing the project 4)balancing existing rights with needs in terms of respecting existing rights and providing compensation, if necessary 5) sound science.

[23] See WWF Document, Interbasin water transfers and water scarcity in a changing world, supra note 4, p.40ff.

[24] Article 95 of the Greek Constitution provides the foundation for the jurisdiction of the Council of State. The Council of State has a general jurisdiction on petitions for the annulment of administrative acts and subsequently also on petitions for annulment of administrative acts on the grounds of violation of environmental legislation. See E. Spiliotopoulos, *The Judicial Review of Administrative Action in Greece* in : E. Spiliotopoulos/A. Makridimitris (Eds), Sakkoulas Publications, Athens 2001, p.7-110.

[25]All things considered, it can be argued that the Council's dynamic jurisprudence is largely positive, as it has set rational standards to a dysfunctional and narrow-minded bureaucracy and has played an important role in increasing awareness of environmental issues in Greek society. The Council's activism, though, has sometimes destabilized the line of demarcation between judicial review and legislative authority. See N. Papaspyrou, *A Farewell to Judicial Passivity: The Environmental Jurisprudence of the Greek Council of State*, Journal of Modern Greek Studies, 1999, Vol. 2, p. 63 (65, 82f).

[26]The Court ruled in similar cases (for example Plenary Decision of the Council of State No. 1847/2008) that modifying a city plan or granting an environmental license by a formal law instead of issuing an administrative act contradicts the constitutionally established principle of separation of powers (Art. 26) and the right to judicial access (Art. 20 par. 1). Furthermore, the Fifth Section of the Council (Decision No. 391/2008) raised serious doubts as to whether the introduction of individual regulations, such a construction license, by a formal law, which does not allow the prior information and participation of the public, could be compatible with the provisions of European environmental legislation (Directives 97/11/EC and 2003/35/EC) providing for public consultation procedures during the authorization procedures.

[27] The Council of State requested the Court of Justice of the European Union to provide an answer to the following questions concerning the interpretation of the

WFD: 1) Does the provision of the Art. 13 par. 6 of the WDF set only an upper time limit or does it set a specific deadline for the transposition of the provisions of the Art. 3,4, 5,6, 9, 13 and 15 of the Directive? In the case that the Court decides that the Directive sets only an upper time limit for the elaboration of the River Basin Management Plans (RBMP), the following question arises: 2) Is the regulative context of a national legislative provision that allows for water transfer from a concrete river basin to another river basin before the elaboration of the relevant RBMP of both river basins, compatible with the provisions of the Art. 2,3,4, 5, 6, 9, 13 and 15 of the WFD?. In the case of a positive answer, the following question arises: 3) Is water transfer from one river basin to a neighbouring river basin acceptable under the regulative concept of the provisions of the Art. 2,3,4, 5, 6, 9, 13 and 15 of the WFD? In the case of a positive answer, the next question that arises is: 4) Is the satisfaction of water needs for sanitation the only acceptable goal for water transfer or could it also serve the purpose of irrigation or energy production? In the case that the ECJ rules that Arr. 13 par. 6 of the WFD sets a specific deadline for the elaboration of the RBMPs and for the transposition of the Art. 2,3,4, 5, 6, 9, 13 and 15 of the WDF the following question arises :5) Does a national legislative provision that allows water transfer endanger the *effect utile* of the WDF or in order to assess this the criteria concerning the scale of the interventions and the goals of water transfer also have to be taken into consideration?. Finally the following question has to be answered by the ECJE : Is a national legal provision which authorizes a RMBP by a formal law, without conducting any public consultation procedures, in harmony with the provisions of the Art. 13,14 and 15 of the WFD that require prior information and public participation during the elaboration of RBMP?.

[28]The Council of State requested of the CJEE to provide an answer to the following question concerning the interpretation of the EIA Directive: Does an EIA Study, which relates to the construction of dams and other technical infrastructure necessary for water transfer and was authorized by the Greek Parliament (Law 3481/2006), fulfill the requirements of the Directive 85/337/EEC, as modified by the Directive 97/11/EU concerning public information and consultation?.

[29] The Council of State requested of the CJEE to provide an answer to the following questions concerning the interpretation of the Strategic Environmental Assessment (SEA) Directive (2001/42/EU) : 1) Does the water plan, on which the transfer scheme is based, fall within the scope of application of the SEA Directive? In the case of a positive answer, the following question arises: 2) Can the relevant acts (environmental licenses, authorization for the construction of dams), which were annulled by the relevant decisions of the Council of State, be regarded as formal preparatory acts, so that there is no obligation for SEA?. In the case of a negative answer, the following question arises: 3) Does the meaning of Art.11 par. 2 of the Directive require a separate SEA for the diversion project or not?.

[30] The Council of State requested of the CJEE to provide an answer to the following questions concerning the interpretation of the Habitats Directive (92/43/ECC) and the Wild Birds Directive (79/409/ECC): 1)Can the national competent authorities, in the meaning of the Art. 3,4 and 6 of the Directive,

authorize a project for water transfer, which is not directly associated with the conservation of a special protected area, when all relevant studies do not include updated data for the Fiona and flora of that area and it is extracted from the EIA Study that its implementation can contribute to the transformation of the physical river eco-system to an anthropogenic riverine system? 2) Can a water transfer project that primarily serves irrigational purposes, be classified as a project of overriding public interest in the meaning of Art.6 par. 4 of the Directive? And in the case of a positive answer, the question that arises is 3) What kind of compensatory measures should be taken?.

[31] See J. Glasson/R. Therivel/A. Chadwick, *Introduction to EIA*, 2nd Ed, Taylor & Francis, 2005, p. 4.

[32] See M. Mardsen, *Strategic Environmental Assessment in international and European Law*, Earthscan, London, 2008.

[33] For the importance of ensuring "environmental flows", see Hydropower Sustainability Assessment Protocol- Key components document, January 2009, p. 32, available at: www.hydropower.org/sustainablehydropowerhydropower/HSAF.html, (last accessed 21 October 2010).

PART V

ECOLOGICAL INTEGRITY AND INDIGENOUS PEOPLES: CASES AND ISSUES

INTRODUCTION

LAURA WESTRA

The concept of ecological integrity finds an accepting "home" in the traditions and lifestyles of indigenous communities everywhere in the world. In this part of the book, the three chapters range from the Maori people in New Zealand, to the Sami people of Norway, to the Onondaga Nation in North America.

Linda Te Aho (Chapter 19) recognizes the connection between ecological integrity and cultural traditions, as she cites the Nuu Cha Nulth, "a first Nation people of Canada...everything is one". For the Maori people, from their creation story on, that "oneness" encompasses both their ancestors, all their different tribes, and their environment. Thus, the Waikato river is much more than a body of water. The tribes, therefore, must protect the river as an ancestor and as a way of life, against attempts of various settlers to use it, without any concern for its integrity. In 2010, the Waikato-Tainui Raupatu Claims (Waikato River) Settlement Act, acknowledges the failure of previous regimes to protect the river, as it proclaims a different vision and strategy for the future, which includes power-sharing between the Waikato River Authority and the local Maori peoples using their traditional knowledge.

In Chapter 20, Mimi Lam and Trude Bosch reaffirm the deep connect between Indigenous peoples' culture and ecological integrity, this time for the Saami Indigenous peoples of Norway. Cultural property or the knowledge and traditions accumulated for generations, appear to reflect the UNESCO category of "intangible goods", deemed to be worthy of the protection of the international community.

Aside from cultural reasons for protection, fisheries are at grave risk globally, whereas small scale, artisanal fisheries are not seriously considered in policy making because larger, industrial operations are supported that often negate marine conservation policies.

The Saami are the indigenous peoples of Norway and other Nordic countries in Europe and Russia; Norway has the largest Saami population where they and their way of life have constitutional protection. In earlier times, strict Norwegian assimilation policies affected both reindeer herders and the Coastal Saami. More recently, there have been new legal instruments introduced to protect Saami rights to salmon fishing. However the latter is an activity with great symbolic value, but does not yield any substantial income to salmon fishers. Hence, the "Sea" or "Coastal" Saami continue their traditional ways of life and place-based subsistence economies. These cultural values not only help preserve the group, but also suggest sustainable environmental policies for all people of the world.

Jack Manno (Chapter 21), extends the theme developed in the two previous chapters. He starts by citing a version of the traditional Haudenosaunee "Thanksgiving Prayer". He emphasizes the close connection between the protection of ecological integrity and all things that support the sacred web of life, such that the struggle of indigenous nations to protect their way of life parallels the GEIG quest.

An example is the resistance of the Onondaga to oppose gas drilling that would have polluted their lands and waters. We note that the "lawsuit" that followed their resistance was dismissed on 22 September 2010, and an appeal is being pursued. The "Haudensaunee Environmental Task Force" also argued that horizontal drilling under Indian territories will be a violation of treaty protected mineral rights. As well, the cultural resources, sacred sites and unmarked burial sites would be affected. Manno appeals to the work of Prudence Taylor, and to Hans Jonas in *The Imperative of Responsibility* (1984), to argue that connections and alliances must be formed to protect not only the local lands, but especially the particular relationship of the Onondaga people to those lands.

CHAPTER NINETEEN

INDIGENOUS ASPIRATIONS AND ECOLOGICAL INTEGRITY: RESTORING AND PROTECTING THE HEALTH AND WELLBEING OF AN ANCESTRAL RIVER FOR FUTURE GENERATIONS IN AOTEAROA NEW ZEALAND

LINDA TE AHO

Aotearoa New Zealand[1] is known throughout the world for its beautiful scenery, but behind those scenes there are many stories of how, over time, human activities have degraded the health and wellbeing of our lands and waterways. The Indigenous Māori have long fought to defend the integrity of natural resources who do not have a voice of their own in the processes of law and policy making. Ecological integrity is an ethical concept that values the state of nature being whole and unimpaired such as might be found in places of wild nature, untouched by human exploitation. This paper seeks to provide a domestic example of the application of the environmental ethic of the indigenous Māori which ensures that humans are kaitiaki (guardians) of their surrounding environment by virtue of shared genealogy, and how this ethic is being applied to restore and protect the health and wellbeing of the Waikato River, New Zealand's longest. This particular river restoration project reflects many of the important aspects of ecological integrity. It serves, too, as a domestic example of the most sophisticated attempt at powersharing to date between Māori and the Crown in relation to natural resources, which may be of interest to other jurisdictions.

Everything is One

As soon as I came across the notion of ecological integrity, it struck a chord with me. The similarities with my own indigenous ways of viewing and valuing the environment were striking. According to a leading proponent of the ecological integrity discourse, Laura Westra, one of the most important aspects of integrity is the self-creative capacity of life to organize, regenerate and evolve over time at a specific location. In relation to water, another leading proponent, James R. Karr, has studied the spatial requirements that are needed to maintain native ecosystems. Ecological integrity, it is said, bridges the concerns of science and public policy. It also acknowledges the global connection between human health and the integrity of nature, and the rights of both. It is argued for instance that human health and function are both directly and indirectly affected by disintegrity.[2]

The indigenous understanding that all things in the physical and spiritual worlds are connected is reflected in the saying from the Nuu Cha Nulth, a first nation people of Canada: *Hishuk is ts'awalk* - everything is one.[3] It is an understanding shared by the Indigenous Māori of Aotearoa New Zealand. According to our creation stories which have been passed down via oral tradition, water originates from the pain of separation of our sky father and earth mother who had been bound together embracing their children in absolute darkness. After much debate and deliberation one of the children, Tāne Mahuta, forced his father skywards to separate his parents, and the siblings emerged into the world of light. They included the revered ancestor of the winds, who disagreed with the separation, and who exacts revenge from time to time via violent storms; the revered ancestor of the seas; and the revered ancestor of the forests and all life within. Māori see ourselves as direct descendants of all of these ancestors who operate in a balancing system, and who are personified and incorporated into everyday life in prayer, in song, and in oratory - reflecting a reverence for creation as a whole entity. While each tribal grouping in Māori society maintains its own particular traditions, we all see ourselves as inextricably bound to the environment, including in particular our waterways, by virtue of genealogy which derives from our creation stories. Consequently we see ourselves as not only 'of the land' but 'as the land'. An oft-quoted saying in relation to water, '*ko au te awa, ko te awa ko au*' 'I am the river and the river is me', articulates this interconnectedness that lies at the heart of the way we view the world and our waterways. It is this worldview that Māori have long fought to have recognised and provided for in the many government policies and programmes that affect the natural world.

Waikato River - An Ancestor, and a Way of Life

The Waikato River is New Zealand's longest River. It winds its way northwards for approximately 425km from its volcanic mountain source in the centre of the North Island and out to the Pacific Ocean at Te Puuaha o Waikato, south-west of Auckland. In 1859 Austrian geologist, Dr Von Hochstetter, spent three months in New Zealand and recorded the following impression:

> ...[T]he sight of the majestic stream is truly grand. It is only with the Danube or the Rhine that I can compare the mighty river, which we had just entered...Its sources spring from the very core of the land; its waters roll through the most fertile and most beautiful fields, populated by numerous and most powerful tribes of the natives, who have taken their name from it...The Waikato is in truth the main artery of the North Island.[4]

At least since the early 1300s, the Waikato peoples have lived in settlements along the banks of our river. The river was everything. The personification of the natural world is a fundamental feature of Māori tradition and the river is conceptualised as a living ancestor by the Waikato-Tainui peoples,[5] and is recognised as having its own life force and spiritual integrity. The river's healing powers are reflected in the words of my elder, Te Kaapo Clark:

> Spiritually the Waikato River is constant, enduring and perpetual. It brings us peace in times of stress, relieves us from illness and pain, cleanses and purifies our bodies and souls from the many problems that surround us...[6]

The river's historical significance is captured in the words of Dr Michael King:[7]

> More than any others in New Zealand, the tribes of the Waikato Valley are a river people. Five centuries of continuous occupation of its banks have embedded the river deep into the group and individual consciousness. The river's associations grew and ripened with the history of the inhabitants until memories of heroes and villains, of battles, significant journeys and natural disasters.

In addition to its spiritual and historical dimensions, the river was also central to our survival. Over generations the Waikato-Tainui people developed tikanga (laws and practices) which reflect a profound respect for the Waikato River and the life within. The environmental ethic of the Māori which ensures that humans are kaitiaki (guardians) of their surrounding environment by virtue of our shared genealogical relationship is not restricted to preservation. It extends to sustainable use. [8] While laws and practices relate to the blessing of children, of cleansing, and of healing, tikanga also recognises that if we care for the river, the river will continue to sustain us as people. Swampy lowlands and the river provided flax for weaving, and water fowl and eels for food.[9] Elderly tribal members recall being taught not to be greedy, to take only enough food for a meal, and not to mistreat the river. Prohibitions on fishing or other activities were imposed in defined areas to prevent fishing for a time to allow for food species to rejuvenate.

Impacts on Integrity and Avenues for Reconciliation

It is our responsibility as tribal members to protect the integrity of the river as an ancestor and as a way of life. That integrity came to be seriously threatened as a result of the familiar patterns of colonisation experienced by indigenous peoples around the world.[10] Exploratory expeditions by the British and French to this country from 1769 and the establishment of a British colony in nearby Australia, culminated in Te Tiriti o Waitangi (The Treaty of Waitangi), signed between many Māori tribal groups and the British Crown in 1840. By the Māori text of the Treaty, Māori gave up 'kawanatanga' - the right to govern - but retained 'tino rangatiratanga' - sovereignty or the right to self-determination in relation to things that Māori valued, such as lands and waterways. Despite these guarantees, the extant indigenous systems of law and government collapsed. Ways of life and landholdings were devastated. Over decades of rapid change, mining, farming, and hydro-electricity development took their toll on the health and wellbeing of the river. Today, as a result of commercial fishing; the introduction of predatory fish; hydro-electric dams disturbing migration; and the inability to survive industrial pollution, river peoples are simply unable to gather important food species from a river once teeming with life. Apart from the tangible loss of food sources, this also means that knowledge about species and fishing practices has not been passed down to the next generations which in turn results in a loss of connection between youth and the elders who possessed such knowledge, and a loss to our language as names of different species, and different stages of their life-cycles, are no longer spoken.

Generations of Māori have tried different avenues over a number of years to have such breaches of Treaty of Waitangi guarantees addressed by the Crown. Their efforts were rewarded in 1975 with the establishment of the Waitangi Tribunal - a forum that has the power to inquire into such claims then make recommendations to the Crown. Such recommendations, while not generally binding, often form the basis of negotiations between the Crown and Māori claimants to settle grievances.

The main source of grievance in relation to the Waikato River is known as raupatu (confiscation). In the Waikato region, the contest for land had become a crucial issue in the 1850s as the British settlers' demand for land grew. A pan-tribal movement known as the Kīngitanga (the King Movement) was established in the 1850s, largely as a unified response by a number of tribes to the upsurge of unauthorised land sales for the ever-growing number of settlers.[11] Warrior chief Pōtatau Te Wherowhero of Waikato was raised up as the first Māori King in 1858 and was soon succeeded by his son, Tāwhiao. It was during Tāwhiao's term as second Māori King that the settler Government, seeing the Kīngitanga as a threat to its stability, sent its forces into the Waikato region in July 1863, labelling the Waikato people as rebels and subsequently confiscating millions of acres of Waikato lands and driving people away from their villages alongside our ancestral river.[12]

In the wake of confiscation the Waikato River became subject to a raft of legislation passed by successive governments.[13] A host of different authorities came to have jurisdiction over the river[14] as the Crown assumed that the English tradition of law applied to rivers in Aotearoa. That law presumed that the Crown owned the beds of tidal rivers as arms of the sea, and that the owners of lands with river frontage owned the beds of non-tidal rivers to the rivers' centre lines.[15] Such distinctions were completely foreign to Māori, who had our own conceptions of rivers, and were imposed without consultation or recognition of prior rights, interests and authority of the indigenous Māori.

In contrast to the way in which Māori tended to view rivers, the settlers seemed to view rivers as either obstacles to be overcome, evidenced by the draining of wetlands;[16] or convenient avenues for waste-disposal.[17] The mainstem waters and those of its streams and tributaries were polluted by sewage, farm run-off, extensive coal mining and other industrial discharges. Whereas, by tradition, Māori saw rivers as central to their survival and would adapt their needs to suit those of the river, the European settlers would adapt rivers to take advantage of opportunities for economic development. The upper reaches of the river were dammed in the 1900s to produce hydro-electric power inundating significant sites and causing ongoing problems with unstable flow regimes. The exploitation of the river was to involve a chain of power stations, mostly in narrow gorges. By the mid 1960s the Waikato River produced nearly half of New Zealand's power. Stories about the land and practices of old may not hold true when the landscape has been altered in such ways. They do not make sense to the younger generations. The words of a Cree elder reflecting upon the impact of the James Bay Hydro-Electric Project could easily be attributed to one of my own elders:

> A river once so strong and deep and that provided for a lot of families is now all dried up and sad to look at. It's so sad that we have lost all our native source of food and our way of life, and all because he wanted power...[18]

In the changing legal and political landscape of New Zealand, Waikato-Tainui has maintained the importance of their unique relationship with the river, and the need to respect the river's integrity and restore its wellbeing. From the 1880s Kīngitanga leaders have fought in political spheres and in the courts to address the deeply held grievances of raupatu.[19] Waikato-Tainui also lodged a claim with the Waitangi Tribunal in relation to the Waikato River. It is telling that claims concerning water resources were the first to be heard by the Tribunal, and although the Tribunal's early reports were important precursors to the Waikato River settlement,[20] Waikato-Tainui decided to settle their claims by negotiating directly with the Crown, rather than via a full inquiry in the Waitangi Tribunal. A settlement was reached in 1995 focusing upon the return of land.[21] The claim in relation to the Waikato River was excluded for future consideration and the ensuing settlement is enshrined in the Waikato-Tainui Raupatu Claims (Waikato River) Settlement Act passed in 2010 (the Act).

The Waikato River Settlement – A Vision to Heal the River

The ecological integrity discourse recognises that, like people, ecosystems or landscapes can be more or less 'ill'. In the words of James R. Karr,[22]

> An ill person may be suffering from a cold or dying of cancer ...an ill river may have ...fish populations depleted by overfishing or no fish at all, or only a few of the river's most tolerant invertebrates may remain after severe chemical pollution.

The Waikato River settlement recognises that as a result of human activities the river is degraded. The river is ill. The preamble of the Waikato-Tainui Raupatu (Waikato River) Settlement Act 2010 records that the deterioration of the health of the Waikato River, while under the authority of the Crown, has been a source of distress for the people of Waikato-Tainui and that the pollution, degradation and development of the river, its lakes, streams and wetlands have caused the decline of once rich fisheries that, for generations, had sustained the people's way of life and their ability to meet cultural obligations.[23]

The settlement's overarching purpose is to restore and protect the health and wellbeing of the Waikato River for future generations.[24] In order to achieve that purpose, the settlement puts forward a vision and a strategy which encompasses an excerpt from a famous lament of the second Māori King, King Tāwhiao in the 1860s. The lament records the significance of the Waikato River as a treasure for all generations and provides a clear vision of what the river once was, and what it could be again:

Tōku awa koiora me ōna pikonga he kura tangihia o te mātāmuri.
The river of life, each curve more beautiful than the last.

Resource Management Act 1991

The Waikato River settlement also serves as an example of how Māori have turned to the Treaty of Waitangi settlement processes to defend the integrity of natural resources since reliance upon New Zealand's principal natural resources statute, the Resource Management Act 1991, has not produced favourable results.[25] The Resource Management Act sets out a comprehensive regime for the sustainable management of land, air and water. Section 5 defines sustainable management of natural and physical resources to mean their use, development, and protection in a way, or at a rate, that enables people and communities to provide for their social, economic, and cultural wellbeing, and for their health and safety while:

(a) sustaining the potential of physical resources (excluding minerals) to meet the reasonably foreseeable needs of future generations;
(b) safeguarding the life-supporting capacity of air, water, soil, and ecosystems; and
(c) avoiding, remedying, or mitigating any adverse effects of activities on the environment.

Central government retains the power to set national environmental standards and issue national policy statements, but it delegates most of its responsibility for day-to-day management to regional councils and territorial authorities who regulate the use of land, air and water.[26] In making decisions under the Resource Management Act, section 6 requires decision makers to recognise five matters of 'national importance'. The first four refer to the protection of coastal marine areas, wetlands, lakes and rivers, outstanding natural features, and indigenous traditions with their ancestral lands, water, sites, significant and sacred sites, and other things treasured by Māori. Those decision makers shall also have regard to eight matters under section 7, the first being 'kaitiakitanga' – the 'exercise of guardianship by the tangata whenua of an area in accordance with tikanga Māori in relation to natural and physical resources; and includes the ethic of stewardship'.[27] Tangata whenua literally translated means people of the land and refers to the relevant indigenous group(s) of certain locations. The seven other matters include the efficient use and development of resources, protecting the heritage value of sites and buildings, and enhancing amenity values and the environment. There is specific provision for protecting the habitat of introduced species, trout and salmon, but no specific reference is made to indigenous fish. In achieving the purpose of the Resource Management Act with regard to the above principles, section 8 requires that those with decision making responsibilities shall 'take into account' the principles of the Treaty of Waitangi.

Though the Resource Management Act 1991 formalised a range of legal rights that were intended have Māori interests in the environment and the different way in which we view the world recognized and provided for, those rights are particularly vulnerable when presented as just one of many other considerations that decision-makers must take into account. In terms of cases brought under the legislation, the results have not been positive for Māori who have lacked political weight.[28]

Waikato-Tainui Raupatu Claims (Waikato River) Settlement Act 2010

The preamble of the Waikato-Tainui Raupatu Claims (Waikato River) Settlement Act 2010, (the Act) records that the Resource Management failed to protect the integrity of the river and the authority of the Waikato Tainui people, and that the Waikato Tainui people have been involved as respondents in many consent hearings, seeking conditions to protect the river.[29] For these reasons, much stronger legislative protection for the

Waikato River was sought via the Waikato River settlement. As a result, section 5 of the Act proclaims the vision and strategy to be the primary direction-setting document for the Waikato River and activities within its catchment affecting the Waikato River, and that it prevails over certain sections of the Resource Management Act, and over national policy statements. There is a list of objectives that will help achieve the vision and that list includes the integrated, holistic, and co-ordinated approach to management of the natural, physical, cultural, and historic resources of the Waikato River; and the adoption of a precautionary approach towards decisions that may result in significant adverse effects on the Waikato River. [30] In addition twelve strategies are to be followed. The first two strategies are to ensure that the highest level of recognition is given to the restoration and protection of the Waikato River; and to establish what the current health status of the Waikato River is by utilising traditional Māori knowledge and knowledge systems together with the latest available scientific methods. This second strategy is has been undertaken in the form of the Waikato River Independent Scoping Study (WRISS) which reports on the current condition of the river; and identifies restoration scenarios, costs and benefits of the restoration scenarios, and priority actions.[31]

There are a host of provisions in the Act which stipulate that decision-makers under statutes such as the Conservation Act 1987 and the National Parks Act 1980 will be required to give effect to the vision and strategy. Decision-makers under other statutes including the Fisheries Act 1996 and the Local Government Act 2002 will be required to have particular regard to the vision and strategy.

Agreeing to Share Power

The Waikato River settlement is a local example of Klaus Bosselmann's global theory that the central problem facing those who aspire to ecological integrity is not that we don't have solutions to the problems of resource depletion, carbon emissions and so on, but that we are failing to agree to implement those solutions.[32] So the problem is not *how* to live in accord with nature, but how to get humans to *agree* on how to live in accord with nature.

While there is a growing number of these sorts of agreements at a domestic level the Waikato River settlement is the most sophisticated example of powersharing in Aotearoa New Zealand.[33] Powersharing enables indigenous peoples and local communities to have greater rights and responsibilities in the governance and management of the landscapes and ecosystems they live in and near on the basis that enhancing peoples' rights will lead to more effective governance and management of resources. The Act provides for co-governance and co-management in relation to the Waikato River. Governance functions are to be carried out by a new entity, the Waikato River Authority, which will be made up of equal numbers of members appointed by the Crown (including regional and local councils) and members appointed by those tribal groups currently recognized by the Crown. The Waikato River Authority is responsible for monitoring and implementing the vision and strategy and will also administer a contestable clean-up fund for restoring and protecting the health and wellbeing of the Waikato River.[34] The authority is to make decisions by consensus. Māori appointed commissioners are to participate in hearing committees and boards of inquiry in respect of applications for resource consents for activities which include taking, using, damming, or diverting water in the Waikato River, and certain discharges to the river.[35]

At a co-management level, joint management agreements are required between Waikato-Tainui and the relevant regional council and between Waikato-Tainui and relevant territorial authorities for specified functions under the Resource Management Act insofar as those functions relate to the Waikato River and activities within its catchment.[36] As well, certain customary activities such as the use of traditional whitebait stands and eel weirs, and the right to continue traditional ceremonies are explicitly recognised.[37]

Positive Practical Effects

How might the new administrative arrangements and policy objectives operate in a practical context? Over time, changes to more intensive land uses have increased the amount of nutrients entering the river from the vast tracts of agricultural land, the municipalities, and the industries within the catchment of the Waikato River. High nutrient concentrations contribute to problems with excessive growth of algae which can be unsightly and damage the ecosystems of streams and shallow lakes. The direct discharge of human waste to water is culturally offensive to Māori. Without wanting to pre-empt the findings of the Waikato River Independent Scoping Study and the review of the vision and strategy that will follow, the management and reduction of nutrient sources from human activities is likely to feature strongly in both documents. There have been significant improvements to the waste treatment systems of municipalities and industries. And, to be fair, much is already being done by landowners to protect the river by fencing streams and removing stock from wetlands and lake margins, and by riparian planting. Unfortunately, these efforts have not been enough to counter increases in nutrient leaching from farmland, so the obvious restoration actions of riparian planting and fencing streams are likely to be highly prioritised by the new Waikato River Authority when making decisions on allocating the publicly contestable clean-up fund. The Authority may also explore ways of assisting research and development of low nitrogen leaching land uses, ways to permanently reduce nitrogen outputs from farming, and more modern treatment systems for nutrient removal. Given the significant contribution that dairy farming makes to the economic wealth of both the country and the region, for short term political reasons, regional councils have refrained from compelling changes to farming practices. However, assuming stronger provisions in the vision and strategy, the regional council will be legally required to give effect to those provisions in their policies and plans which could, in turn, lead to caps on nitrogen outputs from land in the catchment forcing reductions in stock or changes to land use (e.g from pasture to trees). Policies and plan provisions that require land based disposal of sewage would help to meet Māori aspirations.

Enhancing our Future Existence

As well as these anticipated positive effects at a practical level, the Waikato River Settlement also serves to illustrate at a higher level, how we can live wisely on the earth together. Around the world there are many studies relating to river restoration. Some study methods are at a catchment wide scale, some use social science, and some incorporate indigenous environmental knowledge. The Waikato River Independent Scoping Study that identifies restoration scenarios and priority actions has all those features. Its findings will be used by a decision-making body on which indigenous and non-indigenous people have an equal say. I have argued elsewhere that the Waikato River Settlement is far from perfect.[38] It is nonetheless bold and innovative. The settlement aims to restore the health and wellbeing of the Waikato River for future generations. Health and wellbeing is broad enough to include peoples' economic, social, cultural, and spiritual relationships with the river. The settlement provides for the ability to exercise rights and responsibilities of guardianship according to indigenous custom. It weaves together traditional environmental knowledge with western science and heralds a new era of participation for Māori in relation to the catchment. The settlement promotes inclusion rather than exclusion. It promotes mutual respect for Māori and non-Māori values, perceptions and aspirations.

The settlement provides a model for combining the contributions of different knowledge systems which in turn provides hope for the enhanced management of our future existence in Aotearoa New Zealand, hopefully bringing to an end a paradigm of excluding[39] Māori from engaging in decision-making processes that affects the environment of which we form an integral part.

Health and Wellbeing

What is more, in setting out to improve the health and wellbeing of our ancestral river, the settlement enhances human health and wellbeing. As recognised in the discourse of ecological integrity, there is a global connection between human health and the integrity of nature. Human health and function are both directly and indirectly affected by disintegrity.[40] From a broader indigenous perspective, our health and that of our communities is linked to the health of our land and our waters. Land and water must be healthy so that we can be healthy too. A convenient example comes from the western United States where Native American Tribes rely heavily upon salmon. Rachael Paschal Osborn explains that clean, cool, flowing waters are essential to virtually every aspect of the salmon life history. Conversely, the degradation of rivers brought about through post-contact human activities has caused major adverse impacts on salmon abundance and, consequently, on the health and well-being of salmon-dependent tribes.[41]

In the case of the Waikato River, the affirmation and validation of our cultural and traditional knowledge, of our values and ways of viewing the world, and in particular our ancestral river, has had a positive impact on our self-esteem and confidence. We walk a bit taller, we feel a bit stronger. This must have a positive impact on our own health and wellbeing - illustrating that nature is a world wide web of relationships between humans, spirits, and species. Everything is one.

Closing comments

The Waikato River is an ancestor. It is the source of my people's cultural identity and our spiritual and physical wellbeing. It is our duty to guard and protect this gift left to us by our ancestors, and to use its resources wisely and sustainably. Our own health and wellbeing is directly and indirectly linked to the health and wellbeing of the river. For the most part the Waikato River settlement provides a sense of optimism with its overarching purpose of restoring and protecting the health and wellbeing of the river for future generations. The language and purpose of the settlement clearly resonate with the notion of ecological integrity. The Māori ethic of kaitiakitanga will be resoundingly familiar to other indigenous peoples around the world. The resemblance between kaitiakitanga and the notion of ecological integrity leads me to believe that those who aspire to ecological integrity have strong allies in indigenous peoples. There are many lessons to be drawn from traditional knowledge systems built up over generations, which, when combined with contributions from other knowledge systems, provides hope for a shared existence on this earth.

Notes

[1] Aotearoa is a Māori name for New Zealand.
[2] Laura Westra, Ecological Integrity Encyclopedia of Science, Technology, and Ethics. Ed. Carl Mitcham. Vol. 2: D-K. Detroit: Macmillan Reference USA, 2005. p574-578 and the references cited therein, as viewed at www.globalecointegrity.net [date accessed: 7 December 2010].
[3]As recorded in the Museum of Anthropology, Vancouver, Canada.
[4] Ferdinand Von Hochstetter, *New Zealand: its Physical Geography, Geology and Natural History* (JG Cotta, Stuttgart, 1867) pp 294-5; cited in Ann Parsonson, *Waikato River Claim Report* (A confidential working document to assist the negotiating team following the 1995 Settlement).
[5] I descend from the peoples of Waikato-Tainui and Ngāti Korokī Kahukura.
[6] Statement of Evidence of Te Kaapo Clark of Ngāti Korokī Kahukura, prepared on behalf of Waikato-Tainui for the Watercare Hearing before the Franklin District Council, Tuakau, December 1996.
[7] Michael King *Te Puea: A biography* (Hodder and Stoughton, Auckland 1977) 50.
[8] Mere Roberts, et al. "Kaitiakitanga: Māori Perspectives on Conservation" (1995) 2 Pacific Conservation Biology 7 at 7, as cited in Jacinta Ruru, 'Indigenous Peoples' Ownership of Mountains' (2004) 3 Indigenous L.J. 115.
[9] Te Kaapo Clark and Lyn Tairi *Te Ihingarangi A Historyof the Karapiro-Maungatautari Area* (1992, Electricity Corporation of New Zealand).
[10]Patricia Monture-Angus 'At the Boundary: Indigenous Lawyers in the 90s' in Mikaere and Milroy (eds) *Ki te Ao Marama 10th Anniversary Hui a Tau Conference Proceedings*, p5. I make the point here that indigenous peoples must take responsibility for hunting certain species to extinction, such as the moa - a giant flightless bird; and for using fire to clear forests. On this point Jacinta Ruru argues that this does not disqualify the ethic, just as non-indigenous peoples are not disqualified from seeking to improve environmental conditions today. See Ruru, above note 8 at 130 and the references cited therein.
[11] See Michael King, *The Penguin History of New Zealand* (2003) chapter 15, and see www.teara.govt.nz for historical accounts of the King Movement; David McCan, *Whatiwhatihoe The Waikato Raupatu Claim* (2000).

[12] By Orders in council under the New Zealand Settlements Act 1893, the Crown unjustly confiscated approximately 1.2 million acres of land in the Waikato area.
[13] Alan Ward, *National Overview*, Vol 2 Waitangi Tribunal, Rangahaua Whaanui Series (1997), pp 348-353.
[14] Early examples include river boards with powers to control rivers under the River Boards Act 1884 and catchment boards established under the Soil Conservation and Rivers Control Act 1941.
[15] See in particular the Whanganui River Report (Wai 167) 1999; and *Muellar v Taupiri Coalmines Ltd* (1900) 20 NZLR 89 (CA) in respect of the Waikato River.
[16] The Land Drainage Act 1893 enabled wetlands to be drained for agricultural and pastoral production.
[17] The Public Works Amendment Act 1889 vested powers in river boards to declare rivers and streams public drains.
[18] Kreg Ettenger "A River That Was Once So Strong So Deep" Local Reflections on the Eastmain Diversion, James Bay Hydroelectric Project" in John Donahue and Barbara Rose Johnston (eds) *Water, Culture, & Power Local Struggles in a Global Context* (1998) 47.
[19] *Te Kīngitanga The People of the Māori King Movement* from *The Dictionary of New Zealand Biography* (Auckland University Press, 1996) Te Rata, 103. See also *Tainui Māori Trust Board v AG* [1989] 2 NZLR 513 (CA).
[20] Waitangi Tribunal, Manukau Harbour Report (Wai 8) 1985; Kaituna River Report (Wai 4) 1984; Motunui-Waitara Report; (Wai 6), 1983.
[21] The Waikato Raupatu Claims Settlement Act 1995.
[22] James R. Karr Vignette 11.2 Biological Integrity and ecological health, in M.C. Neuman *Fundamentals of Ecotoxicology* (3rd ed) 2010 324-329.
[23] Waikato-Tainui Raupatu Claims (Waikato River) Settlement Act 2010, Preamble, Paragraphs (k) and (l).
[24] Above, s3.
[25] Jacinta Ruru "Undefined and Unresolved: Exploring Indigenous Rights in Aotearoa New Zealand's Freshwater Legal Regime" (2009) 20 JWL 236, 238-241.
[26]Resource Management Act 1991, s2.
[27] Ruru, above note 25 at 237-238.
[28] Prue Kapua "Review of the Role of Māori under the Resource Management Act 1991" [2007] *Resource Management Theory and Practice* 92, 106-108.
[29] Waikato-Tainui Raupatu Claims (Waikato River) Settlement Act 2010, paragraph 14.
[30] Above, Schedule 2 (1) (3)(e) and (f).
[31] See www.river.org.nz for reports relating to the Waikato River Independent Scoping Study. The first major report, The Waikato River – Current Condition and a Framework for Restoration, January 2010 records that over time, human activities have degraded the health and wellbeing of the Waikato River, iv. The WRISS has been completed and is due to be publicly released in 2010.
[32] Klaus Bosselman 'Earth Democray: Institutionalizing Sustainability and Ecological Integrity, in Engel,J.R et al *Democracy, Ecological Integrity and International Law,* Ch6,1.

---

[33]For more information about other examples of co-management and co-governance regimes in relation to natural resources, see Linda Te Aho 'Indigenous Challenges to Enhance Freshwater Governance and Management in Aotearoa New Zealand – The Waikato River Settlement' (2009) 20 JWL 285, 288-290.

[34] Waikato-Tainui Raupatu Claims (Waikato River) Settlement Act 2010, ss22-24.

[35] Above, sections 25-31.

[36] Above, sections 35-55.

[37] Above, sections 56-63.

[38] Linda Te Aho "Contemporary Issues in Māori Law and Society The Tangled Web of Treaty Settlements Emissions Trading, Central North Island Forests, and the Waikato River" (2008) 16 *Waikato Law Review* 229, 242-246.

[39] I have borrowed the expression 'paradigm of exclusion' from the work of Dr Kepa Morgan. See for example "Exploring Knowledge System Synergies for Integrated Decision Making" (2009) 12 *Journal of Australian Indigenous Issues* 299.

[40] Laura Westra, above, note 2.

[41] Rachael Paschal Osborn, 'Native American *Winters* Doctrine and Sevens Treaty Water Rights: Recognition, Quantification, Management' (2009) 20 JWL 224.

CHAPTER TWENTY

CULTURAL VALUING OF FISHERY RESOURCES BY THE NORWEGIAN SAAMI

MIMI E. LAM AND TRUDE BORCH

Introduction

Ecological integrity is characterized by "wild nature," "the autopoietic (self-creative) capacities of life to organize, regenerate, reproduce, sustain, adapt, develop, and evolve," and being "valuable and valued" (Westra et al. 2000). For indigenous people, however, the issue of ecological integrity cannot be separated from cultural integrity, for in indigenous beliefs and practices, people are not separable from nature (Berkes 2008). The interdependence of biological and cultural diversity today may reflect a small-scale ecological phenomenon in the past: the process of coevolution of small human groups with their local environments, modified by humans as they adapted and developed specialized knowledge of it (Maffi 2001). Indigenous ecological knowledge, practices, and ethics can inform modern policies designed to cope with social dilemmas of common-pool resources and environmental challenges. Climate change poignantly exemplifies a modern global-scale ecological phenomenon requiring local-scale cultural adaptations: indigenous and traditional peoples, as communities living in marginal lands and highly dependent on natural resources, are among the most vulnerable groups being impacted by climate change, but they also have the most to offer of tested adaptation and mitigation strategies to environmental changes (Macchi et al. 2008).

Cultural adaptations to changing local environments and policies depend upon cultural knowledge and environmental values. This chapter analyzes natural resource policy from the perspective of cultural property, a way of knowing shared among members of a local community. It values not only access and ownership rights to natural resources, but also the historical relationships of people to places. Identifying cultural property as a distinct property regime in environmental policy may enhance the value and protection of wild living resources and the traditional livelihoods and ecological knowledge of the communities dependent on them. An intergenerational- or *G-index*, proposed to measure the cultural value of natural resources to local communities (Lam, *in review*), is applied to salmon resource use by the Saami, the indigenous people of Norway. The cultural implications of Saami natural resource use within national and regional environmental policies highlight the unique challenges posed of integrated ecological and cultural integrity for indigenous people.

Natural Resource Property Regimes

Property implies a set of social relations between the rights-holders and duty-bearers, recognized by both and sanctioned by social institutions (Kirsch 2001). Four property regimes have been identified in environmental policy (Bromley 1992): 1. open-access (*res nullius*), where resources are owned by no one; 2. private (*res privata*), where resources are owned by individuals or corporate entities; 3. common (*res communes*), where resources are owned by many individuals in a community; and 4. state (*res publica*), where resources are owned by the public, i.e., all citizens of a country, and managed and controlled by the nation state or crown. Common-property regimes and resources, the latter defined by their difficulty of exclusion of access and subtractability of benefits by others' resource use, have been analyzed (Berkes et al. 1989) and show Garrett Hardin's (1968) influential "tragedy of the commons" parable, the dilemma of pursuing individual benefits at the cost of collective ruin, to be overly deterministic.

Cultural property, in contrast, is a way of knowing shared by local community members that recognizes not only possessions and rights to natural resources, but also place attachments, social relationships, and local ecological knowledge (Kirsch 2001). It contains two distinct types of property relations: 1. *alienable property relations*, which are separable from their geographic location and inherently convertible or amenable to economic compensation; and 2.*inalienable property relations*, which are inseparable from the historical relationships of people to places and are characterized by kinship ties, social relations, traditional customs, subsistence production, and livelihood practices. Local ecological knowledge and traditional subsistence ways of living constituted valuable cultural property lost by indigenous communities in the *Exxon Valdez* oil spill (Snyder et al. 2003) and the Marshall Islands atoll radioactive fallout (Kirsch 2001). Local resource-dependent communities also experienced loss of cultural property from the environmental damage caused by the 2010 Gulf of Mexico oil disaster. Recognizing cultural property as a distinct property regime in environmental policy may help value and protect both wild living resources and the traditional livelihoods and ecological knowledge of local and indigenous communities that are still dependent on and attached to their places (Lam, *in review*). Valuing cultural property may also help preserve ecological and cultural integrity.

Cultural Property and Sense of Place

Cultural property, the knowledge and traditions accumulated from generations of living in close contact with a place, can be gauged by the sense of place of the people, as the socially constructed meanings and attachments to place that develop with experience (Stedman 2003). 'Place bonds' of individuals strengthen as the rootedness, associated with residential status in the place, and bondedness, or social belonging and attachment to the community, increase (Hay 1998). Sense of place and its associated feelings of security, belonging, and stability develop with the length and timing in an individual's life history of residence, insider status, and local ancestry. Hay (1998) proposed five developmental stages of connection to place: 1. superficial (tourists and transients); 2. partial (campers, cottagers, and resident children); 3. personal (new residents and immigrants without roots in the place); 4. ancestral (residents with roots); and 5. cultural (indigenous residents with roots and spiritual ties, as affirmed by culture). An intense sense of place was found after forty years in a place for those who moved to the community and after thirty years for those raised in it, while new migrants began to feel committed to the place only after about ten years, as they developed local knowledge, a social network, and some community standing (Hay 1998).

Cultural indicators measure the resilience or ability of an individual, family or community to adapt, by shared accumulated knowledge and lived experiences, to fluctuations in natural resources and their policies. As a cultural indicator for non-market resource valuation, Lam (*in review*) proposed the number of continuous lived generations in a place as a proxy for sense of place, with the human generation interval set at thirty years. The ensuing intergenerational-index, *G-index*, assesses cultural resilience, a signature of the value or relative impacts of resource access or loss to the overall livelihood or vitality of the organizational unit. As a basic measure of the length of the relationship with the resources, *G-index* estimates the historical local resource investment and exploitation by the resource users. As more generations are spent in a locale, the cultural resilience to transfer to an alternative place lowers, compared with transients or migrants with a lower sense of place and local ecological knowledge (Lam, *in review*). Loss of access to traditional resources may exceed the cultural resilience for an individual, leading to a loss of livelihood or sense of self; for a family, a loss of tradition; and for a community, a loss of a way of life. The identities of all would be affected, as the resources reflect more than a source of income, but a way of knowing and being, i.e., valuable cultural property in relationships with places sustained over multiple generations.

Human Dimensions of Fisheries Management

Human impacts from degradation of natural resources are threatening human wellbeing and global sustainability (MEA 2005, Rockström 2009). Modern industrial fisheries, using highly mobile fleets, sophisticated gear and technology and accumulated ecological knowledge, have expanded both the range and depth of fishing. Millennial exploitation of marine resources (Pitcher 2001, Jackson et al. 2001) has intensified human appetite for seafood, fishing pressure, and dependence on marine resources. This precipitated a global crisis in the 1990s, when major fisheries collapsed and catches declined from serial depletions of commercial target species (Pauly et al. 2002). In contrast, indigenous and subsistence fisheries based on traditional and local ecological knowledge sustained historical relationships with marine ecosystems, intimately woven into the social fabric and cultural identity of coastal communities (Haggan et al. 2007).

Policies for rebuilding fisheries and restoring ecosystems (Pitcher and Lam 2010) recognize the benefits of ecosystem-based management (Pikitch et al. 2004) and responsible governance (Sissenwine and Mace 2003). However, lacking tangible benefits or appropriate incentives (Grafton et al. 2006), public apathy and private resistance often negate marine conservation policies (Mangel et al. 1996). Current natural resource property regimes encourage fishing activities geared towards short-term private gain, with nefarious consequences on the long-term carrying capacities of global marine ecosystems. Small-scale, artisanal, place-based fisheries traditionally have had less ecological impact and involved more individuals (Pauly and Maclean 2003), but their economic inefficiency is often used politically in favour of developing large-scale, industrial fisheries.

Valuing the cultural property of coastal communities may help to sustain marine resources and preserve the human dimensions of fisheries (de Young et al. 2008), by focusing on the socio-cultural drivers and impacts affecting local communities and global society. This has led to calls for new social contracts defining reciprocal rights, duties, and responsibilities to evolve ethical fisheries globally (Lam and Pauly 2010) and improve resilience to climate change impacts (O'Brien et al. 2009). O'Brien et al. (2009) write: "Within the context of a changing climate, social contracts must recognize the interests, values, and knowledge systems of local communities." This sentiment runs counter to most modern fisheries policies, which view local, place-based fisheries as being economically unimportant and therefore not important for cultural preservation. In this chapter, we argue fishery resources have both economic and cultural value vital to maintaining the traditional life ways and places for the indigenous people of Norway, with likely parallels the situation for indigenous people globally.

The Indigenous People of Norway: The Saami

The Saami are an indigenous people who form an ethnic minority in Norway, Sweden, Finland and Russia's Kola Peninsula, which, together with Karelia, form a geographically and culturally distinct region known as Fennoscandia (Figure 20-1). The Saami population is estimated at 60,000 - 100,000, with 40,000 - 45,000 Saami in Norway, concentrated in Finnmark, 17,000 in Sweden, 5,700 in Finland and 2,000 in Russia. Official censuses have not given reliable counts, as not all Saami wish to acknowledge or declare their ethnic identity because of historical assimilation policies that eroded Saami cultural traditions and practices. The traditional Saami region in Norway extends from Finnmark County in the north to Hedmark County in the south. The coastal regions of northern Norway, including the fjord areas in Finnmark County, the north of Troms County, and some sea-bordering municipalities in Nordland County, have been defined by the Norwegian government as the Coastal Saami area (Figure 20-1). The Saami population in these coastal and fjord communities, known as the Sea Saami or Coastal Saami, have a tradition for adaptation to the sea and sustaining themselves from aquatic resources (Nilsen 2003). Saami coastal societies traditionally sustained themselves by harvesting marine resources combined with small-scale agriculture.

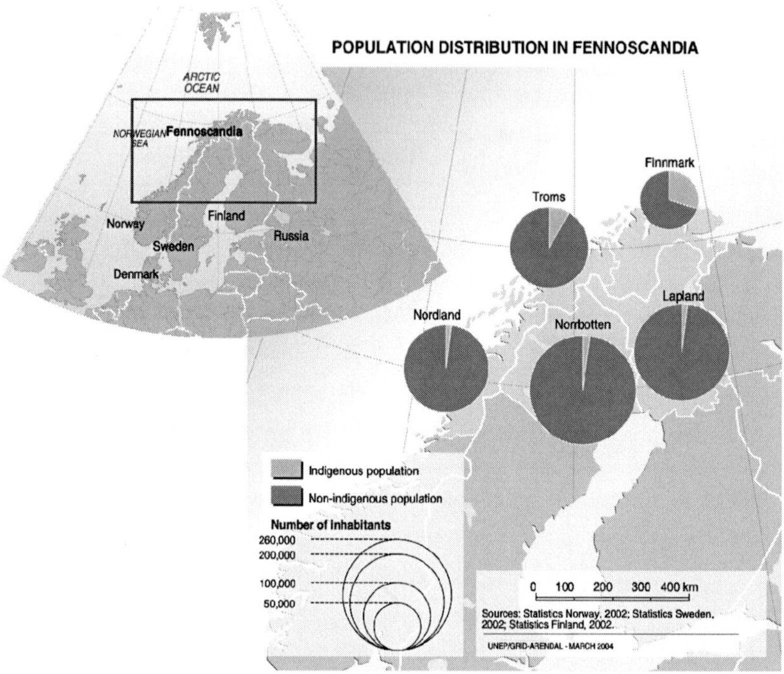

Figure 20-1. Distribution of Saami population in Fennoscandia (see inset), showing concentration of indigenous population in Norway, and within Norway, in its northernmost county, Finnmark (modified from UNEP/GRID-Arendal).

The Saami have a special right to cultural protection, but Norway's policies have only recently shifted from cultural assimilation to preservation of its indigenous people. Saami cultural identity is now protected by Article 110a of the Norwegian Constitution, ratified in 1988, giving rights to the Saami population and creating duties and obligations to the Norwegian state: "It is the responsibility of the authorities of the State to create conditions enabling the Sami people – to preserve and develop its language, culture and way of life." It was based on the Saami Act of 1987, which sets out legal regulations for the Saami and the main guidelines for the Saami Parliament of Norway, the representative body for Norwegians of Saami heritage. Norway has also ratified international conventions, declarations and agreements applicable to the Saami, such as Article 27 in the UN Convention of 1966 on civil and political rights and the International Labour Organization (ILO) Convention No. 169 of 1989 on the rights of indigenous and tribal peoples in independent countries.

The Saami have a traditional economy based on a direct relationship to nature in the arctic and subarctic areas, with local adaptations to reindeer herding, hunting and fishing, and agriculture. They historically sustained themselves from different types of natural resources, exhibiting a resilient group adaptation to a variable local resource base (Nilsen 2003, 165):

> "An abundant supply of fish from the sea, lakes and rivers, together with wild, extensive wilderness pastures and water power have been a blessing, because these resources have provided a basis for life and periodic abundance in an otherwise meagre and hard existence. They have been a foundation for self-sufficiency in the traditional combined adaptations in households and *siida* arrangements that are so typical of Saami culture."

Siidas were the former organizational units of Saami societies, where several family groups cooperated in the practical management and sharing of natural resources and game. The individual *siida* had a collective right to hunting and fishing within its area. The head of the *siida* oversaw regulations for natural resource use and ensured that hunting and distribution followed its rules and traditions, while labor and economic burdens were shared among all members.

Following the First World War, the Coastal Saami became dependent on "the Pomor trade" with the Russians and fish was their most important commodity. When this trade was stopped in 1917, the Coastal Saami had to adjust to Norwegian society. With less opportunity for trading fish, it became important for the Coastal Saami to acquire land for farming, but Norwegian legislation stated that only citizens mastering the Norwegian language could purchase land, forcing the Saami to learn Norwegian. After the Second World War and the closeness between the Saami in the north and Russia during the cold war, the Norwegian government worked hard to assimilate the Saami population into Norwegian society.

This assimilation or "Norwegianization" policy affected both the nomadic reindeer herders and the Coastal Saami. For reindeer herders, the granting of exclusive indigenous rights has been less contested than for the Coastal Saami, as the rights to reindeer herding are inherited and the majority Norwegian population does not participate. The Coastal Saami, on the other hand, utilize the same resources and have the same livelihood as other coastal Norwegians, and the rights to fishing have to be granted, leaving the Saami struggling for exclusive rights to some marine resources, such as cod. Also contributing to the assimilation of the Saami in coastal areas were their integration into the Norwegian welfare community and its developing prosperity (Nilsen 2003, 164):

> "An especially important part of the Norwegian policy of distribution for the coastal population was the adoption of a law concerning unprocessed fish in 1938 (the Raw Fish Act). Through this arrangement, fishermen in Norway are assured a large part of the fisheries' income. So on the one hand, integration into Norwegian society has created a pressure as far as giving up one's own identity..."

This assimilation policy dramatically altered Saami social and economic life, with Saami culture now sustained through business activities developed along with traditional livelihoods. The marginal resources in Saami areas are seldom sufficient to support single occupations, so combinations of livelihoods are used as an economic adaptation to give a more balanced and resilient use of natural resources. Saami ownership rights in fishing are assigned by residency status in Saami areas (Borch et al. 2009), not to the individual, as in the individual transferable quota systems widely advocated in fisheries management today (for critique, see Pitcher and Lam 2010). Since indigenous rights are secured through places designated as Saami areas, i.e., communities in areas with strong Saami traditions, Saami culture and identity are strongly coupled to natural resources and places. This has strong implications to the impact of Norwegian fisheries policy on the maintenance of Saami culture.

Norwegian Sea Salmon Fisheries Policy

In recent years, the Norwegian government has increased efforts to protect its wild salmon population, heightening regulation of net fishing for salmon in the sea to reduce fishing on mixed-stocks and preserve the genetic composition of its wild stocks. The government has also argued that the economic contribution from sea salmon fishing is marginal today. Through the Saami Parliament, the Saami have opposed this argument, claiming that even though the economic impact of sea salmon fishing may be declining in Saami areas, the activity remains of cultural importance. They argue that sea salmon fishing is part of a cluster of cultural practices that contribute to preserve Coastal Saami culture. In an assessment process for implementing licenses for net sea salmon fishing, the Norwegian Ministry of the Environment commissioned a study of sea salmon fishing in Coastal Saami areas (Borch et al. 2009) to review both its historical and current importance, as well as Coastal Saami rights to sea salmon fishing. This is the cultural and political background to the indigenous case study presented in this chapter.

The Norwegian government's Norwegianisation policy, in tandem with the centralisation process of the 1950s and 1960s, led to a higher density of Saami in the coastal communities of northern Norway and their designation as Coastal Saami areas. Sea salmon fishing here historically contributed to food supply, trade and combination income (income from several sources). The combination approach is not as widespread as before for various reasons, such as fisheries and agriculture policies affecting all coastal adaptation in Norway. Today, only a small fraction of the sea salmon fishers in Saami areas has substantial income from the activity. In spite of this, sea salmon fishing continues to play a role in combination adaptation, settlement, and activity through traditions for food gathering (berry picking and fishing) and "being in nature" in Coastal Saami areas.

Coastal Saami culture, the knowledge, values and traditions associated with a way of living in coastal areas with a Saami history, is maintained not only as part of work, but also leisure. Most inhabitants of Saami areas now make a living from sources other than traditional nature-based industries, leading to a change in economic adaptation, industrial structure and outdoor recreation (Pederson 2002). Many youth now come in contact with nature only via outdoor recreational activities (Pedersen 2002, 179):

> "A challenge following from this is how traditional activities in nature can be forwarded as outdoor recreational activities in ways that are considered as meaningful to young people, parallel with a strengthening of their knowledge about Saami language and culture as well as their feeling of belonging to the Saami community." (Translated from Norwegian)

The exclusive rights of indigenous groups to use and manage natural resources are typically granted based on their importance to their cultural preservation. This cultural value of harvest rights need not be related to the economic importance of an ecological adaptation, but may contribute to the economy of a group and its culture in a multi-dimensional adaptation. The activity may have great symbolic value or value as an identity marker for ethnic groups whose cultural borders are challenged (Cohen 1985, 44):

> "...the greater the pressure on communities to modify their structural forms to comply more with those elsewhere, the more are they inclined to reassert their boundaries symbolically by imbuing these modified forms with meaning and significance which belies their appearance."

present struggle of the Coastal Saami population to protect their rights to harvest ocean resources, such as in sea salmon fishing.

Saami Sea Salmon Fishing

The study of sea salmon fishing (Borch et al. 2009) surveyed all sea salmon fishers in the Coastal Saami areas, not exclusively those identified or registered as Saami, as there was no practical or ethical way to check for ethnic identity among respondents (n = 200, response rate 48%), and indigenous rights are assigned based on residency status in Saami areas. Traditionally, Coastal Saami would engage in salmon fishing, in both fresh water and the sea. In a Stone-Age study in the Varanger Fjord, Odner (1964) describes how historical Coastal Saami settlements were localized to the inner parts of fjords, but would move during spring to the outer parts to fish for salmon, collect eggs, and hunt for seal and sea birds. In another study from the Pasvik Valley, Tanner (1929) describes how women and children would fish for salmon in the inner parts of the fjords, while men would go further out in the sea to fish for cod and other marine resources.

The natural adaptation, industrial structure and demography in Coastal Saami areas have changed significantly (Borch et al. 2009). From 1980 to 2007, the overall population in Coastal Saami areas decreased by 16%, while in Alta, a municipality where a new college opened, it increased by 36%. Out-migration of younger age groups has triggered the population decline, closely connected to declines in the primary industries. For example, individuals with fishing as their main occupation halved, dropping from 2,763 in 1990 to 1,320 in 2007, and similarly for fishing in combination adaptation. Sea salmon fishers in Finnmark and Troms parallel these trends, with an average age of 62 years, 92% aged 40 or over, and low recruitment of new fishers With an aging population profile, opportunities to share and retain cultural knowledge, practices and values are fading in Saami communities. Intergenerational transmission of knowledge of sea salmon fishing does continue from grandfathers (22%), fathers (68%) and uncles (17%) to children and grandchildren. Sea salmon fishing also continues to contribute to the total income, though often in combination with other income sources. For Finnmark County, 52.7% of sea salmon fishers are on pension or other types of governmental support, 22.8% have income from other marine fisheries, 4.7% have income from agriculture and 15.7% have wages from the private or public sector. For these fishers, 41.1% of catches is sold for income, 46.5% contributes to the household consumption and 12.4% is shared with family and friends.

When the sea salmon fishers from Finnmark and north Troms were asked about their motivations for sea salmon fishing (Figure 20-2), to "assure coastal culture", "safeguard rights based on use" and "continue living in area" were ranked 1st, 2nd, and 3rd out of 10 motivations, while for "own consumption" was ranked 4th, "income" as 8th, and "share catch with family and friends" as 9th. The motivation to "assure Coastal Saami culture" was 5th, with a wide standard deviation, reflecting the divisiveness associated with Coastal Saami culture. This contrasts with the first-ranked "assure coastal culture," which had the smallest standard deviation.

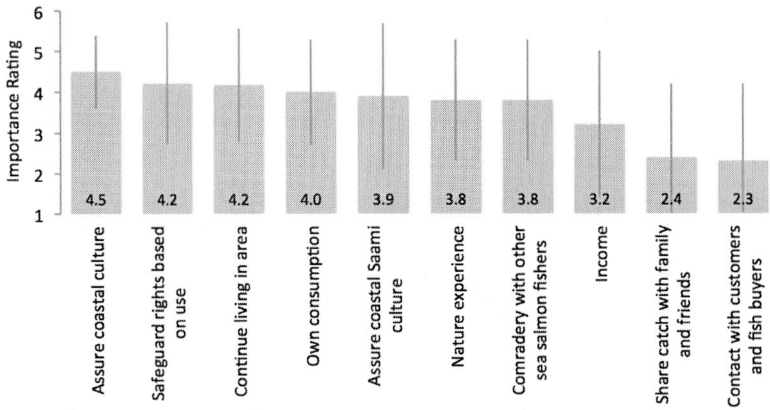

Figure 20-2. Responses of survey respondents in Norway's Coastal Saami areas to the question: How important are the following reasons for you to continue sea salmon fishing? Vertical bars denote standard deviations about response means.

Where sea salmon fishers in Coastal Saami areas lived when fishing was probed (Figure 20-3) to elucidate the motivation to "continue living in area" as a driver to continue sea salmon fishing. During the fishing season, 44.1% of fishers in Saami areas lived in their primary homes, 46.3% in secondary or transient homes, 1.5% in rental facilities, and 8.1% lived elsewhere. In comparison, for Norway, 69.0% of fishers fish from their primary homes and only 25.9% from secondary homes, 1.2% from rental homes and 4.0% from homes elsewhere. We propose two explanations for these differences: first, the Coastal Saami tradition of staying while fishing in small, primitive cabins (identified as secondary homes in the survey); and second, of individuals who grew up in the area returning during the fishing season to live in their former family home, now a secondary home. Data for fishers in Finnmark County are intermediate of the Saami areas and Norway, likely reflecting a transition between traditional and urban lifestyles and mentalities in this northernmost county. We see these trends in residency while fishing as a gradation from stronger to weaker sense of place dependence and attachment in Saami areas, Finnmark, and Norway.

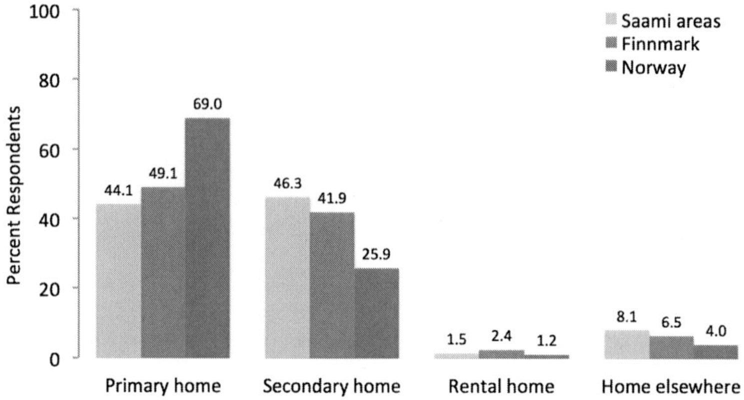

Figure 20-3. Responses of survey respondents in Norway's Coastal Saami areas to the question: Where are you living when you fish?

Conclusions

The indigenous people of Norway have lived in its coastal regions for at least 2000 years (based on written records from 95 AD), accumulating locally adapted ecological knowledge and resource use practices. While this was disrupted by European colonization in the 1600s to 1700s, Saami ecological knowledge of local fishery resources remains strong and distinct from that of Europeans, despite both groups fishing commercially. The longer historical relationship of the Saami to the local place suggests that the resources have not only ecological and economic value, but also cultural value, manifest in both Saami attachment and dependence on the place and its resources. The layered lenses of cultural value of resources, sense of place, and an intergenerational-index may help bring into focus natural resource policies appropriate for local indigenous communities.

Sea salmon fishing contributes to continued settlement and activity in Saami areas and thus to Coastal Saami culture. Its cultural value originates from the places where the Saami traditionally lived, which are embedded with cultural meaning and attachment. In spite of losses of ways of living in and from nature, a strong Saami tradition persists for hunting and gathering. A strong motivation for sea salmon fishing is to continue living in the place, sustaining a population necessary to replicate Saami culture. Measuring cultural value by the number of continuous lived generations (*G-index* = 1, taken as 30 years) in a place (Lam, *in* review), we estimate *G-index* = 64 for the Saami and *G-index* = 14 for Europeans and conclude that Saami ecological knowledge, management practices and cultural values should be weighted with economic value in just fisheries policies.

Traditional lifeways and subsistence economies based on local natural resources cannot compete with global markets and government regulations favoring industrialization. To preserve these ways of living, the cultural value and meaning of natural resources must be assessed in context. For the Saami, sea salmon fishing allows them to sustain their traditional ways of life, while also providing income, by preserving traditional practices while living in traditional places. Use of resources in such communities is embedded within a cultural cluster that has evolved over many generations of resource use in places with strictly codified rules and traditions of use. Valuing both the cultural and economic significance of local knowledge and practices protects the local resources and people dependent upon those resources. In Norway, indigenous rights are assigned to all residents, regardless of heritage, of areas with strong Saami traditions, emphasizing the protection of places by people living in and attached to those places.

The Saami, by their status as indigenous peoples, have special rights to participation in both the management and practice of fishing. The cultural value of sea salmon fishing to the Saami should be factored in to fisheries policy in Norway, given its domestic and international mandates to preserve Saami culture. An indigenous right to sea salmon fishing is not tantamount to a right of use, however, if it is unsustainable. It is incumbent on Norwegian salmon management to ensure that the exercise of rights to fishing comes neither at the expense of the government's international obligations to the conservation of biological diversity nor its obligations to protect and uphold the rights of its indigenous peoples. That is, both ecological and cultural integrity must be preserved, particularly for local indigenous and resource-based communities, if environmental policies are to be just and sustainable for all people of the world.

Acknowledgements

MEL thanks Bill Rees and Laura Westra, respectively, for inviting her to give a presentation at the GEIG 2010 conference in Vancouver, BC and to write this book chapter. TB thanks the Norwegian Research Council for funding a visit to Simon Fraser University, which led to this collaboration and Norway's Directorate for Nature Management for commissioning the study on the Saami sea salmon fishing. We are indebted to Tony Pitcher for the figures and to the survey respondents for sharing their information.

References

Berkes, F. 2008. *Sacred ecology.* 2nd ed. New York: Routledge.

Berkes, F., D. Feeny, B. J. McCay, and J. M. Acheson. 1989. Commentary: The benefits of the commons. *Nature* 340:91-93.

Borch, T., A. Buanes, G. R. Karlsen, and F. Olsen. 2009. Sjølaksefiske i sjøsamiske områder – Rettigheter og kulturell betydning. Rapport nr.: 1/2009. Tromsø, Norway: Norut.

Bromley, D. W. 1992. The commons, common property, and environmental policy. *Environmental and Resource Economics* 2:1-17.

Cohen, A. 1985. *The symbolic construction of community.* London: Routledge.

De Young, C., A. Charles, and A. Hjort. 2008. Human dimensions of the ecosystem approach to fisheries: An overview of context, concepts, tools and methods. Food and Agriculture Organization of the United Nations (FAO) Fisheries Technical Paper. No. 489. Rome: FAO.

Grafton, R. Q., R. Arnason, T. Bjørndal, D. Campbell, H. F. Campbell, C. W. Clark, R. Connor, et al. 2006. Perspective: Incentive-based approaches to sustainable fisheries. *Canadian Journal of Fisheries and Aquatic Sciences* 63: 699-710.

Haggan, N., B. Neis, and I. G. Baird, eds. 2007. *Fishers' knowledge in fisheries science and management.* Coastal Management Sourcebooks 4. Paris: UNESCO Publishing.

Hardin, G. 1968. The tragedy of the commons. *Science* 162:1243-1248.

Hay, R. 1998. Sense of place in developmental context. *Journal of Environmental Psychology* 18:5-29.

Jackson, J. B., M. X. Kirby, W. H. Berger, K. A. Bjorndal, L. W. Botsford, B. J. Bourque, R. Bradbury, et al. 2001. Historical overfishing and the recent collapse of coastal ecosystems. *Science* 293:629-638.

Kirsch, S. 2001. Lost worlds: environmental disaster, "culture loss," and the law. *Current Anthropology* 42(2):167-198.

Lam, M. E. Valuing cultural property in fisheries: A generational-index for ecosystem-based governance. *Human Ecology Review* (*in review*).

Lam, M. E., and D. Pauly. 2010. Who is right to fish? Evolving a social contract for ethical fisheries. *Ecology and Society* 15(3): 16. http://www.ecologyandsociety.org/vol15/iss3/art16/.

Macchi, M., G. Oviedo, S. Gotheil, K. Cross, A. Boedhihartono, C. Wolfangel, and M. Howell. 2008. Indigenous and traditional peoples and climate change IUCN Issues paper.

Maffi, L. 2001. Introduction: On the interdependence of biological and cultural diversity. In *On biocultural diversity: Linking language, knowledge, and the environment*, ed. L. Maffi, 1-50. Washington, D.C.: Smithsonian Institution.

Mangel, M., L. M. Talbot, G. K. Meffe, M. T. Agardy, D. L. Alverson, J. Barlow, D. B. Botkin, et al. 1996. Principles for the conservation of wild living resources. *Ecological Applications* 6(2):338-362.

Millennium Ecosystem Assessment. 2005. Ecosystems and human well-being: Synthesis. Washington, D.C.: Island Press.

Nilsen, R. 2003. From Norwegianization to Coastal Saami uprising. In *Indigenous peoples: Resource management and global rights*, ed. S. Jentoft, H. Minde and R. Nilsen, 163-184. Delft, The Netherlands: Eburon Academic Publishers.

O'Brien, K., B. Hayward, and F. Berkes. 2009. Rethinking social contracts: Building resilience in a changing climate. *Ecology and Society* 14(2):12. http://www.ecologyandsociety.org/vol14/ iss2/art12/.

Odner, K. 1964. Erverv og bosetning i Komsakulturen. *Viking* XXVIII:117-128.

Pauly, D., and J. Maclean. 2003. *In a perfect ocean: The state of fisheries and ecosystems in the North Atlantic Ocean.* Washington, D.C.: Island Press.

Pauly, D., V. Christensen, S. Guénette, T. J. Pitcher, U. R. Sumaila, C. J. Walters, R. Watson, and D. Zeller. 2002. Towards sustainability in world fisheries. *Nature* 418:689-695.

Pedersen, K. 2002. Mellom multer og mygg, rock og rein. Natur og turliv blant ungdom i samiske og samisk relaterte miljøer. I *Individ, identitet og kulturell erfaring*, red. S. Bjørkås, 161-190. Kristiansand, Norway: Høgskoleforlaget.

Pikitch, E. K., C. Santora, E. A. Babcock, A. Bakun, R. Bonfil, D. O. Conover, P. Dayton, et al. 2004. Policy Forum Ecology: Ecosystem-based fishery management. *Science* 305:346-347.

Pitcher, T. J. 2001. Fisheries managed to rebuild ecosystems: Reconstructing the past to salvage the future. *Ecological Applications* 11(2):601-617.

Pitcher, T. J., and M. E. Lam. 2010. Fishful thinking: rhetoric, reality, and the sea before us. *Ecology and Society* **15**(2):12. http://www.ecologyandsociety.org/vol15/iss2/art12/.

Rockström, J., W. Steffen, K. Noone, Ã. Persson, F. S. Chapin, E. F. Lambin, T. M. Lenton, et al. 2009. A safe operating space for humanity. *Nature* 461:472-475.

Sissenwine, M. P., and P. M. Mace. 2003. Governance for responsible fisheries: an ecosystem approach. In *Responsible Fisheries in the Marine Ecosystem,* ed. M. Sinclair and G. Valdimarsson, 363-390. Rome: FAO and CABI Publishing.

Snyder, R., D. Williams, and G. Peterson. 2003. Culture loss and sense of place in resource valuation: Economics, anthropology and indigenous cultures. In *Indigenous peoples: Resource management and global rights*, ed. S. Jentoft, H. Minde and R. Nilsen, 107–123. Delft, The Netherlands: Eburon Academic Publishers.

Stedman, R. C. 2002. Toward a social psychology of place: Predicting behavior from place-based cognitions, attitude, and identity. *Environment and Behavior* 34(5):561-581.

Tanner, V. 1929. Antropogeografiska studier inom Petsamo-området. 1. Skolt-Lapparna. *Fennia* 49(4).

UNEP/GRID-Arendal, Distribution of Saami dialects and Nordic countries, Arctic indigenous population, *UNEP/GRID-Arendal Maps and Graphics Library*.

Westra, L., P. Miller, J. R. Karr, W. E. Rees, and R. E. Ulanowicz. 2000. Ecological integrity and the aims of the Global Integrity project. In *Ecological integrity: Integrating environment, conservation, and health*, ed. D. Pimentel, L. Westra, and R. F. Noss, 19-41. Washington, D.C.: Island Press.

CHAPTER TWENTY-ONE

PROMOTING ECOLOGICAL INTEGRITY THROUGH AN ALLIANCE OF AN INDIGENOUS NATION AND ITS NEIGHBOURS: A CASE STUDY OF OPPOSITION TO INDUSTRIAL SCALE GAS DRILLING ON HAUDENOSAUNEE LAND

JACK P. MANNO

Giving Thanks

At the 2010 Vancouver meeting of the Global Ecological Integrity Group on which this book is based, I opened my talk with the following:

> Let us bring our minds together as one to greet and thank our Mother the Earth. This morning when we awoke we could see and appreciate that the earth continues to perform her duties as given by the Creator. The waters flowed. The plants soaked up the light. The trees breathed. The critters did what critters do. The birds sang. The breezes rustled. The sun warmed. The moon tugged at the sea. The stars guided. And Creation at this place was fully alive as we came together as the Global Ecological Integrity Group to notice and give greetings and thanks, love and respect to our Mother the Earth.

The idea and tone of this opening derive from the Thanksgiving Address of the traditional Haudenosaunee, People of the Longhouse, otherwise known as the Iroquois Confederacy (HETF 1996). Before and following every important occasion, council meeting or ceremony, the people are asked to bring their minds together to acknowledge with greetings and thanks, love and respect the special gifts provided by each of the elements of Creation. The ceremonial oration is known in the Mohawk language as,

Ohen:ton Karihwatehkwen (The Words That Come Before All Else). It is not a prescribed speech that is always the same. It is shorter or longer, more florid or direct, depending on the occasion. But the elements always remain the same. It is meant to remind everyone that there is a web of life made up of beings that are intrinsically and perpetually dependent on each other. Life is fragile. In the Haudenosaunee Creation story the continuance of life is the result of a game of chance and the odds were heavily against it. Each day that the world goes on is a day for gratitude for each element of Creation, each fellow being who continues to carry out its particular way of life, its intrinsic duties and powers that are considered by the Haudenosaunee to be Original Instructions from the Creator. Among the obligations that human beings have is to give thanks together which, in turn, encourages life to continue.

The Onondaga Nation, the central fire, or capital of the Confederacy, is at the center of New York State. My school, the State University of New York College of Environmental Science and Forestry, sits on a hill overlooking the Onondaga Valley, an ecological and historically rich part of the ancestral homeland of the Onondaga Nation. As a scholar of environmental science, policy and governance it was perhaps inevitable that my thinking would be influenced by the culture and perspective of my Indigenous neighbors. It was assured in the mid-1990s when Onondaga Nation leaders asked me for assistance in responding to an environmental crisis caused by a leaking underground gasoline storage tank contaminating their groundwater. Subsequently, I worked with the Onondaga Nation on a variety of environmental concerns. Through my work and deepening friendships I have come to respect their philosophies, values and perspectives on the natural world. Even before this influence I had long been interested in reforming environmental institutions and policies to be responsive to our best understanding of the workings of ecological systems. The questions I asked were: How can we build communities that flourish without undermining the ecological systems on which all flourishing depend? How can we build institutions that learn from Nature rather than try to tame Nature? How do we make sure our economic cycles mimic natural cycles so that the output of one life-sustaining process feeds the next? These questions brought me to the discipline of Ecological Economics and led me also to the Global Ecological Integrity Group. GEIG asked important questions about how to define, characterize and preserve the integrity of life-sustaining ecological systems and include ecological integrity as a goal of environmental law and policy. As I have come to know more about Haudenosaunee philosophies, ways of life, forms of government and processes for policy making I have begun to explore how they can provide important models for the personal changes and policy reforms required by our environmental crises.

Forming Alliances, Communicating a Haudenosaunee Perspective

In the mid-1970s, recognizing the gathering environmental storm, the Grand Council of Chiefs of the Haudenosaunee Confederacy decided that it was time to speak more publicly about their perspectives and concerns. They authorized a number of their leaders to join with spokespeople from other Indigenous Nations to deliver a message to Non-Governmental Organizations (NGOs) credentialed by the United Nations. The position papers they prepared for that meeting in Geneva were later gathered in a book titled, *Basic Call to Consciousness* (Akwesasne Notes 1977). Arguing that while the national liberation struggles and human rights struggles of the time were important, "What is needed is the liberation of all the things that support life—the air, the waters, the trees—all the things that support the sacred web of life." They argued that "The traditional Native peoples hold the key," and they called for alliances "with the other peoples of the world to assist in our struggle to regain and maintain our ancestral lands and to protect the way of life we follow."

A number of alliances were formed and some continue. In order to understand how alliances between Indigenous Nations and non-Indigenous allies might promote ecological integrity I recently initiated a long-term research project to study groups that have organized as allies to Indigenous Peoples. With some of my students I have begun with a case study about an organization known as the Neighbors of Onondaga Nation (NOON). In the case study we are looking at the role of NOON as allies in the effort to bring a Haudenosaunee Perspective to the debate over hydraulic fracturing (hydrofracking) in the Marcellus Shale that is the subject of the remainder of this chapter.

Natural Gas in the Marcellus Shale

The Marcellus Shale gets its name from Marcellus, New York, a small community in Onondaga County where rare outcroppings of the vast shale formation appear. Marcellus shale contains large amounts of fossil natural gas in microscopic seams deep underground. Innovations in drilling known as slick water high volume horizontal hydraulic fracturing, commonly referred to as hydrofracking, have facilitated the extraction of natural gas from deep shale formations by injecting a mixture of water, sand and petroleum distillates and other potentially toxic compounds under high pressure to form cracks in the shale thereby releasing natural gas (NYSDEC 2009; Horwitt 2010).

A large portion of the Marcellus Shale in New York State lies below the ancestral lands of the Onondaga Nation. This land is the subject of a legal action filed in federal court by the Onondaga Nation and the Confederacy well before hydrofracking became an issue of concern. The lawsuit asks the court to declare that New York State violated federal laws and treaties when it took Onondaga land and the Onondaga therefore continue to be the rightful title holders. According to the Onondaga Nation's leaders the lawsuit is the latest stage of a centuries-long effort to maintain their culture and way of life and to protect and restore the lands and waters of their territory for all its inhabitants, human and non-human alike. Although not a consideration in the original lawsuit, the potential opening of the Marcellus Shale to high volume, high-pressure hydrofracking, the Onondaga Nation's response has been clearly placed within the context of the Land Rights Action[1].

The Onondaga Nation website contains analysis titled, "You Can't Drink Gas!! THE ONONDAGA NATION & H.E.T.F. Oppose Gas Drilling By Hyrdo-Fracking." It begins with language that also forms the preface to the land rights action:

> "The Nation and its people have a unique spiritual, cultural and historic relationship with the land, which is embodied in the *Gayanashagowa*, the Great Law of Peace. . . . The people are one with the land and consider themselves stewards of it. It is the duty of the Nation's leaders to work for a healing of this land, to protect it and to pass it on to future generations."

It continues with,

> "This obligation to act as stewards of the land and the waters is being fulfilled in the Nation's work to stop the dangerous natural gas drilling method known as hydrofracking. Around 1600 leases have already been signed by landowners in Onondaga County to gas drilling companies and some are just outside the Nation's currently recognized territory. The Nation is very concerned with the dangers that will be created to the waters in general and to the Nation's drinking water system in particular" (Onondaga Nation 2010(a)).

NOON and Onondaga Nation Ally Against Hydrofracking

Before the Onondaga land rights action was formally filed, NOON had formed as a project of a local peace organization in hopes of avoiding the sort of White backlash seen in neighboring regions subject to Indian land claims. NOON published materials explaining the Onondaga land rights action, the history behind it, and reasons communities within the boundaries of the Onondaga ancestral homeland should welcome and support it.

The goals of NOON are to:

- Promote understanding of, and respect for, the Onondaga people, history and culture within the Central New York community;
- Provide accurate information about the Onondaga Nation's current issues of concern, such as their Land Rights Action;
- Challenge racism towards the Onondaga people through education, building relationships, and encouraging shared experiences between the people of our Nations;
- Support, and collaborate with, the Onondaga Nation in their initiatives to promote environmental healing and restore respectful relationships between the governments of our Nations; and,
- Advocate for just and fair treatment of the Onondaga people at all levels of our own government" (NOON 2010).

Given these goals it was natural for NOON and the Onondaga Nation to collaborate in opposing hydrofracking in the Onondaga ancestral homeland. The southern tier of New York State borders on parts of northern Pennsylvania where the boom in shale gas was underway. In 2009, activists from New York's Southern Tier concerned about the effects of rapid rural energy development they were witnessing invited New York environmentalists and the HETF to a tour of drilling pads and homes where domestic water wells had been contaminated by the drilling. Representatives of the HETF, and others from Onondaga Nation, went to Radford Pennsylvania on a fact-finding mission to visit with homeowners who were affected by hydrofracking. NOON came to see hydrofracking in the context of its support for the Onondaga Land Rights Action. Members of NOON perused County Court records and documented over 1300 gas leases in Onondaga County. They produced a map showing the distribution of the leases that was published on the front page of the local newspaper, The Syracuse Post-Standard (Goldberg 2009). Lyndsay Speer (2010), the Onondaga legal counsel's environmental organizer, has written that, "suddenly it [hydrofracking] was no longer a 'Southern Tier' issue. People began to pay attention. Combined with the hundreds of grassroots groups that have formed downstate, on the Southern Tier, and in

Pennsylvania, and with the assistance of the big environmental groups such as Environmental Advocates of NY, hydrofracking has become a major issue."

In March 2009, the Haudenosaunee Environmental Task Force (HETF) released a statement on hydrofracking that stated, "every part of the natural world is important and interrelated; when humans tinker more and more with the natural balance, we do so at the peril of our grandchildren." The statement closes by asking NYSDEC to partner with the HETF to find alternative energy sources that, "do not destroy our grandchildren's ability to live long and healthy lives." In November, the traditional Chiefs of the Haudenosaunee Council banned hydrofracking on their lands:

> "The Haudenosaunee will not allow hydrofracking on or near their aboriginal territory, and call on the Government of New York State to similarly ban hydrofracking and other unconventional gas drilling methods within New York State... we do so for the future of all our relations" (Onondaga Nation 2009).

The statement expresses several components of the Haudenosaunee worldview relevant to the controversies surrounding hydrofracking that can be summarized as including:

- Native sovereignty and treaty rights require Nation to Nation consultation and agreement,
- The territory under consideration is peopled by many non-human beings that are considered relatives,
- Water is sacred, "the first law of life," and humans have special duties given by the Creator to protect water.
- Environmental decisions are moral decisions that require deliberation over a long period of time to consider the unintended consequences of similar past decisions and the potential consequences for future descendants at least to the seventh generation.
- While New York State officials believe that gas companies have a legal right to obtain natural gas and the NY DEC has a legal duty to permit and regulate the practices, the Haudenosaunee have a legal right through treaties to carry out their sacred duties to care for and protect the land and waters.

In addition to these broad principles, the Onondaga Nation press release states that "two major concerns were raised (by the Haudenosaunee Environmental Task Force) that are unique to the Indian Nations: (a) horizontal drilling under Indian territories will be a violation of treaty protected mineral rights; and (b) the proposed regulatory system will not have any mechanism for the protection of cultural resources of the Nations: sacred sites, unmarked burial sites and former village and other archeological sites" (Onondaga Nation 2010). The HETF concluded that despite the best efforts of the state environmental agency to regulate the gas industry's practices, unintended consequences from the hydrofracking process would inevitably damage the environment long into the future (HETF 2009).

The HETF statement concluded that a meeting between the Haudenosaunee delegation and representatives of the state environmental agency was successful in that their concerns were heard and treated with respect but the state agency, despite its good intentions, is institutionally incapable of considering and acting on the HETF's specific issues and their overall perspective on unconventional gas drilling. The press release reported that the delegation believed, "the DEC lacks the authority to protect the environment or its people by identifying an industry as inherently detrimental and banning it. The best they can do is regulate and fine polluters when the inevitable accidents occur; their hands are tied by policy set higher up in the government" (Onondaga Nation 2009). According to HETF members present, the Commissioner told them that, "The Creator put riches above the earth, and he put them below the earth for the benefit of the people" (personal communication). While clearly the Commissioner was attempting to communicate in a language he believed he shared with the Haudenosaunee, the idea that the riches of the earth were placed here to benefit people contrasts sharply with a responsibility-based worldview common to Indigenous perspectives (Corntassel 2008; Manno and Powless 2008).

In its communication with NY State concerning hydrofracking the Haudenosaunee are contesting the state's right and asserting its own duties under Haudenosaunee law, the Great Law of Peace (*Gayanashagow).* The Great Law is a comprehensive orally transmitted Constitution involving all aspects of Haudenosaunee collective life, it specifically includes and reinforces an ethic of responsible resource management, a perspective of respect and gratitude toward the natural world, a requirement to consider the impacts of decisions on future generations (those "whose faces are coming from beneath the earth") and the assignment of caretaking responsibilities.

The global movement for sustainable forms of community and economic development requires a similar institutionalization of environmental responsibility and accountability into all government decision-making

(Walker et al 2009). It was also the original intent of the US Environment Protection Act that broadly required impact assessments for all decisions affecting the environment.

Among the tasks of environmental law is to balance individual rights to use, enjoy and benefit from one's own property with positive responsibilities toward collective wellbeing, which must include the ecological foundations on which all wellbeing ultimately depends. Environmental law thus understood requires "the fulfillment of responsibilities that go far beyond the reciprocal legal duties owed to other individuals," as Prue Taylor (2010:200) insists in her essay on the German philosopher, Hans Jonas', "The Imperative of Responsibility" (1984). This degree of responsibility would necessarily set limits and standards on individual or corporate resource entitlements.

New York State's process for determining whether hydrofracking in the Marcellus Shale will be permitted and how it will be regulated is taking place within the context of an environmental legal system where the only countervailing duties to property rights are those required to protect the rights of others. Often under such a system, environmental agencies see their task as finding a balance, a kind of Pareto optimization, among the rights claims of competing interests in the context of a responsibility, a duty even, to promote economic growth. The resulting wholesale commoditization of land and resources has been devastating to Indigenous ways of life (Manno 2010). Under such a system, the interest of the Haudenosaunee in carrying out its responsibilities to protect future generations and other forms of life by sustaining healthy and resilient ecosystems will inevitably receive short shrift, unless either the Haudenosaunee Nations can successfully require the state to acknowledge and respect their treaty rights, which they understand as being their right to carry out their duties under their own laws and under Natural Law, or NOON and other allies are successful in reforming the dominant society's environmental laws to perpetuate and enforce environmental responsibility.

Taylor (2010) categorizes three approaches to mitigating the current neglect of responsibilities in environmental law. First, expand the range of activities to which environmental laws apply; second, grant specific rights to species or ecosystems or, to Nature as a whole, and third extend moral consideration to Nature. Taylor argues that the first two extend the "rights" framework to include new claims to rights and as a result can lead simply to ever more complex environmental law and inevitably to inclusion of the environment as another interest competing for resources against alternative uses. Taylor imagines a legal system based on the third approach, of expanding the range of moral responsibilities by creating positive duties in law requiring individuals and organizations to guard and protect ecological integrity. Such a mandate could be satisfied through the creation of multiple forms of trusteeship or guardianship that recognize a direct relationship of reciprocal care between humanity and nature, as the Haudenosaunee insist. Concepts of trusteeship and guardianship that create obligations to care and duties to protect the well being and future capabilities of those under one's care already exist in Western legal systems, most relevantly, under US Indian law. Recognizing legally that ecosystems are under our care also creates a relationship of reciprocity if it is combined with a recognition that we are equally under the care of Nature, that we are dependent on the balance and integrity of the ecosystems of which we are a part and our fates our inextricably intertwined.

Zoltan Grossman (2005:24) reports in his study of Native/Non-Native alliances that alliances sometimes grow out of a mutual recognition that some outside threat exists. These alliances sometimes also mature into "more equal and stable political, economic and cultural links between the communities when the allies are successful in developing a shared contemporary sense that the territory on which they reside is also a place they share as a common home." Grossman refers to the building of a shared territoriality as "geographies of inclusion" in which "different groups might mutually include each other within a territorially defined place they view as a common home."

It remains to be seen whether the controversies over hydrofracking will continue the pattern of Native and non-Native pro-environmental alliances or whether the relationship with the surrounding community will become one of conflict over resources and money, and disagreement among environmentalists. The Onondaga leadership and NOON have taken a strong stand against the new drilling technology. The County has banned such drilling on land it owns. There have been numerous public meetings to discuss the threat of hydrofracking, including as part of NOON's educational series on the Onondaga Land Rights action. NOON issued a press release in November 2009 calling for a ban on permitting gas wells that use horizontal drilling and hydraulic fracturing. NOON stated that it took this position "after hearing from friends at the Onondaga Nation that their concerns on this issue went unheeded by the NYS Department of Environmental Conservation."

Landowners have signed thousands of leases; several near the currently recognized Onondaga Nation territory. The potential for conflict appears considerable and may either strengthen or threaten the geographies of inclusion that NOON and the Onondaga Nation have begun to co-create. Whether the alliance between NOON and the Onondaga Nation is successful in its attempt to prevent hydrofracking remains to be seen. Meanwhile, alliances between Indigenous Peoples and their allies appear to be growing worldwide. They have the potential to introduce a powerful imperative of responsibility into decision-making processes that affect the environment.

References

Akwesasne Notes. 1977. *Basic Call to Consciousness.* Summertown, Tennessee: Native Voices.

Corntassel, Jeff. 2008 *Toward Sustainable Self-Determination: Rethinking the Contemporary Indigenous-Rights Discourse* Alternatives 33, pp 105–132.

Goldberg, Delon 2009. "Land agents scour Central New York seeking rights to drill for natural gas." *Syracuse Post-Standard*, July 18. Retrieved April 20, 2010 (www.syracuse.com/news/index.ssf/2009/07/land_agents_scour_centra l_ new.html).

HETF (Haudenosaunee Environmental Task Force) 1996. *Words That Come Before All Else; Environmental Philosophies of the Haudenosaunee* . Turtle Island: North American Traveling College.

Horwitt, Dusty. 2010. "Drilling Around the Law." *Environmental Working Group Drilling Around the Law Report.* Retrieved April 15, 2010 (www.ewg.org/drillingaroundthelaw).

Kimmerer, Robin W. 2008. "The Rights of the Land: The Onondaga Nation of Central New York proposes a radical new vision of property rights," *Orion* November/December 2008.

Manno, Jack 2010 "Commoditization and Oppression: A Systems Approach to Understanding the Economic Dynamics of Modes of Oppression." *Ecological Economics Reviews, Annals of the New York Academy of Sciences*, Volume 1185 pp. 164-178.

—. 2010. "Haudenosaunee Great Law of Peace: A Model for Global Environmental Governance?, Ch. 9 in Democracy, Ecological Integrity and International Law, eds. J. R. Engel, L. Westra and K. Bosselman. Cambridge Scholars Publishing.

Manno, Jack P. and Chief Irving Powless, Jr. 2008. "Brightening the Covenant Chain: The Onondaga Land Rights Action and Neighbors of Onondaga Nation." Pp. 149-158 in *Governance for Sustainability - Issues, Challenges, Successes,* edited by. K. Bosselman, R. Engel, and P. Taylor. Gland, Switzerland: IUCN.

NOON (Neighbors of the Onondaga Nation) 2010. *"Neighbors of the Onondaga Nation: Working in Solidarity with the Onondaga Nation."* Retrieved January 15, 2010 (www.peacecouncil.net/NOON/articles/culture1.html).

NYSDEC (New York State Department of Environmental Conservation. 2009. "Draft Supplemental Generic Environmental Impact Statement (DSGEIS) Relating to Drilling for Natural Gas in New York State Using Horizontal and Hydraulic Fracturing" Retrieved January 15, 2010. (www.dec.ny.gov/energy/58440.html).

Onondaga Nation 2010. "You Can't Drink Gas!! THE ONONDAGA NATION & H.E.T.F. Oppose Gas Drilling By Hyrdo-Fracking," *Onondaga Nation Website.* Retrieved May 8, 2010. (www.onondaganation.org/news/ 2010/2010_0122.html).

Sibley, David. 1995. *Geographies of Exclusion: Society and Difference in the West*. London: Routledge.

Speer, Lindsay. 2010. "Haudenosaunee Leadership" ONONDA?GEH ONGWAWENNA? *(Onondaga Nation Community Newsletter).* Retrieved October 10, 2010 (https://docs.google.com/fileview?id=0B8Kc5AyHrUwoNj kxZjUyOTItOGE1Zi00Y2MwLTgxNzgtMjgwOWJmNjlhY2Ni&hl=en&authkey=CM7xqJoB).

Taylor, Prue. 2010. "'The Imperative of Responsibility' in a Legal Context: Reconciling Responsibilities and Rights" pp. 198-225 in Engel, J.R, Westra L, and K. Bosselman (eds), *Democracy, Ecological Integrity and International Law,* Cambridge, UK: Cambridge Scholars.

Walker, Brian. et al. 2009. "Looming Global-Scale Failures and Missing Institutions," *Science,* Vol. 325, 11 September 2009, pp.1345-1346.

Notes

[1] The lawsuit was dismissed by the 5th circuit court in Albany, NY on September 22, 2010. The decision of Judge Kahn can be read at www.peacecouncil.net/NOON/OnonLRtsDism9-22-10.pdf.

According to Onondaga leaders (personal communication) they are preparing an appeal.

PART VI

GLOBALISATION AND PUBLIC HEALTH: ON SCIENCE AND THE HUMAN RIGHT TO HEALTH

INTRODUCTION

COLIN L. SOSKOLNE

Science provides a systematic body of evidence intended to serve as the rational basis for policy formulation. But, when ideological forces drive the global health agenda, the rightful role that evidence can play is thwarted. Unless scientists are true to their mission of pursuing the public interest over any other interest, the scientific enterprise can be marginalized and the interests of those bent on maintaining the *status quo* will be permitted to dominate.

Two examples are presented in the first two chapters in this part of the book. These show how failing to act works to the detriment of the health and well-being of both present and future generations. These chapters highlight questions of the human right to health and inter-generational equity. In the final of the three chapters in this part of the book, new technologies and approaches in the applied science of cancer epidemiology could serve to untangle the connections between environmental determinants and epigenetic phenomena to better protect human health in a degrading world.

First, Donald Spady, in his thoughtful essay (Chapter 22), anticipates the nature of life for people as global ecological integrity declines and Earth's resources, particularly fossil fuels, decline with costs climbing out of reach of most individuals, even in the developed world. He argues that the environmental, physical, economic, and social infrastructure essential to today's society will become increasingly fragile and prone to failure. As a result, public health will be challenged to perform its preventive functions and, on the primary care front, hospitals will be hard-pressed to remain viable. Spady appeals to governments and business leaders to implement policies that will help societies globally to adapt to a new reality that acknowledges their dependence upon ecological integrity as well as on the human constructs of society. He points to a collective denial that is not allowing leaders to act responsibly in protecting the health and well-being of both present and future generations. The human right to health is being undermined by a societal disregard and inability to implement actions that science has been suggesting for decades.

Second, in a case study out of Sydney, Nova Scotia, Canada, Colin Soskolne and Shira Kramer (Chapter 23) reveal the wanton disregard for environmental safeguards to the detriment of environmental and community health over many decades. Sydney, Nova Scotia, remains one of Canada's worst industrially contaminated sites. The ownership of coal mining, coking and steel industries has been held in more recent decades by provincial and federal authorities whose jurisdiction also includes public health. This chapter exposes conflicting interests and raises profound questions of suppression bias and social injustice in Canada. Justice now is being sought in Nova Scotia's first environmental class action law suit, a process that has taken some seven years thus far and, at time of writing, the class action certification has yet to be allowed. The latter is needed in order for the case to proceed on its merits. Meanwhile, justice in determining the community's right to a healthy environment remains in the balance.

Third, in Chapter 24, Vladimir Bencko presents a relatively more positivist view of the future. Recognizing the challenges faced through declining ecological integrity, Bencko identifies hopeful new approaches in cancer epidemiology research that may help this science in fulfilling its role in the prevention of community health harms. He points to study designs that are prospective and multi-centred that would have the statistical power needed to untangle the connections between environmental determinants and epigenetic phenomena that permit disease processes to advance. By discovering such connections, the sustainability of human health will be able to be more assured through the application of evidence-based interventions that could protect human health. In this way, epidemiology, as the science that drives public health decision-making, will be more able to contribute to assuring the public of its right to health in a degrading world.

As globalisation persists, this Part of the book underscores the systemic flaws in governance that undermine the right to health, not only for present generations, but also of future generations. As we adapt to ecological declines, we are encouraged that the applied science of cancer epidemiology could generate the knowledge needed to better ensure the sustainability of human health into the future, at least with respect to cancer prevention and control.

CHAPTER TWENTY-TWO

PUBLIC HEALTH IN A TIME OF RESOURCE DEPLETION AND ECOLOGICAL DISINTEGRITY

DONALD W. SPADY

Introduction

Earlier chapters in this book have described the importance of ecological integrity to living systems and how disordered ecological integrity threatens Earth's life support systems. Disordered ecological integrity also threatens human health. In this chapter I explore how the environment plays a key role as part of the infrastructure of a public health service, and how ecological disintegrity threatens the ability of public health services to meet their obligations. I define Public Health and briefly outline its core roles. I argue that the successful implementation of these roles depends on a complex social, cultural, physical, and environmental infrastructure, all based on a series of assumptions. However, because of ecological disintegrity, the depletion of fossil fuels and Earth's resources, and the social, economic, and physical consequences that arise as a result, the infrastructure upon which we build and run our societies may well fail, and many assumptions we use to plan and live our personal and professional lives are false. To minimize the effects of the inevitable change that will arise from these problems, public health must make immediate plans for a transition to a simpler, more local, independent existence.

Many people think of the environment in terms of mountains, seashores, rivers, or lakes; as sources of rest and recreation. The environment as a determining force in their lives may become apparent after events like the 2010 heat wave in Russia that killed thousands of people, or a human induced disaster that seriously harms the natural world, such as the Macondo oil-well blowout in the Gulf of Mexico. Only rarely do we consider the environment as the primary provider for all life of shelter, clean air, clean water, fertile soils and other essentials. Instead of thinking of the natural world as the maintainer of all life on Earth, we mine its "endless" resources and exhaust its services. But nothing is truly endless, and now those services and resources nature provides are beginning to fail; and our survival as an advanced human society is thus threatened.

This threat takes many forms: as an economic threat, as a threat to our culture, our institutions, our security, our ability to grow food, to fish, to build safe, warm, shelter, our ability to maintain our health and to treat our illnesses. Because our ability to treat illness is threatened, and will become even more compromised, we will need to work actively to maintain our health. This is the task of Public Health.

Public Health Defined and Described

John Last defines public health as "*An organized activity of society to promote, protect, improve, and, when necessary, restore the health of individuals, specified groups, or the entire population. ... It is a way of thinking, a set of disciplines, an institution of society, and a manner of practice*" (Last 2007). This definition emphasizes populations, not individuals, and the actions of public health are directed to maintaining health, not treating illness.

Table 22-1. Core Functions of Public Health (Chambers & Sullivan 2007)

Function	Examples of Task
Health Promotion	Addresses public education and health advocacy issues, maternal & child health, chronic disease
Health Protection	Food safety, enforcement of laws and regulations, occupational health
Disease & Injury Prevention	Immunization programs, Advertising campaigns (bicycle helmets), workplace safety
Health Surveillance	Monitoring prevalence and incidence of specific diseases (TB, AIDS, sexually transmitted disease), air & water quality, pollution
Population Health Assessment	Monitors vulnerable populations, immigrants, homeless, health surveys.

The core functions of public health are summarized in Table 22-1. They are not mutually exclusive, and they subsume many other tasks. Public health maintains and monitors the health of the population, detects hazards, advises governments and private citizens, and plans, implements, and evaluates appropriate responses to threats. As they act, public health services integrate their actions with other agencies and institutions so that affected individuals can receive appropriate care.

Like many ecosystem services, public health works in the background to ensure human health. Unless there is a public health emergency, such as an epidemic, when the public health service takes a higher profile, it functions in the background of society, and success for a public health service is when nothing happens. This differs from the medical care system which treats the ill and injured individual, costs more in money and resources, and is often in the public eye.

The Requirements of a Public Health System

An effective public health system depends on an infrastructure composed of people, materials, attitudes, and services. Infrastructure generally means physical structures, like roads and buildings, but this can be broadened; thus Baker et al. refer to a public health "infrastructure of workforce, informational systems, and organizational capacity" (Baker et al. 2005). The definition used here is even broader; stressing also the importance of society and culture and the environment. Although they differ in form, structure, and function, each component is necessary to maintain the public's health, and together they form a structure upon which an effective public health system must depend to do its task.

A. Social and Cultural Infrastructure

People are always the key component of any human system, and a public health service requires motivated, well qualified professionals and support personnel. It needs doctors, nurses, epidemiologists, statisticians, and investigators, plus lawyers, accountants, and administrators. Support staff are needed to maintain records, communication and information systems, buildings, vehicles, and equipment. It is clear that while much of the above, and more, is essential to an effective public health system, it is not necessarily specific to public health.

Good public health also depends on an effective political system forming a supportive stable government that provides the necessary funds and creates the necessary legislation to enable public health action. It also depends on an educated populace that has good general health practices and understands why public health is important. Public health also depends on an economy that can generate the tax revenue needed by the government to fund adequately the public health system and its other obligations. This economy also has the financial and productive capacity to supply the goods and services needed by public health and at the same time can meet society's basic needs for food, shelter and clothing. Ideally it also meets the needs of education, jobs, entertainment, and culture, factors giving people a sense of purpose and self-worth.

All this happens in the context of the cultures, attitudes, and expectations of the members of society. These abstract qualities provide the perspective and dictate the behaviours that allow a system to function; their absence cripples the remaining infrastructure. In a simpler, more local, and less technically based future, these factors will be critical infrastructure components of any public health system, and of any society.

B. Physical Infrastructure

The physical infrastructure is the technical and physical fabric that defines today's modern societies. Almost everything we ride or fly, enter, or live or work in, constitutes part of this fabric. As well it includes the delivery services for energy in the forms of electrical grids, pipelines, and generating plants. It includes many of the material goods used in industry, in laboratories, in hospitals and in universities, plus many services, such as communication networks and information management.

This description does not describe the specific infrastructure of a public health system, because, with few exceptions, much of this infrastructure is generic. A public health information system uses computers and software, paper, electricity, oil and coal, wood and water; these are not specific to public health, but they are essential to public health. This non-specificity is a strength in that shortages can be resolved by finding another supplier. However, a systemic threat, for example, an energy or petroleum shortage, which would affect a large part of a society's infrastructure, may also threaten the effectiveness of the public health care system.

Finally, infrastructure is not benign, it has many effects on personal and population health. Infrastructure creation can cause migration in or out of an area, can create or destroy employment, can clean or pollute the environment, can alter a community's ability to respond to disaster or opportunity, and can beautify an area or totally destroy it.

C. Environmental Infrastructure

C1. Service Provision

Some readers may question the inclusion of the environment as infrastructure, given that it is inherently natural in origin and, apart from that which man has inferred, has no apparent affection for the human species. However, the fact that many natural resources and services have been appropriated by humanity for its own benefit and economic value is well recognized (Costanza et al. 1997; Daily et al. 1997), thus I think its infrastructure credentials are legitimate. Ecosystems act in an integrated manner within nature to provide critical provisioning, regulating, supporting, and cultural services (Pollard et al. 2010); although we may look at these services through a human eye, they are for the benefit of all life, not just humans. Like public health, ecosystems function in relative anonymity; when things go well, no-one notices, but when they fail, the repercussions can be widespread, dramatic, and even catastrophic.

C2. Resource Provision

Earth provides us what we need to live; animals and plants, water, minerals and their important subset, metals. With these resources humanity has created cultures, civilizations, developed theologies, philosophies, and sciences, built universities and institutions and armies, and has increased its numbers. Some resources were so valuable as to have the ages in which they were developed named after them, thus we had the Bronze Age and Iron Age.

For the past 200 years we have experienced the Carbon Age, an age defined by the use of the fossil fuels coal, oil, and natural gas. Before this time energy was derived mainly from wood, which was burned for heat and for cooking, and used to build ships and buildings. Today, fossil fuels pervade our lives. Their attributes of being concentrated and cheap energy sources enabled the industrial revolution and many of the improvements in living standards that followed (Cottrell 1955). Fossil fuels give us gas and diesel fuel for our cars and kerosene for airplanes. They drive our industries, fuel our power plants, kindle our economy, and provide chemicals and products used daily by virtually everyone, everywhere. Fossil fuels provide the fertilizers, pesticides, and the fuel for agriculture. Global food production, and hence population growth, has increased dramatically, largely due to fossil fuel. Fossil fuels enabled the development of a society urged to consume more and more, and not to repair, but to throw away (Lebow 1955). Cheap transport by air and ship facilitated globalization and hence the practice of outsourcing manufacture to lands with cheap labour and marginal environmental protection.

Comment

This discussion about infrastructure outlines the dependencies of a public health care system, from the obvious ones of a suitable workforce to those less obvious, such as intact physical infrastructure, a stable economy and government, to the rarely acknowledged dependencies of ecosystem health and the availability of resources. All of these enable a public health system to function, but they also are important determinants of individual and population health. When considering the needs of public health, one can not consider only its primary requirements; public health functions within a broader context, and the healthy functioning of this broader context is as important a determinant of the ultimate ability of the public health system to function as are its immediate requirements.

Some Common Assumptions that Guide Society

As well as infrastructure, public health, and also much of society, bases its actions and plans on various assumptions and beliefs regarding the way things are. I discuss some of the more salient of these below.

A. Life is only going to get better

One common assumption is that life will only get better. Our lives are better than our parents' lives were, and we expect our children to live better lives than we do. This has not always been the case, but it has been generally true for the past two hundred years or so, and life before that was so different that any assumptions based on earlier experience probably do not apply.

B. Humans are exempt from nature

Humans commonly believe that we are special, that we are exempt from the laws of nature and physics and that "*Because I can; because I want to*" (a quote I saw on a t-shirt) is true because human ingenuity, creativity, and technology will make it so (Cairns 2009). This attitude is exemplified in Genesis 1:26 where man is created "*...Let us make man in our image: after our likeness,*" and Genesis 1:28, where man is told to "*Increase and multiply, and fill the earth and subdue it and rule over the fishes of the sea and the fowls of the air and all living creatures that move upon the earth.*" The Genesis statements are heavenly edicts; but they have been expanded upon and used to encourage Western civilization to exploit nature. Thomas Berry discusses this interesting assumption in an address to Harvard University (Berry 1996).

C. Business-As-Usual Prevails, indeed, it MUST prevail

The assumption that Business-As-Usual (BAU) will prevail codifies the attitude that the physical, economic, and social infrastructure essential to today's society will be intact and will work as expected. Thus, goods will arrive just-in-time, food will be safe to eat, clean water will flow from the tap, waste will be treated appropriately, electricity will flow from the wire, oil and gas will flow from our pipelines, busses and trains will run on schedule, as will airlines, and the mail will finally deliver that cheque. Businesses, cultural institutions, and hospitals will be open and accessible to the public, and the public will be able to pay for the goods and services provided by these institutions. Security of country, person, and possessions will be reasonably ensured and justice will be effective and fair. All this will function within a society that works through an effective public service directed by a stable, elected government.

Perhaps the most important component of BAU is that the economy must continue to grow. The survival of our debt based economy, and our lifestyle, means BAU **must** prevail; failure leads to recession, unemployment, and hardship. But such a model depends on more and more consumption of goods and services, which derive ultimately from Earth's resources. Hence, along with the BAU assumption is the unstated requirement that the necessary resources be available to meet demand.

D. Nature-As-Usual Will Prevail

We tend to assume that nature will continue as it has throughout the history of mankind; i.e. as well as Business-As-Usual it will be Nature-As-Usual, and the more we distance our lives from nature, the stronger we believe this assumption. The seasons will come and go, the rains will fall, the crops will grow, the waters flow, the oceans provide, plants will flower and insects mature and birds and animals will migrate, all in synchronized rhythm. We plant our crops and harvest them, we hunt and fish, and we enjoy our festivals and celebrations and our holidays, based on the unchanging cycles of nature. With some allowance for the odd natural disaster, we don't really think about nature. We know there will be blizzards, floods, severe storms, and drought, but experience tells us that they will resolve in time and our lives will return to normal. Nature never *really* changes, we can depend on it, can't we?

E. We Know What Can Happen

The final assumption is that our problem sheet includes all the possibilities. In life the unexpected happens, sometimes its good, like winning a small lottery, sometimes we are not sure, like winning a huge lottery, and sometimes it is very bad, like Hurricane Katrina. A big part of a successful Public Health service is planning for the problems, looking at all the possibilities, and working out solutions. It also plans for expected problems with predictable timeframes, such as for heat waves in the summer. It plans for events that are expected, but their timing is uncertain, such as an epidemic of food poisoning. An important caveat to these plans is that the problem is in our lexicon of possibility; if it is totally unforeseen, or denied, we cannot plan for it and our response will probably be inadequate.

"Unforeseen" and "Denial" are not that far apart. For example, many responsible officials said Hurricane Katrina was unforeseen, yet the very real risk New Orleans faced from a hurricane had been predicted in a national magazine at least a year earlier. The ozone hole above Antarctica is an example of an unforeseen but major public health problem. The hole was detected, but ignored because the statistical program analyzing the satellite data had been modified to ignore certain values of ozone which were considered too low. It was 'found' when a theoretical prediction of ozone depletion was being verified using different measurement tools, and the conflicting satellite data were reanalyzed and the hole was confirmed (Rowland & Molina 2000). Today, climate change denial is common and this denial can have profound political and social consequences. For example, climate change denial played a pivotal role in the United States 2010 mid-term elections, and it appears that denial may dictate the direction of American climate change policy for years. However, the science is clear. Climate change is real. We let arrogance and denial trump reality at our collective peril.

To summarize, we live our lives and make our plans based on various assumptions: our children's lives will be better than ours, nature is man's servant, the lights will always turn on, the economy will grow forever, the rains will come, and we know where the wild beasts are. Public Health also makes its plans using these assumptions. However, they are all wrong.

Why are the Assumptions Not Valid?

A. Our natural systems are failing

As we developed and created our civilizations, we grew too fast, extracting too much of Earth's resources, damaging Earth's life-support systems, and throwing too much waste away. We persist today and seem to have acquired the attitude that the Earth is infinite in size and resources and resilience, and that humanity cannot "significantly" affect it; this is good because the leaders of BAU say we cannot afford to change.

But Earth is singular, and finite: it has a finite size, finite resources, finite resilience. This year we are using the resources and services of 1.5 Earths, and by 2030 we will need two Earths to meet our needs (Pollard et al. 2010). In 1900 there were 1.6 billion people on Earth, today, there are close to 7 billion, and there will be billions more by 2050 (Cohen 1995). It is not just the numbers; it is what people consume. My city, Edmonton, Alberta, Canada, in 2008 had an ecological footprint 3.2 times the global average (Anielski 2010); if the world lived as Edmontonians do, we would need 4 Earths to meet our needs. Most population growth is occurring in the third world, and the ecological footprint is much smaller there; but the expectations of a better life, and thus a bigger footprint, are omnipresent. The first world also will grow, but children born there will likely adopt the western "multi-planet" footprint. This unsustainable consumption cannot continue.

The abundance of evidence of man's assault on the Earth boggles the mind. Much of this evidence is summarized in Part I of this book and reiterating it here proves nothing. Suffice to say that our actions have imperiled the fabric of life essential to all living organisms. We are exceeding the planetary boundaries of Earth and the ecosystems and resources that sustain life are failing (Rockstrom et al. 2009; Pollard et al. 2010). Our exploitation of the Earth has triggered the sixth major extinction event in the history of life, the only one due largely to human activity (Chapin et al. 2000). We have exploited our fisheries (Jackson 2008), forests (Gibbs et al. 2010), soil (Bai, Dent, Olsson & Schaepman 2008), aquifers (van Beek, van Kempen, Reckman, Vasak & Bierkens 2010), and rivers and lakes (Gleick & Palaniappan 2010). We continue to pollute our water and soil with minerals, organic matter, and chemicals that persist and act in nature in ways both known and unknown (Hayes et al. 2006; Kelly et al. 2009). We have released foreign species into new areas with no natural predators (Gallaugher & Wood 2003) and in doing so have caused great harm to local ecosystems and economies.

A critical economic and environmental problem relates to the use of fossil fuels. These fuels are essential to today's economy, but the CO_2 released when they are burned has led to the cascade of ecosystem failures generally called global climate change (Stone et al. 2009). Effects of climate change are apparent now, but they will become more obvious and severe as the century proceeds and global temperatures rise even more (Sokolov et al. 2009; Allison et al. 2009). The rate, nature, and extent of the consequences of global change, or of ecosystem failure, are hard to predict; one reason is that they manifest non-linear characteristics and changes can be sudden and catastrophic (Scheffer, Carpenter, Foley, Folke & Walker 2001).

The economic problems associated with fossil fuel consumption relate more to the pending decline in their use, a decline partly due to the need to slow or stop CO_2 release, but mainly because fossil fuels are non-renewable energy sources that are approaching peak production and then will decline. This is most urgent with respect to oil and the phenomenon is called Peak Oil, the situation when global oil production reaches a maximum, plateaus for a time and then declines to relative insignificance (Aleklett & Campbell 2003; Hirsch, Bezdek & Wendling 2006). When oil will peak is uncertain, some think it has already occurred, most of the rest say within ten years. The world is not running out of fossil fuels, but extraction is becoming too hard technically, too expensive, too energy intensive, and too polluting; by 2100, fossil fuels will be effectively marginal sources in global energy supply.

Because of its pervasive economic impact, this loss of fossil fuel energy will cause much economic, social, and human hardship over the coming decades. But, fossil fuels are the single most important source of CO_2 and their continued use is done at significant risk to humanity, so their phase-out is necessary (Kharecha & Hansen 2008). We must adapt to this loss.

Humanity's effects on the Earth are so profound that some people call today's times the Anthropocene (Zalasiewicz, Williams, Steffen & Crutzen 2010), a name that summarizes the evidence of how much humanity has affected the Earth, and diminished its resilience and its ecological integrity. This evidence is a clarion call from Earth to all humans that if they are to sustain themselves as societies, and not just as isolated groups of the species *Homo sapiens,* they must change their lifestyles and their attitudes towards nature. This clarion call also tells humanity that it faces immediate threats to life and health, that in addition to any long-term changes in lifestyles and attitudes, it must prepare for immediate problems that are arising because of ecological disintegrity.

B. The Built Infrastructure is Also Failing

Most of our current infrastructure was built over the last century and much of it is old and fragile and its integrity is compromised due in part to age, overuse, neglect, too little money and too few resources to maintain it or rebuild it. Even today we do not adequately maintain our infrastructure. In 2007, Canada had an overall municipal infrastructure deficit estimated at $123 billion, and a further $115 billion was needed for new infrastructure (Mirza 2007). In the United States in 2007, total infrastructure costs were estimated to be $1.6 trillion; Europe's costs are even more (Thompson, Cullian, Bloomsbury & Englebretson 2008).

Climate change will affect infrastructure integrity through extreme climate conditions and events, such as floods, ice-storms, sea level rise, heat, and drought. Transportation, electricity generation and distribution systems, water and sewage systems, and seaports are examples of major physical infrastructure that will also be affected by climate change (Lemmen, Warren, Lacroix & Bush 2007; CCSP 2008; Karl, Melillo & Peterson 2009).

Reinforcing, repairing, and replacing infrastructure is costly in terms of money and resources. The current global economic situation inhibits great expenditure, even if necessary, and many essential minerals are approaching depletion and others will face this reality within the century (Diederen 2009). As well, we are running out of fossil fuels, which have kindled much of the world's growth, but will soon not be available in the amounts needed, and climate change considerations preclude their continued use anyway. The depletion of these and other resources means that they cannot be used to create or repair infrastructure, which is troubling, since we all must cross bridges, drink clean water, and stay warm or cool. The economy may suffer too, spiraling into depression. Government revenues will drop, and thus the services provided by governments, such as social-security, health-care, research monies, education, defense, may fail or be restricted. Maintaining steady food supplies will be hindered due to the lack of energy for industrial agriculture, climate inconsistency, high transport costs, an increasing population, and peak globalization (Curtis 2009).

When we consider these issues, their enormity, and their consequences for how we must live in the future, all the assumptions we explored earlier fall apart. New assumptions are needed, new value systems required, perspectives changed, economic theories revisited, and desires diminished. For mankind to continue as a society, life will have to become simpler, less demanding of nature, and more caring for nature. Our perspective must change to that of man being a part of nature and not apart from nature.

So, What about Public Health?

This chapter has raised several critical issues. The first issue is that the stresses placed upon Earth by the actions of humanity are unsustainable and, while Earth will survive this crisis, as it has always done, humanity's survival depends on how effectively and quickly it can create and change to a sustainable way of life. A second issue is that the infrastructure that Public Health is so dependent on is at risk of failure, and the ability of Public Health to carry out its mandate could be crippled. Therefore, Public Health should revisit its core functions and the assumptions upon which it makes its plans.

Society is facing a transition. This can be a transition forced upon it as the laws of physics and biology play themselves out, or it can be a planned transition. John Michael Greer, who described one potential course of transition in *The Long Descent* (Greer 2008), more recently wrote that "*...the future after peak oil is a pathless land...*" (Greer 2010). Paths give us direction and hope, and paths can broaden into roads. "*Hope ... is just like roads across the earth. For actually the earth had no roads to begin with, but when many pass one way, a road is made*" (Lu Xun 1881-1936). We need to start making roads.

Each road will be different, governed by local concerns, goals, and resources. In each case, however, building these roads needs individuals with an expert knowledge of what Public Health can do, how it fits into and depends upon society and the environment. Skills such as these are not found in one person, but in teams.

The destination is unclear, but a suitable target may be that described in Galbraith's *The Good Society* (Galbraith 1996), a vision placing primacy on community, not on economic power, inequity, and inequality. Wilkinson and Pickett, in *The Spirit Level* (Wilkinson & Pickett 2009) validate many of Galbraith's arguments by showing that education, employment, food, and reduced inequality can profoundly influence the health of a population. These determinants of health, already at risk, will become more endangered during the transition, and Public Health must ensure that this risk is minimized. Although some would argue that these are political and economic goals, not Public Health goals, they are also critically important determinants of health. Thus Public Health can and should be their potent and vocal advocate; given the infrastructure constraints anticipated, such advocacy may be one of few things it can do effectively.

There will be few resources available to create these roads, thus our expectations as to how they meet the needs of the coming decades will have to be conservative, the destinations tentative and amenable to change, and based on the needs of humanity in harmony with nature. Public Health has functioned in the past on ethical and functional paradigms that are too anthropocentric for tomorrow's needs. Soskolne and Bertollini (Soskolne & Bertollini 1999) stress that Public Health, and health care in general, must create a new view of human relationships with living systems, and incorporate it into current perspectives of health. Difficult ethical decisions may have to be made as to resource use, especially deciding what should be done as opposed to what can be done. Finally, the resources used must be renewable, non-polluting, and in harmony with nature, as we plan and live for the centuries ahead, and not just for today.

Building these roads has no fixed schedule, but we must start now. There will be detours and delays, and these will affect our health. We will face injury and illness as the economy slips, jobs wither away, money needed for food, shelter, and clothing shrinks, our resources deplete or become too dear, and natural disasters occur. The transition will likely wax and wane, with shock interspersed by stability, but that is just a guess. Society will have to adapt to these cycles at the same time as it relearns and learns both old and new but simpler ways of living, and provides for its physical and mental needs. It may face the angst of knowing "what was", and maybe even have a systemic—unwarranted—sense of guilt for wasting and destroying so many resources and leaving so little to its children.

Some will view this picture with horror, others with a sense of challenge; challenge is more fruitful, more fun, and the only viable option. We have a rich cultural and scientific legacy, much of which we can still use. And, while diminished, technology is not lost. It can help us develop renewable energy that can let us maintain some aspects of our lifestyles. It can deliver some services (including critical aspects of healthcare), and help maintain our culture, but it is not a panacea and we will have to consume carefully and conserve well.

As we build these roads, time will pass, climate change will grip the world, resource depletion will squeeze society, and the infrastructure will be increasingly tested, thus Public Health will find meeting its core tasks incrementally difficult. The structure of problems will be more systemic, acting not only locally, but nationally or even internationally, and maybe all at the same time, thus there may be fewer places to turn for help, and the assistance scant. While anticipated, many problems will be unprecedented, such as dealing with the consequences of sea level rise, or a food crisis secondary to a failing agriculture infrastructure. Most importantly, although some will be more affected than others, all of society will feel the effects. So, with declining energy and natural resources, failing infrastructure, reduced economic support, and facing climate change, social transition, refugees from climate change or other disasters, possibly a demoralized public, and likely a politically unstable society, Public Health will have more demands placed upon it and less ability to deal with them. What can it do? I don't know.

But, humans are the only species that can anticipate the future, that can think through a problem, and can plan a complex response that takes decades to implement and to see the expected effect; we are also the only species that has been so destructive to Earth. Let us use our unique abilities to increase our chances of societal survival by being courageous enough to acknowledge that we face a major Public Health emergency. Then let us define the components of this emergency adequately enough to start to respond. Let us determine what knowledge and resources we need and the likelihood of their being available to address the problems we expect, while maintaining ecological integrity.

Having defined the problems, we must educate our politicians that a healthy populace is much more resilient and productive than a sick one, that maintaining population health is cheaper than providing acute health care, and thus the actions of Public Health must be enabled and supported as a high priority within government. We must also educate the public as to why they are experiencing a transition which to many will be unexpected, that life will be difficult, but that in the end, life could be better than today; different for sure, but not necessarily worse. If we do these things, then we will likely traverse this transition in far better shape than if we just continue with Business-As-Usual, which, with few exceptions, unfortunately seems to be the current plan.

This planning cannot be postponed. Global climate is changing rapidly and unpredictably. A "wait and see" attitude is irresponsible and the apparent denial, dithering, posturing, and procrastinating of national governments demoralizing. It is frustrating to realize that much of what I have written is old news to leading politicians and business leaders, even if they do not like the message. It is the kind of news that is hard to accept or to act upon because it implies the need for major changes in the lives and businesses at all levels of society. For the past 200 years or so, every generation has felt that their way of life and level of comfort will be better than that of their forebears, and this has been because of the incremental access to cheap, abundant, portable, energy. Now it needs to change. We have no choice, we must grow up and accept the challenge of our time.

References

Aleklett, K., & Campbell, C. J. (2003). The peak and decline of world oil and gas production. *Minerals and Energy - Raw Materials Report Minerals and Energy - Raw Materials Report*, *18*(1), 5-20.

Allison, I., Bindoff, N. L., Bindschadler, R. A., Cox, P. M., De Noblet, N., England, M. H. et al. (2009). *The Copenhagen diagnosis. Updating the world on the latest climate science.* Sydney, NSW, Australia: University of New South Wales Climate Change Research Centre (CCRC).

Anielski, M. (2010). *Edmonton's ecological footprint* (12). Edmonton, AB: The City of Edmonton. Retrieved from http://thewaywegreen.ca/discussion-papers/.

Bai, Z. G., Dent, D. L., Olsson, L., & Schaepman, M. E. (2008). *Global assessment of land degradation and improvement 1. Identification by remote sensing* (2008/01). Rome/Wageningen: FAO/ISRIC. www.fao.org.

Baker, E. L., Potter, M. A., Jones, D. L., Mercer, S. L., Cioffi, J. P., Green, L. W. et al. (2005). The public health infrastructure and our nation's health. *Annual Review of Public Health*, *26*(1), 303-318.

Berry, T. (1996, 04/9/1996). *Ethics and Ecology*. Proceedings from Harvard Seminar on Environmental Values, Cambridge, MA.

Cairns, J. (2009). Sustainability, exceptionalism, and exemptionalism. *Ecosystem Health*, *7*(3), 147-154.

CCSP. (2008). *Impacts of Climate Change and Variability on Transportation Systems and Infrastructure: Gulf Coast Study, Phase I. A Report by the U.S. Climate Change Science Program and the Subcommittee on Global Change Research.* (4.7). Washington, DC: Department of Transportation.

Chambers, L. W., & Sullivan, S. M. (2007). Reflections on Canada's public health enterprise in the 21st century. *Healthcare Papers*, *7*(3), 22-30.

Chapin, I. I. I., F. Stuart, Zavaleta, E. S., Eviner, V. T., Naylor, R. L., Vitousek, P. M., Reynolds, H. L. et al. (2000). Consequences of changing biodiversity. *Nature*, *405*(6783), 234-242.

Cohen, J. E. (1995). *How Many People Can the Earth Support.* New York, London: W.W. Norton & Company.

Costanza, R., d'Arge, R., de, G., Rudolf, Farber, S., Grasso, M., Hannon, B. et al. (1997). The value of the world's ecosystem services and natural capital. *Nature*, *387*, 253-260.

Cottrell, F. (1955). *Energy and society. The relation between energy, social change, and economic development.* Westport, Connecticut: Greenwood Press, Publishers.

Curtis, F. (2009). Peak globalization: Climate change, oil depletion and global trade. *Ecological Economics, 69*, 427-434.

Daily, G. C., Alexander, S., Ehrlich, P. R., Goulder, L., Lubchenco, J., Matson, P. A. et al. (1997). Ecosystem services: Benefits supplied to human systems by natural ecosystems. *Issues in Ecology, 2*(Spring 1997), 1-16.

Diederen, A. (2009). Minerals scarcity: A call for managed austerity and the elements of hope. http;//europe.theoildrum.com/node/5239.

Galbraith, J. K. (1996). *The Good Society: The Humane Agenda.* New York, NY: Houghton Mifflin Co.

Gallaugher, P., & Wood, L., editors. (2003). *Water and the future of life on earth* (05/22-24/2002, Trans.). Vancouver, BC: Simon Fraser University.

Gibbs, H. K., Ruesch, A. S., Achard, F., Clayton, M. K., Holmgren, P., Ramankutty, N. et al. (2010). Tropical forests were the primary sources of new agricultural land in the 1980s and 1990s. *Proc Natl Acad Sci, 107 online 08/31/2010*, doi:10.1073/pnas.0910275107.

Gleick, P. H., & Palaniappan, M. (2010). Peak water limits to freshwater withdrawal and use. *Proc Natl Acad Sci, 107*(25), 11155-11162.

Greer, J. M. (2010). A pathless land. Retrieved 06/23/2010, 2010 from http://thearchdruidreport.blogspot.com/2010/06/pathless-land.html.

Greer, J. M. (2008). *The Long Descent. A User's Guide to the End of the Industrial Age.* Gabriola Island, VC: New Society Publishers.

Hayes, T. B., Case, P., Chui, S., Chung, D., Haeffele, C., Haston, K. et al. (2006). Pesticide mixtures, endocrine disruption, and amphibian declines: Are we underestimating the impact? *Environ Health Perspect, 114 Suppl 1*, 40-50.

Hirsch, R. L., Bezdek, R., & Wendling, R. (2006). Peaking of world oil production and its mitigation. *Aiche Journal, 52*(1), 2-8.

Jackson, J. B. C. (2008). Ecological extinction and evolution in the brave new ocean. *Proc Natl Acad Sci U S A, 105, Suppl 1*, 11458-11465.

Karl, T. R., Melillo, J. M., & Peterson, T. C. (2009). *Global climate change impacts in the United States.* New York, NY: Cambridge University Press. www.globalchange.gov/usimpacts.

Kelly, E. N., Short, J. W., Schindler, D. W., Hodson, P. V., Ma, M., Kwan, A. K. et al. (2009). Oil sands development contributes polycyclic aromatic compounds to the Athabasca river and its tributaries. *Proc Natl Acad Sci, 106*(online). doi:10.1073'pnas.0912050106.

Kharecha, P. A., & Hansen, J. E. (2008). Implications of "peak oil" for atmospheric co2 and climate. *Global Biogeochemical Cycles*, *22*(GB3012), 1-12.

Last, J. M. (2007). *A dictionary of public health.* New York: Oxford University Press, Inc.

Lebow, V. (1955). Price competition in 1955. *Journal of Retailing*, *31*(1), 5-11.

Lemmen, D. S., Warren, F. J., Lacroix, J., & Bush, E. e. (2007). *From Impacts to Adaptation: Canada in a Changing Climate 2007.* Ottawa: Natural Resources Canada. http://adaptation2007.nrcan.gc.ca.

Mirza, S. (2007). *Danger ahead: The coming collapse of Canada's municipal infrastructure. A report for the Federation of Canadian Municipalities.* Ottawa, ON: Federation of Canadian Municipalities. www.fcm.ca.

Pollard, D., Almond, R., Duncan, E., Grooten, M., Hadeed, L., Jeffries, B. et al. (2010). *WWF Living Planet Report 2010* (2010). Gland, Switz: Worldwide Fund For Nature. www.panda.org.

Rockstrom, J., Steffen, W., Noone, K., Persson, A., Chapin III, F. S., Lambin, E. et al. (2009). Planetary boundaries: Exploring the safe operating space for humanity. *Ecology and Society*, *14*(2), 32(online).

Rowland, F. S., & Molina, M. J. (2000, 12/07/2000). *The CFC-Ozone Puzzle: Environmental Science in the Global Arena*. Proceedings from John H. Chafee Memorial Lecture on Science and the Environment, National Academy of Sciences, Washington, DC.

Scheffer, M., Carpenter, S., Foley, J. A., Folke, C., & Walker, B. (2001). Catastrophic shifts in ecosystems. *Nature*, *413*(6856), 591-596.

Sokolov, A. P., Stone, P. H., Forest, C. E., Prinn, R., Sarofim, M. C., Webster, M. et al. (2009). Probabilistic forecast for twenty-first-century climate based on uncertainties in emissions (without policy) and climate parameters. *Journal of Climate*, *22*(19), 5175-5204.

Soskolne, C.L. & Bertollini, R. (1999). Global Ecological Integrity and 'Sustainable Development': Cornerstones of Public Health: A Discussion Document. World Health Organization, European Centre for Environment and Health, Rome Division, Italy. 74 pages.

Stone, D. A., Allen, M. R., Stott, P. A., Pall, P., Min, S.-K., Nozawa, T. et al. (2009). The detection and attribution of human influence on climate. *Annual Review of Environment and Resources*, *34*, 1-16.

Thompson, C., Cullian, J., Bloomsbury, A., & Englebretson, R. (2008). *The problem is more than money. Global infrastructure crisis.* Deloitte LLP.

van Beek, L. P. H., van Kempen, C. M., Reckman, J. W. T. M., Vasak, S., & Bierkens, M. (2010). Global depletion of groundwater resources.

Geophysical Research Letters, *in press*, 17.
Zalasiewicz, J., Williams, M., Steffen, W., & Crutzen, P. (2010). The new world of the Anthropocene. *Environmental Science & Technology*, *44*(7), 2228 - 2231.

CHAPTER TWENTY-THREE

CANADA'S SYDNEY TAR PONDS: CONFLICTING INTERESTS RESULTING IN A CASE OF SUPPRESSION BIAS AND SOCIAL INJUSTICE?

COLIN L. SOSKOLNE AND SHIRA KRAMER

"Industry's offensive against the regulation of health and safety hazards uses academics to downplay or deny the seriousness of the hazards..."
—Clayson and Halpern, 1983

The Problem

In this case study from Sydney, Cape Breton Island, Nova Scotia, Canada, the wanton disregard for environmental safeguards — through coal mining, coking operations and steel production effluents discarded over many decades into the Sydney area and harbour — has served to create the Sydney Tar Ponds. For decades, Sydney has remained one of Canada's worst industrially contaminated sites.

The ownership of these industrial operations has been held in more recent decades by provincial and federal authorities whose jurisdiction also includes public health.

This chapter exposes conflicting interests and raises profound questions of social injustice in Canada. Despite knowledge and awareness of carcinogenic hazards to both workers and the local community, no analytical health studies — normative at that time — were ever undertaken to our knowledge, thereby suppressing the opportunity to develop compelling evidence in response to expressed community health concerns.

Justice now is being sought in Nova Scotia's first environmental class action law suit, a process that has taken some seven years thus far, but, at time of this writing, remains at the class action certification stage.

Historical Context

Why is Context Important in a Case Study?

Coal, coking and steel manufacturing are large-scale industrial undertakings that can bring great prosperity. However, as with any economic activity, near-sighted planning can bring untoward consequences. Profit-seeking and political kowtowing can obscure the vision and well-being of a whole province, its leaders and its citizens. Lessons learned from past experiences can serve to pave the way for preventing future harms, especially if we can influence change at the policy level.

In this ethical-legal analysis of the Sydney Tar Ponds, consideration of context is essential for understanding normative frameworks prevailing in the period under review, stakeholder viewpoints and actions, and the present status of the case.

The Historical Context of Cape Breton's Steel Industry, According to Abbass (2006)

Cape Breton Island, re-annexed to the province of Nova Scotia in 1820, was recognised in 1672 as the first place in North America for its coal reserves. With Cape Breton's trade in coal beginning as a regular, authorised and systematic business in 1766, General Mining Association Limited (**GMA**) was organised in 1825. By 1827, GMA held possession of all of Nova Scotia's mines and minerals, including Cape Breton and, as a company, pioneered coal mining, not only in Cape Breton, but in all of Canada. Coal became the basis for Nova Scotia's prosperity, and the basis for successful manufacturing enterprises in both Nova Scotia and throughout Canada.

Between 1830 and 1834, GMA opened coal mines, built a foundry and built a railway three miles in length, from the Sydney-area coal mines to their newly-built, coal-shipping pier in North Sydney. In 1858, the final and complete transfer of authority over the mineral wealth of the province of Nova Scotia was transferred from the Imperial Government to the Provincial Government of Nova Scotia, and the mines were again leased to the GMA.

After Cape Breton's first economic boom, the town of Sydney was incorporated in 1885 (population 2,300). The opening of a coal mine in Reserve, and building a railway to Sydney, provided a port for shipments of coal abroad; the town further developed through the exploitation of the broader region's coal reserves for the production of steel in Sydney. At the turn of the century (1900), industrial developments continued, with Sydney becoming a major provider of its products to North America and Great Britain, and also to the West Indies, South America and Germany.

Incorporation of the Dominion Coal Company (**Domco**) was effected in 1893 by Boston financier Henry Whitney with a syndicate of American and Canadian investors. Domco constructed the Sydney & Louisbourg Railway, connecting all of the various mines to piers at Sydney and Louisbourg, Nova Scotia.

In 1900, Nova Scotia Steel Company purchased the Sydney Mines assets of the GMA. They enlarged their name to Nova Scotia Steel and Coal Company Limited (**Scotia**) and continued with major industrial expansion as the hub of economic activity and growth in the region, coal and steel production being the major single employer. By 1900, a steel plant was being built in Sydney with close ties to business interests in Boston, Massachusetts, USA.

With the rapid development of expanding coal and steel manufacturing operations around the Sydney industrial base and harbour, the first blast furnace of the newly-established Dominion Iron & Steel Company (**Disco**) was lit on December 19, 1900. Disco created employment at good wages that led to an enormous influx of carpenters, masons, bricklayers, machinists and labourers. Nova Scotians who had sought work in the USA were able to return home, and the eyes of the world were on Sydney. On February 3, 1903, Sydney was incorporated as a city with a population of about 16,000. By 1904, technical innovations led to tar being used for fuel in the open hearth furnaces.

With the backdrop of feedstock quality challenges, business conflicts and financial investment opportunities, a new corporation was formed by 1907, the **Dominion Steel Corporation**, which was by no means independent of Disco and Domco interests. Indeed, Disco was producing almost half of all steel used across Canada. Based upon investor interest in 1917, control was taken of Scotia and a merger with Disco was attempted.

Until 1920, the Sydney steel plant was the focus of attention for American, British and Canadian investors. Disco, owned by the Dominion Steel Company (which also controlled Domco), had become a huge conglomerate. In April 1921, British Empire Steel Corporation (**Besco**) was incorporated, uniting Disco and Scotia to become the largest consolidation in the British Empire, and second only to US Steel Corporation on the American continent. This was an all-British corporation including English, Canadian, Australian, Indian and South African investors.

The post-World War I years brought falling prices for coal, ore and steel, with intense competition for Canada from the USA and Europe.

Attempts to reduce the workers' wages led to industrial strife in Nova Scotia until 1925. Provincial and federal police in strike actions were aligned with Besco. In 1923 and 1925, pitched battles raged between workers and police.

Workers died in Canada's first coal mine disaster in Nova Scotia in 1873; between 1917 and 1969, 234 men died in the Pictou County pits alone.

Fiscal turbulence describes the steel markets and the questionable economic viability of the Nova Scotia steel industry from the end of World War I through 1928, when control of Dominion Steel Corporation was purchased by Toronto investors. In 1930, investors from central Canada reorganised Disco as the Dominion Steel & Coal Company Limited (**Dosco**), implementing significant expansion to its production facilities over the next several years.

In 1942, the Federal Government of Canada provided funds for further expansion to the facilities; by 1949, 50 years into the plant's existence, still more spending was required to modernise it. In 1951, all steel-makers in Canada initiated big expansion programmes to meet the rising demand for steel. Throughout, the production capacity was significantly increased.

By 1957, with a payroll including some 4,500 employees, the profitable Sydney steel plant, Dosco, was about to set a yearly production record. Then, **Hawker Siddeley**, a company based in the United Kingdom and one of Canada's fastest growing holding companies, was preparing to acquire control of Dosco through its Canadian subsidiary, **A.V. Roe (Canada)**. Sydney was then the eastern Canadian steel capital. A.V. Roe (Canada), in its 12 years of operations, never paid any dividends to parent company Hawker Siddeley and, instead, reinvested to grow the company. Blast furnaces, spewing an average of 150 tons of dust each month on parts of the Whitney Pier area, painted rainbows on windows, ruined hanged clothes and stripped paint off houses (Abbass, 2006). Had a $6 million cindering plant been installed, these effects could have been mitigated. The Sydney steel plant occupied 546 acres and consumed on average 60,000 net tons of coal each month. By 1960, Dosco had emerged as a great industrial empire.

The More Recent Context: Decline and Demise of an Industry, According to Abbass (2006)

Since 1960, no new technology was introduced into the steel plant, and no new diversification took place. By 1966, with a new steel plant in Quebec employing some 2,500 workers, layoffs took place in Sydney, reducing the workforce to some 3,000. The Sydney plant was left producing only semi-finished steel. In April 1966, Dosco announced the closure of all coal mining operations. In response, the federal government created the Cape Breton Development Corporation (**Devco**) to assume control of the coal mines. When the coal mines closed later that year in Newfoundland, the provincial government took them over, but did not operate them.

In October 1967, Hawker Siddeley announced the formal shutdown of the Dosco Sydney steel plant, which had been the engine of Cape Breton's economy since 1900. Hawker Siddeley was under no obligation to further develop the industry or the region in the national interest, or for the common good. It thus made no effort to take the economic development of Nova Scotia into account. Faced with the prospect of seeing the local area become a ghost town, the Province of Nova Scotia took ownership of the Sydney steel plant in 1967, with the federal government providing an operating grant. A **crown corporation** was formed, and in 1968, all ties between Hawker Siddeley and the now named Sydney Steel Corporation (**Sysco**) were severed. New markets were found in Mexico, Chile, South Korea and the USA, and an operating profit was announced that year (and through 1970), resulting in a government grant for further expansion despite production levels being lower than what they had been four years previously.

From 1972-1975, Sysco underwent a multi-million dollar modernization. Sysco invested in an oxygen plant, a lime burning plant, installed vacuum degassing equipment and bought two new casters, among other expensive investments in equipment. It turned out that the performance efficiency of any new environmental equipment was undermined by the absence of basic oxygen furnaces which had been recommended by consultants. Millions of dollars had been wasted along with a continuing absence of control over industrial emissions.

The Nova Scotia government tried to further facilitate expansion of the steel industry in the region after the permanent workforce was reduced to 2,150 in 1980. With the federal government investing 80% and the province 20%, phase one of a 10-year modernization plan began in 1981, but by 1986, with market uncertainties, the future of Sysco came into doubt. However, the plan went ahead and operations were in full cycle in 1990. At this time, Sysco employed 1,350 full-time workers. However, beset with design and management issues, the Sydney steel pant was closed in 2000. It is claimed that Sysco cost the taxpayers $3 billion over 33 years.

Despite community appeals to keep a reduced-sized plant in operation, the province liquidated Sysco's capital assets between 2001 and 2004. The plant required $10 million to tear down most of the remaining structures, and to clean up chemicals left on the grounds of the 450 hectare site in the heart of the city. Taxpayers had to contribute some $23 million to the clean-up. Workers removed 50,000 kg of PCBs placed for containment in the plant's brick house, 600,000 kg of asbestos-tainted material, two bunker fuel storage tanks, 3 km of underground fuel and coke oven tar lines, and 1.4 million kg of other potentially toxic chemicals.

Omissions of Note in Abbass (2006)

First, while alluded to in Abbass (2006), early stakeholders were mostly preoccupied with profit and were given relative free rein by the provincial government to operate. Then, when profits began to fluctuate, the government took over as the main stakeholder for political reasons. At no point in time did anyone appear to entertain a global or long-term vision of the impacts of this industry on the province or its people.

Second, in addition to deaths from mining accidents, morbidity and mortality associated with exposures to toxic pollutants from the region's industrial activities were not accounted for in considering the human toll of economic growth.

Third, no mention of which we are aware is made of any likely community health concerns (which we now know were prevalent, including concerns about risks of cancer, adverse reproductive outcomes, respiratory disease and other adverse health outcomes), except to note the role of government in protecting environmental and human health and that this ought to have been on record. Indeed, community health started to be formally investigated in the 1970s, which must have been triggered by some expressed concerns. Although formal exposure assessments and air pollution studies started in the late 1950s, human health research was not initiated until the late 1970s. It was only later in the 1990s and early 2000s that more intensive enquiry into both exposure assessment and human health effects took place.

Fourth, and finally, various immigrant communities of all ethnicities and cultures from around the world came to work in Sydney. These groups were economically disadvantaged, and came seeking employment and opportunities for their families. Most egregious is the fact that there was no regard for environmental protections from the disposal of industrial wastes and, exposures long-established as definitive human carcinogens, were permitted to saturate these communities over decades.

Background to Scientific Considerations

The discipline of epidemiology is the science basic to informing policy relating to the protection of public health. Epidemiology is the field that investigates the causes of health problems, and provides the evidence for polices aimed at reducing harms by preventing disease and premature mortality at the community level. However, policy is influenced by many forces/interests that the epidemiologist must recognise. In particular, it is the role of applied sciences, such as epidemiology, to submit the best possible replicable assessments into the policy debate.

Environmental epidemiology is a sub-specialty of epidemiology in which the health effects on populations from exposures to physical, chemical and biological agents external to the human body are studied. We try to clarify the relationships between exogenous factors and host susceptibility factors.

Any organisation, group or individual whose primary interests are not consistent with those of public health can obstruct the advancement of knowledge. Forces that support the *status quo* can operate to influence whether a research question is pursued, or the nature of the questions that scientists ask, leading to Suppression/Oppression Bias. Research Suppression (or Oppression) is the act of obstructing the conduct of a study or the release of its findings, an unethical activity when imposed for reasons other than concern for scientific validity and objectivity (Porta, 2008). "Suppression Bias" undermines public health because it distorts and delays the discovery of scientific information on health risks.

Research suppression operates at the interest group level such as from government, industry, the scientific community, or by any other group or individual with a vested interest in maintaining the *status quo*. Suppression can be motivated, for instance, by financial or ideological interests that a topic not be researched or reported on, or even by professional jealousy.

Such acts deny the fundamental principle of advancing scientific knowledge, namely that of research in pursuit of the public interest. Research suppression can lead to:

- an absence of evidence;
- bias in a purported risk factor-disease association in the literature; and
- a decrease in the precision of that association's estimate of risk.

Certainly as taught since the 1960s, epidemiological leads may be generated from numerous sources, including anecdotal evidence, case studies and descriptive analyses, which then can be verified through the rigorous application of analytic epidemiological methods (Figure 23-1).

Figure 23-1. Hierarchy of Epidemiology Study Designs*

Descriptive/Observational Studies (hypothesis generating)

- Population-based (correlation studies)
- Individual-based
    - Case/anecdotal reports
    - Case series
    - Cross-sectional/prevalence surveys

Analytic Studies (hypothesis testing)

- Controlled observational studies
    - Case-control studies
    - Cohort studies (retrospective/prospective)
- Controlled intervention studies
    - Randomised controlled trials (RCTs)

* The lower down on the chart, the greater the complexity and cost. Most often, RCTs are unethical in environmental epidemiology.

The epidemiological method requires prudent approaches to research by first taking advantage of existing data sources resulting in more descriptive research. These methods are generally cheaper and quicker than more sophisticated research, warranted when the index of suspicion about potential contaminants is further confirmed by descriptive studies. Where further research is called for to test hypotheses, this can take the form of more complex study designs such as case-control or cohort studies.

Figure 1, however, does present an early paradigmatic approach for reducing uncertainties in building a body of evidence about cause and effect. Since then, our body of knowledge about disease mechanisms has grown such that the need to replicate the findings from studies done elsewhere on local populations subjected to similar exposure circumstances is less justified for informing health policy. Certainly, more research is almost always beneficial to advance scientific understanding and to reduce uncertainties, but it has no place in public health policy. For policy, prudence requires that we err on the side of caution, particularly with established carcinogens, when there is no justification to believe that local populations would be immune from what has impacted populations elsewhere. Hence, even in the absence of epidemiological data at the local level, but in the presence of knowledge about the exposure circumstance of a population, action based on knowledge gleaned elsewhere is the responsible approach to be taken locally. To do otherwise is to delay action at the expense of both local environmental and community health and well-being.

This more robust approach to applying the epidemiological method brings into focus even more starkly the suppression of evidence. Consideration of the epidemiological tools available to environmental epidemiologists is relevant to this analysis of the Sydney Tar Ponds in two ways: (1) the types of studies that should have been conducted, given the nature of the contaminants and the extent of contamination; and (2) the data that were evaluated in our weight of evidence analysis of the relationship between exposures and disease risk in this community.

Method

The authors were retained as expert witnesses/consultants on behalf of the community (plaintiffs) to evaluate and opine about the community's position that it has suffered harm due to exposures sustained from the Sydney Tar Ponds. We accessed and reviewed the published and grey literature pertaining to health effects of exposure among occupational cohorts and communities to the hazardous substances that have polluted the area of Sydney for the past some 100 years. In addition, we reviewed exposure assessment reports, epidemiological studies and all relevant health research relating specifically to the population of Sydney. The exposures that were sustained by workers in the industrial facilities in Sydney, as well as by the general population in the community, have been recognised for decades — and in some cases for over 200 years — to be highly toxic and carcinogenic. These contaminants include soots, tars, petroleum by-products, polycyclic aromatic hydrocarbons, particulate matter, polychlorinated biphenyls, heavy metals and volatile organic compounds.

In the evaluation of reports or studies specific to the Sydney community, each accessible report was evaluated according to its methodological strengths and weaknesses, and its place in the hierarchy of epidemiological study designs, ranging from descriptive (hypothesis generating) to more complex analytical (hypothesis testing) study designs. In considering the weight of the evidence, we were concerned with the following questions: (1) What do we know from the general epidemiological and toxicological literature about the health risks of exposure to the contaminants of concern? (2) What should the entities with authority over industrial and public health activities have done in order to protect the health of the public, evaluate health status/risks and monitor the community, in light of the exposures sustained? (3) What is the state of evidence, including presence/absence of data from the affected community, about health risks?

More so in the past, the epidemiological framework of study designs, from anecdotal evidence, to observational studies, to experimental studies, as noted above, provides a framework for advancing knowledge about exposures that could be damaging to public health. Epidemiologists play an important role in communicating to public health agencies and other authorities about the need to conduct more sophisticated studies, if justified from findings derived from designs lower in the hierarchy (Fig.1).

Analysis and Results

Despite the large body of general epidemiological evidence for human health harms due to exposures from the Sydney Tar Ponds, long-known to be carcinogenic, risk evaluations conducted by governmental agencies, consultants hired by the government and academic researchers over decades generally carried out only anecdotal and descriptive assessments, all designed at the lower end of the hierarchy of epidemiology enquiry. Based upon the failure to conduct more definitive analytical epidemiological assessments of worker cohorts, or of community health impacts, despite having been proposed, our review revealed that the scientific body of evidence to demonstrate health effects in this specific community remains unnecessarily incomplete.

A total of six epidemiological studies relating to Sydney and Cape Breton County, Nova Scotia, were conducted over the period 1985 through 2006 (Band et al, 1998; Burra et al, 2006, Dodds and Seviour, 2001; Veugelers and Guernsey, 1999; Guernsey et al, 2000; and Mao et al, 1985). Three of the studies were ecologic mortality studies (Band et al, 1998; Veugelers and Guernsey, 1999; and Mao et al, 1985); there were two studies of reproductive outcomes (Burra et al, 2006; and Dodds and Seviour, 2001); and there was one ecologic study of cancer incidence rates (Guernsey et al, 2000). All six studies demonstrated increased risk of disease in Sydney/Cape Breton County compared to reference populations (Nova Scotia or Canada), and four studies showed that closer proximity to the Tar Ponds was associated with increased risk of disease. Despite the consistency of these finding, and the voluminous epidemiological literature about the health risks of exposure to the contaminants in the community, the findings of elevated risks of disease were dismissed by public health authorities as inconclusive, or were attributed to other factors.

Applied in the context of environmental determinants of health, epidemiology has been subjected to pressures that result in work that is not in the public interest; and, also in the failure to conduct research that ought to have been done to address a public health concern. These pressures can emerge from "Suppression Bias" which results from actions aimed at obstructing the conduct or publication of research. A bias is thereby produced in the available evidence on the relationships between exposures and health outcomes. An organization, a group or an individual whose priority interests are not consistent with those of public health can be obstructive to advancing knowledge. Suppression bias may, in turn, lead to publication bias. It undermines public health because it distorts and delays the discovery of scientific knowledge on health risks, and it compromises credibility in science and administrative processes for assessing and preventing exposure to hazards or risks. In the realm of litigation, such as toxic and environmental tort cases, the absence or paucity of epidemiological evidence may preclude the establishment of a causal link between exposure and disease, thereby enabling the continuation of harmful exposures or obstructing fair compensation for harm.

Today, it is recognised that descriptive studies in the context of the full body of knowledge can, indeed, carry significant weight in deliberating the evidence for health harms. At the time of writing, and some seven years after its initiation, the Sydney Tar Ponds case continues to seek judicial approval to be heard as a class action law suit. In fact, in our view this process needed to be decided much more expeditiously, and thus raises questions of both conflicting interests and social injustice.

Conclusions and Implications

Any organization, group or individual whose primary interests are not consistent with those of public health can exercise influence by obstructing the advancement of knowledge. "Suppression Bias" can result. This undermines public health because it distorts and delays the discovery of scientific information on health risks.

The scientific evidence base for the relationship between exposure(s) and outcome(s) is often contested by industries or other actors that have vested interests in causal links between harmful exposures and outcomes not being established, sufficiently doubted or deemed irrelevant (Davis, 2007; Davis, 2010; Lewis et al, 1992; Markowitz and Rosner, 2002; McCulloch and Tweedale, 2008; McGarity and Wagner, 2008; Mehlman,

2009). The history of the tobacco industry since the middle of the 20th century provides a classic case of a prolonged effort to discredit the science which, even several decades ago, was demonstrating that smoking causes lung cancer. However, other industries have employed similar tactics of "manufacturing uncertainty" (Michaels, 2008); these include the lead industry, the chemicals and the plastics industries, and the asbestos industry (Michaels and Monforton, 2007).

Numerous tactics are used to discredit the science of determining exposure-effect relationships and/or to affect policy or regulation that attempts to limit or mitigate the exposure and its putative or actually established outcome(s). These tactics include claiming that cause-and-effect relationships have not been established; claiming that statistical data do not provide the answers; and affirming that more research is needed. Often the scientific base or individual studies are labelled "junk science" when no clear definition of this term is provided. Early advice to the tobacco industry by the company Hill and Knowlton is generally characteristic of these tactics (Oreskes and Conway, 2010).

At a time when descriptive studies were deemed less compelling than analytical studies, by funding only descriptive epidemiological research and by not facilitating the conduct of higher-level analytical studies of exposed workers and citizens, epidemiological evidence was denied to the Sydney community. This suppression of research in the exposed community occurred during a time when there was awareness among public health and industry officials of the dangers of exposure to the contaminants of concern, and of health concerns in the exposed population. Exposure assessments, descriptive and ecological studies, primarily carried out in the 1980s through the mid-2000s, demonstrated excess rates of various cancers and other health outcomes, and increased levels of dangerous pollutants, including particulates and benzo(a)pyrene, compared to control communities with coke ovens (Atwell et al, 1984; Lambert and Lane, 2004; Lambert et al, 2006). Health risks in the community were evident in the studies that were conducted, despite the low power of the studies, inappropriate control groups and other methodological weaknesses that diminished the ability of the studies to detect statistically significantly elevated rates of disease.

The fact that no occupational studies are known to have been undertaken is particularly indicative of suppression bias, especially in light of the known toxicity of the exposures, and the findings of Mao et al (1985) of excess mortality from pneumoconiosis among men in Cape Breton County during the period 1971-1983, using provincial mortality rates as the standard.

Furthermore, there was no regard for past, current or future environmental protection in the disposal of waste, which contained well-known human carcinogens. The Sydney harbour became an industrial dump site, infamously known as, if not the worst, one of Canada's most severely contaminated sites by cancer-causing substances recognised as such since Percival Pott's study of chimney sweeps over 200 years ago. Perhaps the motivation for technical innovations that might have respected the environment would have been stronger had the economic paradigm of the day required consideration, for instance, of the need to internalise costs related to the consequences of an environmental clean-up associated with the disposal of wastes into, among other locations, the Sydney Harbour, in the pricing of steel.

Continuing exposures, even after the plant was shut down, were well recognised even by the provincial and federal authorities. These hazardous substances could not possibly be limited to the Tar Ponds alone. Indeed, Band and Camus (1998) established extensive community exposure, not only in proximity to the Tar Ponds, but also through stack emissions to the broader Sydney community. Indeed, particulate pollution was dispersed over a wide geographical area.

Did conflicting interests interfere with governments' responsibility to protect public health? The provincial and federal governments misinformed workers and communities about the health effects associated with the high levels of industrial pollution that rained down on these communities for decades. Excess rates of disease have been dismissed by government authorities as attributable to "lifestyle factors" in the absence of adequately designed, controlled studies. By funding only descriptive epidemiological research and not facilitating/enabling the conduct of higher-level studies, epidemiological evidence was denied to the community. The Sydney community has been comprised of vulnerable immigrant populations seeking employment and safe haven. The question of environmental justice in this community must be considered, given the disadvantaged status of the population and workforce.

A critical aspect of conflicting interests in this case is the dual role of the government as owner of the industrial site, as well as the responsible authority regarding public health of Canadian citizens. The powerful interests and forces at work in the suppression/oppression of research are difficult to overcome, and great vigilance as well as personal integrity are required in order to follow ethical principles. This notion of integrity underlies the conduct of science, be it basic or applied. Because applied science has a more proximate relationship to the community in which the work has been done, there is a significant risk to the public when these conflicts lead to biases, such as the suppression of evidence within this realm of research. Regardless of the nature of the conflicting interests, suppression bias threatens the validity of scientific evidence, the integrity of public health practitioners, and it undermines the credibility of public health policy.

References

Abbass, F. DISCO DOSCO SYSCO (2006). 1672-2005: Chronology of Leading Events & The Pioneers of Coal Mining and Steel Making in Nova Scotia. Sunday Novel Publishing, Sydney, Cape Breton. Atwell, J., Henningar, P., Kozak, J., Morin, M., Oldreive, C. (1984). Ambient air polynuclear aromatic hydrocarbons study, Sydney, Nova Scotia. Report PS-5-AR-84-7. Environmental Protection Service, Environment Canada.

Band, P. et al (1998). Mortality rates within Sydney, Nova Scotia by exposure areas to airborne coke ovens and steel mill emissions: 1961-1988 (Ottawa: Environmental Health Directorate).

Burra et al (2006). Effects of residential exposure to steel mills and coking works on birth weight and preterm births among residents of Sydney, Nova Scotia. The Canadian Geographer 50(2): 242-255.

Clayson, Z.E., Halpern, J.L. (1983). Changes in the Workplace: Implications for Occupational Safety and Health. Journal of Public Health Policy 4(3): 279-297.

Davis, D. (2007) The Secret History of The War on Cancer. Basic Books, New York, NY.

—. (2010). Disconnect: The Truth About Cell Phone Radiation, What the Industry Has Done to Hide It, and How to Protect Your Family. Dutton Press, New York, NY.

Dodds, L. and Seviour, R. (2001). Congenital anomalies and other birth outcomes among infants born to women living near a hazardous waste site in Sydney, Nova Scotia. Canadian J Publ Hlth 92 (5): 331-334.

Guernsey, J.R., Dewar, R., Weerasinghe, S., Kirkland, S., Veugelers, P.J. (2000). Incidence of Cancer in Sydney and Cape Breton County, Nova Scotia, 1979-1997. Canadian J Publ Hlth 91(4): 285-291.

Lambert T.W., Lane, S. (2004). Lead, Arsenic, and Polycyclic Aromatic Hydrocarbons in Soil and House Dust in the Communities Surrounding the Sydney, Nova Scotia, Tar Ponds. Environmental Health Perspectives 112(1): 35-41.

Lambert T.W., Guyn L, and Lane S.E. (2006). Development of local knowledge of environmental contamination in Sydney, Nova Scotia:

Environmental health practice from an environmental justice perspective. Science of the Total Environment 368: 471-484.

Lewis, S., Keating, B., Russell, D. (1992). Inconclusive by design: Waste, fraud and abuse in federal environmental health research. Environmental Health Network, National Toxics Campaign Fund. 71 pages. www.csu.edu/cerc/researchreports/humanhealth.htm

Mao et al (1985). Mortality in Cape Breton, Nova Scotia 1971-1983. (Ottawa: Bureau of Epidemiology, Laboratory Centre for Disease Control, Health and Welfare Canada).

Markowitz, G., Rosner, D. (2002). Deceit and Denial: The Deadly Politics of Industrial Pollution. University of California Press, Berkeley, CA.

McCulloch, J., Tweedale G. (2008). Defending the Indefensible: The Global Asbestos Industry and its Fight for Survival. Oxford University Press, Oxford, England.

McGarity, T.O., Wagner, W.E. (2008). Bending Science: How Special Interests Corrupt Public Health Research. Harvard University Press, Cambridge, MA.

Mehlman, M.A. (2009).The power of the industrial product lobby: controversy, doubt, contradiction, lies and deceit and attacks on the Ramazzini Institute. European J Oncology 14(4): 223-229.

Michaels, D. (2008). Doubt is their product: How industry's assault on science threatens your health. Oxford University Press, New York, NY.

Michaels, D., Monforton, C. (2007). How litigation shapes the scientific literature: asbestos and disease among automobile mechanics. J Law Policy 5(3): 1137-1169.

Oreskes, N., Conway, E.M. (2010). Merchants of Doubt: How a Handful of Scientists Obscured the Truth on Issues from Tobacco Smoke to Global Warming. Bloomsbury Press, New York, NY.

Porta, M. (Ed). (2008). A Dictionary of Epidemiology (Fifth Edition). Oxford University Press, New York, NY.

Veugelers, P.J., Guernsey, J.R. (1999). Health Deficiencies in Cape Breton County, Nova Scotia, Canada. 1950-1995. Epidemiology 10(5): 495-499.

CHAPTER TWENTY-FOUR

PUBLIC HEALTH, ENVIRONMENTAL CANCER EPIDEMIOLOGY, AND SUSTAINABILITY

VLADIMÍR BENCKO

Introduction

Our day-to-day experience of health ailments easily misconstrues health as a local and individual-level issue. We thus fail to recognize that the health profile of a population is the real "bottom line" indicator of prevailing environmental and social conditions.

For man, environmental sustainability must be ultimately about sustaining health-supporting environmental conditions (McMichael 2001). Many civilizations have faltered or collapsed along the way as local environmental resources became depleted. But, it is only now, in the 21st century, that striving for the betterment of human living conditions and social experience must, for the first time, take account of constraints at the global and major regional levels; constraints imposed as a result of humankind having reached, indeed exceeded, many of the natural environment's limits to its carrying capacity for human populations. That poses a new, huge challenge to modern societies (McMichael 2001; Raven 2002; WHO 2009; Soskolne, Westra, et al. 2008). What role can there be for sustaining human life from the field of cancer epidemiology as we adapt to global environmental change?

Epidemiology is a branch of biomedical science, dedicated to the study of health in populations. The objective of this chapter is to explicate some issues around sustainability through a state of the art example from cancer epidemiology. I will describe some methodological obstacles and challenges for the future (Bencko 2007) in the broader context of "risk assessment" and the "precautionary principle" (Bencko 2010).

The field of epidemiology has reached a crucial point with both challenges and opportunities. On the one hand, it seems that most of the major occupational carcinogens have already been identified. Many of the chemicals classified as carcinogens by the International Agency for Research on Cancer (IARC) were first evaluated in the workplace. Occupational exposure to known human carcinogens has diminished in many countries over the past decades, and awareness of their hazards has increased. At the same time, we are still confronted with a long list of substances for which epidemiological data are lacking or inconclusive. Estimates of the number of chemicals in commerce range from tens of thousands to over 140,000 (Stayner, Boffetta and Vainio 2006); for most of them, relevant toxicological information needed for setting up regulatory standards is still lacking (Gemignani, Landi, et al. 2007; Hung, Hashibe and McKay 2007).

We are now at an important junction: advances in the interrelated disciplines on which health risk assessment depends hold promise for a more comprehensive understanding of the influence of environmental stressors on human health. Since the 1960s, there have been major developments in the field of cancer risk assessment. Also, there have been remarkable advances in the broad area of cancer epidemiology, including research not only on human exposures to major cancer risk factors in environmental and occupational settings, but also on lifestyle and nutrition-related risks.

Traditional approaches to study designs in cancer epidemiology have not been successful in identifying and evaluating these potential risk and/or protective factors. Two main reasons for this failure are often due to a small sample size resulting in under-powered studies, and inadequate exposure assessment resulting in exposure misclassification. In this paper, I discuss issues and approaches relevant to these two challenges, and the opportunity for using emerging genomics information in epidemiology.

Increasing Sample Size

An important characteristic of research over the last decade is the increasing number of collaborative studies involving various countries, and thus, as a consequence, sample size is greatly increased. By increasing the sample size, the power of a study to identify statistically significant associations between exposures and a disease endpoints is enhanced. For example, considering that the lifetime prevalence of occupation-related exposures in the population is low (typically below 10%), and the associated risk can be small as well (e.g. a relative risk of 2), the sample size for a community-based study must be large in order to identify statistically significant associations. This is even more crucial if exposure or disease outcome is misclassified in a study population. A sample size of at least 1,000 cases and controls has been recommended for a community-based case-control study on cancer. However, it is often the case that one centre or one country can not provide such numbers within a reasonable amount of time. Therefore, multi-centre studies provide one practical solution. In addition to increased statistical power, multi-centre studies can provide additional advantages that include more heterogenous exposure experiences in the study population, so useful for exposure-response analysis, and also provides the opportunity to study differences in exposure and disease patterns between countries.

Multi-centre studies can be realized in two different ways. Collaboration between centres can be initiated after the completion of each individual study (retrospectively planned multi-centre studies), or before the studies have actually started (prospectively planned multi-centre studies). Meta- and pooled analyses are examples of retrospectively planned multi-centre studies, and these have been applied in many areas of epidemiology.

Prospectively planned multi-centre studies have only recently become possible since international organizations and institutions such as the European Commission started to offer funding for these costly operations. These studies offer the advantage of identical protocols for data collection in each centre involved, avoiding the loss of information at the stage of data pooling. From the perspective of the Central and East European Countries (CEEC), an important example of this approach is the CEEC Multi-Centre Lung Cancer Study that was organized by IARC/WHO/Lyon, France about 15 years ago, and was supported by EC Inco Copernicus FP4. It represented the collaboration of centres in the Czech Republic, Hungary, Poland, Romania, The Russian Federation, and Slovakia (Brennan, Hsu, Moullan et al. 2005; Brennan, McKay, Moore, et al. 2009; Moore, Brennan, Karami, et al. 2007; Karami, Brennan, Hung, et al. 2008). Later on, the same organizational principle was applied in the Kidney Cancer Study supported by NCI/Bethesda, USA (Chalmers 1991; Thompson and Pocock 1991).

An advantage of multi-centre studies is that the potentially serious problems associated with the use of meta-analysis in observational studies can now be substantially mitigated. The merit of applying meta-analysis to observational studies has been questioned and controversial (Blethner, Saurbret, et al. 1999; Stroup, Berlin, et al. 2000) because errors and biases can be easily introduced when studies with different designs, methods, and population characteristics are combined. It would not be a problem if the study designs were comparable, and if the methods for data collection were coordinated prior to the initiation of the study among different centres. There are many factors that make the meta-analysis of individually conducted studies less appropriate. Consider, for example, meta-analysis of odd ratios or relative risk estimates that require auxiliary information such as age, sex, smoking status, and sample size. If all auxiliary variables are homogenous across studies, it would be appropriate to conduct meta-analyses of these studies (Benke, Sim, et.al. 1997; Mannetje, Fevotte, et al. 2003). In reality, however, these conditions have never been satisfied. Therefore, the concept of the multi-centre (prospective) study offers an attractive alternative to the traditional single-centre epidemiological study design.

Improving Exposure Assessment

Optimizing methods of exposure assessment and reducing exposure misclassification represent the most difficult components of epidemiology studies. A perfect exposure assessment for long latency diseases such as cancer would consist of quantitative measurements of internal dose, over the whole exposure period, for each person in the study. This is an ideal goal, but is unlikely to be attained in the near future. The availability of group-based external exposure measurements at selected points in time is presently the best that epidemiologists trying to estimate the exposure of study participants can achieve. More often, one has to fall back on subjective methods of exposure assessment, such as individual recall, proxy measures, and approximations.

Exposure assessment largely depends on the design of the epidemiological study, with community-based and industry-based studies imposing their own specific limitations. In community-based case-control studies, the exposures of interest often cannot be measured directly, and have to be estimated retrospectively. As a consequence, exposure estimates are often based directly on the job-information provided by the research participants (or their proxies), or inferred from the job-information through job-exposure matrices or by expert assessment on a case-by-case basis. The subjectivity and the limited use of data-driven quantitative estimates of exposure used in case-control studies are considered important shortcomings that can lead to substantial exposure misclassification. How to improve retrospective exposure assessment methods is a topic of frequent debate. The difficulty in evaluating the validity of all retrospective exposure assessment methods in the absence of a gold standard is considered a major shortcoming.

Reliability studies indicate that a considerable amount of misclassification can occur in all retrospective methods (Blair, Stewart, et al. 2006; Bencko and Symon 1977). There is, therefore, a compelling need to quantify levels of misclassification expected from each method in order to anticipate the attenuation of the resultant odds ratio (OR) risk estimates. A major area for improvement requires a departure from crude exposure indicators such as "never/ever exposed" to a more quantitative (ordinal or ratio) exposure assessment covering the entire exposure period. Quantitative exposure data that permit the investigation of the exposure-response relationship is an important consideration in establishing causation. The availability of more quantitative exposure data can facilitate valid comparisons of risk and of exposure-response relationships among studies, countries and industries, and can provide a solid basis for risk assessment and standard setting.

Biomarkers and Genomics

A potentially useful approach for improving exposure estimates is to rely on biomarkers of exposures. Biomarkers for exposures in health risk assessment have been available for a long time, but their use in risk assessment does not seem to have been fully exploited. The utility of biomarkers was greatly enhanced when body-burden was a result of exposure from multiple sources, or when the source of exposure was difficult to identify (such as in exposure to second hand smoke).

There are situations where biomarkers of exposure can provide valuable information in addition to the measurement of exposures in traditional epidemiological studies. For instance, when data obtained from epidemiological studies of a community with high arsenic contamination in drinking water are used for risk assessment, a contentious issue is: what is the total arsenic intake for a person? Assessment of exposure is often based on the measured concentrations in the drinking water and estimations of the amount of water consumed and used for cooking, as an estimate of total dietary intake.

Under this range of exposure sources, biological monitoring provides data on the absorbed dose for each individual studied and offers a useful solution to the problem. Biomarkers may include arsenic in urine, in blood, in hair, and in nails. However, the suitability of the various biomarkers to serve as indicators of acute or chronic exposure to inorganic arsenic under the different exposure pathways needs to be carefully evaluated. Although the objective here is not about arsenic risk assessment, I use it as an example to illustrate the potential utility of biomarkers (Uchino, Roychowdhury, et al. 2006; Herceg 2007). In actual risk assessment, one could use the estimated total arsenic body burden coupled with the incidence of a health endpoint to assess the risk associated with the exposure, and to compare the result with the same endpoints observed in other community-based studies.

In the past, the field of cancer genetics has enjoyed a great deal of attention among cancer researchers (Bencko 2010; Gemignani, Landi, et al. 2007). Appreciation of cancer epigenetics is more recent. The study of the role of epigenetic changes induced by environmental, dietary and lifestyle factors is rapidly emerging, but still is in its infancy. Little is known about the precise contribution of epigenetic mechanisms to different types of human health effects induced by adverse stimuli in the environment and diet. While there is accumulating evidence showing that aberrant DNA methylation may result from adverse exposures to epimutagens, there is a paucity of evidence regarding the effects of stimuli causing heritable changes in epigenetic information stored in histones, owing to the fact that this is a new and largely unexplored field. Although it seems inevitable that perturbations in histone modifications are induced by dietary and environmental factors that contribute to the development of human disease such as cancers, rigorous proof of such a relationship remains to be established (Herceg 2007).

Multi-centre and large cohort or case–control studies offer some of the most exciting opportunities to study the contribution of epigenetic events induced by the diet and environment to human cancer. Such examples are the objectives of the European Prospective Investigation into Cancer and Nutrition, a large prospective cohort study designed to investigate the relationship between diet, lifestyle, and the incidence of cancer in 10 European countries (Riboli and Kaaks 1997); and a case–control study on lung and upper-aero-digestive tract cancers in Central and Eastern Europe (Hung, Brennan, et al. 2005; Hung, van der Hel, et al. 2005; Scelo, Constantinescu, et al. 2004). These multi-centre studies boast a large sample size of several thousand participants, and provides the unique possibility to identify which dietary/lifestyle practices, and environment stimuli, may exert risk and/or benefit effects through epigenomic changes.

Epigenetic alterations in comparison with genetic changes are reversible and are typically acquired over many years. These features offer important intervention opportunities for prevention (Herceg 2007). In particular, if these methods prove themselves, they could well play a crucial role in helping with sustaining human life into the future with the adaptations that will be needed under global ecological change.

Risk Assessment and the Precautionary Principle

Risk assessment is an evolving science. The methods of conducting risk assessment are still in their infancy and are undergoing rapid development. One of the important aspects of a recent discussion is about "cessation lag" and "effect lingering" and their potential applications to dose-response analysis in risk assessment (Chen and Gibb 2003; Chen 2008). In addition to providing insight about biological modes of action, the concept of cessation lag is useful for cost-benefit analysis. Effect lingering can be used to analyze epidemiological data by uncovering the hidden biological implications related to disease endpoints, thereby advancing current efforts to characterize and reduce the uncertainties associated with risk assessment.

Controversies abound concerning the appropriate methods and data to use and they are likely to persist, giving the great uncertainties involved in extrapolating beyond the range of available data, the underlying biases, and the other limitations of observational data (Bencko 2003; Bencko, Rames, et al. 2006; Sasco 2007; Bencko, Rames, et al. 2009), and the political and societal implications of these analyses (Stayner, Boffetta, et al. 2006). Skeptics have argued that risk assessment, at least as it is currently practiced, has not been a useful tool for addressing societal concerns about exposures to environmental and occupational hazards (Silbergeld 1993). Their primary concern is that the increasingly intense debates concerning risk assessments may come to be used as an excuse for delay in the development of appropriate regulatory and other responses to environmental and occupational hazards. For example, it has taken the U.S. EPA more than 20 years to finalize its risk assessment for exposure to diesel exhaust particulates (Stayner 1999).

A spirited debate has emerged over the use of the "precautionary principle" as an alternative basis for public health decision-making, and this approach has recently been incorporated in some environmental legislation of the European Union (Commission of the European Communities 2000). The precautionary principle has been defined as the need to take some precautionary measures to prevent threats to human health even when a cause-and-effect relationship has not been fully established (Kriebel and Tickner 2001).

The story of how John Snow convinced the authorities in England to remove the Broad Street pump well before the cause of the cholera epidemic in London was properly understood is well known (Bencko 2010; Stayner, Boffetta, et al. 2006). In my view, risk assessment and the precautionary principle should not be viewed as conflicting paradigms but, rather, as complementary approaches for developing appropriate policies to address risks posed by exposure to carcinogens and other hazards.

Identification and quantification of risks is clearly a useful tool for informed decision-making. Appropriately, this is also the underlying principle for the current U.S. EPA's default approach for cancer risk assessment. This is not a new principle for epidemiologists. Risk assessments are inherently uncertain and should, as the NAS (1996) (NRC Report 1999) suggested, be viewed as an iterative process in need of continual improvement through research targeted to fill the gaps in our evidence-base.

Tremendous advances are taking place in our understanding of basic epigenetic mechanisms. Rapid progress is being made in developing new and powerful technologies such as those for the sensitive and quantitative detection of epigenetic changes, as well as for genome-wide analysis (epigenomics). These advances hold great promise in that the role of epidemiology for sustaining public health may be even more fully addressed in the near future (Herceg 2007; Bencko 2010). It may well be embraced as we make adaptations in the face of global ecological change.

Conclusions

Epidemiological data play a crucial role in the evolving evidence-base for the classification of human carcinogens. With rapidly emerging genomic and molecular data, and with the adoption of multi-centre study designs for their exploration, there is hope that, in the foreseeable future, long-awaited answers to the problems we encounter today will be forthcoming. Improved knowledge about the impact of aberrant genetic and epigenetic interactions with both the environment and diet should soon be available. Epigenetic alterations, in contrast with genetic changes, are reversible, and are typically acquired over decades. These gains in knowledge will offer the important potential for primary prevention strategies as a high priority for both medicine and public health in 21st century.

Acknowledgements

This paper was written under the umbrella of activities supported by the League Against Cancer, Prague, and also under the Follow-up to an EC-supported grant on Lung Cancer [CEEC and ARCAGE P-05/DZ-30/99/2000 IARC/WHO study GEE/09/03] for the dissemination of experience gained in project-related activities.

References

Bencko, V. and Symon, K., Health aspects of burning coal with a high arsenic content: arsenic in hair, urine, and blood in children residing in a polluted area, Environ Research 1977; 13:378-385.

Bencko, V., Environmental epidemiology, present changes and challenges for future. Cent Eur J Public Health 2007; Suppl. 15:S6-S8.

Bencko, V., Rames, J., Ondrusova, M, Plesko, I.., et al., Human exposure to plyhalogenated hydrocarbons and incidence of selected malignancies – central European experience. Neoplasma 2009; 56(4):359-366.

Bencko, V., Rameš, J., Van den Berg, M., Pleško, I. and Trnovec, T., Human exposure to polyhaloganated hydrocarbons and incidence of selected malignancies. In:Environmental Health in Central and Eastern Europe.K.C.Donnley and Leslie H. Cizmas (Eds.) Springer, Dordrecht 2006; 31-37.

Bencko, V., Risk assessment and human exposure to endocrine disrupters. In: Molecular Epidemiology in Preventive Medicine. Jedrychowski, W.A., Perera,F.P., Maugeri,U.(eds.) International Centre for Studies and Research in Biomedicine in Luxemburg 2003; 315-327.

—. Use of epidemiology data for cancer risk assessment: approaches, concepts and issues. Open Epidemiology J. 2010; (3):147-151.

Benke, G., Sim, M., Forbes, F. and Salzberg, M., Retrospective Assessment of Occupational Exposure to Chemicals in Community-Based Studies: Validity and Repeatability of Industrial Hygiene Panel Ratings, International Journal of Epidemiology 1997; 26(3):635-642.

Blair, A., Stewart, P., Lubin, J., Forastiere, F., Methodological issues regarding confounding and exposure misclassification in epidemiological studies of occupational exposures, Amer J. Ind Med 2006; 50(3): 199-207.

Bletner, M., Sauerbret, W.,Schlehofer, B., Scheuchenpflug, T. And Friedenreich, C. Traditional reviews, meta-analyses and pooled analyses in epidemiology. Int J. Epidemiol 1999; 28:1-9.

Brennan, P., Hsu Ch. C. Moullan, N, Szeszenia-Dabrowska, N, Lissowska, J, et al., Effect of cruciferous vegetables on lung cancer in patiens stratified by genetic status: a mendelian randomization approach. Lancet 2005; 366(4):1558-1560.

Brennan, P., McKay, J., Moore, L, Zaridze, D, et al., Obesity and cancer: Mendelian randomization approach utilizing the FTO genotype. Internat J Epidemiology 2009; 38:971-975.

Chalmers, T., Problems induced by meta-analysis. Stat Med 1991; 10:971-980.

Chen, C., and Gibb, H., Procedures for calculating cessation lag. *Regul. Toxicol. Pharmacol.* 2003; 38:157-165.

Chen, C.W., Lingering Effect: Epidemiological information useful for risk assessment. Regul. Toxicol. Pharmacol.2008; 52:242-247.

Commission of the European Communities. Communique on the Precautionary Principle. Brussels: European Commission 2000.

Gemignani, F., Landi, S. and Szeszenia-Dabrowska, N., Development of lung cancer before the age of 50: the role of xenobiotic metabolizing genes. Carcinogenesis 2007; 28:1287-1293.

Hartung, T. and Rovida, C., Chemical regulators have overreached. Nature 2009; 460(7259):1080-1081.

Herceg, Z., Epigenetics and cancer: towards an evaluation of the impact of environmental and dietary factors Mutagenesis 2007; 22(2):91-103.

Hung RJ, Hashibe M. and McKay J., Folate-related genes and the risk of tabacco-related cancers in Central Europe. Carcinogenesis 2007; 28:1334-1340.

Hung, R.J., Brennan, P., Canzian, F. et al., Large-scale investigation of base excision repair genetic polymorphisms and lung cancer risk in a multicenter study. J. Natl. Cancer Inst. 2005; 97:567–576.

Hung, R.J., van der Hel, O., Tavtigian, S.V., Brennan, P., Boffetta, P. and Hashibe, M., Perspectives on the molecular epidemiology of aerodigestive tract cancers. Mutat. Res. 2005; 592:102–118.

Karami, S., Brennan, P., Hung, J.R., Boffetta, J., Toro, J., et al., Vitamin D receptor polymorphism and renal cancer risk in Central and Eastern Europe. J Toxicol Environ Hlth 2008; Part A ,71:367-372.

Kriebel, D. and Tickner, J., Reenergizing public health through precaution. Am J Public Health 2001; 91:1351—1361.

Mannetje, A., Fevotte, J., Fletcher, T., et al., Assessing Exposure Misclassification by Expert Assessment in Multicenter Occupational Studies, Epidemiology 2003; 14(5):585-592.

McMichael, A.J., Human Frontiers, Environment and Disease: Past Patterns Future Uncertainties, Cambridge University Press, 2001.

—. Human population health: sentinel criterion of environmental sustainability. Current Opinion in Environmental Sustainability 2009; 1:101-106.

Moore, L., Brennan, P., Karami, S., Hung, R.J, Hsu, C., Boffetta P. Glutathione S-transferase polymorphisms, cruciferous vegetable intake and cancer risk in the Central and Eastern European Kidney Cancer Study. Carcinogenesis 2007; 28(9):1960-1964.

NRC Report---Subcommittee on Arsenic in Drinking Water, National Research Council (NRC) 1999.

Raven, P.H., Science, sustainability, and the human prospect. Science 2002, 297:954-957.

Riboli, E., and Kaaks, R., The EPIC Project: rationale and study design. European Prospective Investigation into Cancer and Nutrition. Int. J. Epidemiol. 1997; 26(Suppl. 1), S6–S14.

Sasco, A.J., Cancer and globalization. Biomed Pharmacother 2007; 62:110-121.

Scelo, G., Constantinescu, V., Csiki, I., et al., Occupational exposure to vinyl chloride, acrylonitrile and styrene and lung cancer risk (Europe). Cancer Causes Control 2004; 15:445–452.

Silbergeld, EK., Risk assessment: the perspective and experience of U.S. environmentalists. Environ Health Perspect 1993; 101:100-104.

Soskolne, C.L., Westra, L., Kotzé, L.J., Mackey, B., Rees, W.E., Westra, R. (Eds.): Sustaining Life on Earth: Environmental and Human Health Through Global Governance. Lexington Books, Lanham, Maryland, 2008.

Stayner, L., Protecting public health in the face of uncertain risks: the example of diesel exhaust. Am J Public Health 1999; 89:991-993.

Stayner, LT, Boffetta, P. and Vainio, H., Risk Assessment of Carcinogenic Hazards. In: Cancer Epidemiology and Prevention (Third Edition), Schottenfeld, D., Fraumeni, JF, Jr., M.D. (Eds.), Oxford University Press 2006; 65-69.

Stroup, D., Berlin, J., Morton, S.; et al., Meta-analysis of observational studies in epidemiology: A proposal for reporting. JAMA 2000; 283(15):2008-2012.

Thompson, S., Pocock, S., Can meta-analysis be trusted? Lancet 1991; 338:1127-1130.

Uchino, T., Roychowdhury, T., Ando, M. and Tokunaga, H., Intake of arsenic from water, food composites and excretion through urine, hair from a studied population in West Bengal, India, Food and Chemical Technology 2006; 44:455-461.

World Health Organization: Protecting Health from Climate Change: Global research Priorities (Executive Summary" p ii) Geneva, WHO, 2009 978 92 4 159818 7.

PART VII

GLOBAL GOVERNANCE AND THE ROLE OF MEDIA AND COMMUNICATIONS TECHNOLOGIES: POWERFUL TOOLS IN RELATION TO ECOLOGICAL INTEGRITY

Introduction

Colin L. Soskolne

For its reach and access by the masses — influencing values, culture and public opinion — the control of and influence in communications has long been essential to those in pursuit of power. Indeed, in countries where there has been an overthrow of an existing regime, forces leading change first seize control of the broadcast media. The primary reason for such control is for access to the masses, to "inform" and "control" them.

As a powerful form of communications, mainstream television and film (movies) have tended to reinforce dominant ways of thinking. They have appealed to normative social standards associated with the dominant paradigm, whether is has to do with normalising behaviours (e.g., the use of cigarettes through glamour appeal throughout the 20th century), to controlling and, indeed, censoring newspaper reports and editorial content to support moneyed interests allied with the dominant paradigm. The latter is ever-present, particularly as control of the media has become dominated by few powerful individuals whose primary concern is to entrench the *status quo*, ensuring business-as-usual and bolstering their dominance.

Since the advent of the Internet and associated communications technologies, the democratization of information has been at hand. Left and right of the political spectrum, information is accessible to all with computer access. Despite the benefits of access to alternative points of view, some government authorities persist in seizing control and ensuring their access to Internet communications. Laws in most developed countries, even those with advanced industrial economies, permit access to Internet communications and thereby retain ultimate control, all said to be in the so-called public interest for preserving peace and security.

In this part of the book, we are called to recognize through some case studies that all advances, whether in communications technology or not, can be used for the public good, or harm. Vigilance on the part of public interest groups is needed for ensuring that the line is clearly drawn between good and evil uses, and everything in between.

In Chapter 25, Rose Dyson's enduring challenge, one that she has been driving for almost two decades, is made all the more urgent because of a global culture so influenced by powerful interests that fuel what and how messages are portrayed to the public. Her concerns expand in this Chapter to the global arena. She calls for even greater vigilance in recognising the impact that media used for promoting business-as-usual interests is having on planet-wide ecological integrity. Propaganda and myth-building are rooted in cultural norms. A compliant public then permits a level of manipulation that is very damaging to global life-supporting ecosystems. Having tolerated the deterioration of our cultural environment, despite countless studies warning us of harmful effects, Dyson identifies several strategies to correct this decline. First, we must break the code of silence and fear on the subject for a transformative change in consumer-driven lifestyles fuelled by billions of advertising dollars and government subsidies. Second, the need for ministries of education to work, inter-disciplinarily, with ministries of culture will help to determine effects of changes in policy on both the cultural and the natural environments. Third, health care providers need to become involved in the issue of advertising targeted at children because it has implications for the health and well-being of both present and future generations. Finally, the need to ensure an arms-length distance between media regulatory bodies and powerful interests to avoid corporate-friendly plans toward further deregulation is critical. Only then will we begin to move toward real systemic change. Rose Dyson, as President of Canadians Concerned About Violence in Entertainment (C-CAVE), believes that the public has a right to know that the overwhelming weight of research points toward harmful effects. While C-CAVE is a strong advocate of free speech, it does not believe that the commercial and sexual exploitation of children through the media should be protected along with political expression.

Robert Rattle, in Chapter 26, discusses how the power of Internet and communication technologies (ICTs) could be harnessed to transform global cultural trajectories to better protect and enhance ecological integrity. Despite their remarkable power to support the human enterprise, one unintended consequence of the ICTs is that they are proving quite energy, resource and waste intensive. The energy consumption of the Internet and the associated greenhouse gas emissions, mountains of e-wastes, a stew of toxic chemical inputs, and high-precision production refinement requirements and operational tolerances are but a few of the challenges that tarnish the potential benefits of ICTs. Despite these impacts, the social consequences from ICTs may prove even more disastrous. The ownership and control of media extends it's tentacles of power and persuasion to all media, and those with the greatest power continue to exert the loudest messages. ICT access is remarkably distinct along numerous lines of inequality, and media messages are overwhelmingly normalized to global trajectories of cultural evolution. These trajectories appear to covet unsustainable consumption practices including economic growth. Further, ICTs endow the global commons with incredibly powerful tools to - often unintentionally - contribute to those practices. It is from this perspective that ICTs prove to present an incredibly powerful social tool that simultaneously expands the human footprint and inflicts enormous ecological devastation.

The last, but not least important chapter in this Part of the book (Chapter 27) is from Doug Daigle. He analyzes two case studies that were brought to closure in 2010, revealing the motives of powerful interests in misdirecting public opinion to support the interests of the forces of climate denial through the "Climategate" scandal. Daigle demonstrates how deliberate and relentless powerful interests can be, and he reveals how systemic the infiltration of these forces has become in their ability to influence public opinion through various kinds of media and communications technologies. Ideology is shown to influence, distort and marginalise scientific knowledge through the misuse of the media. In his second case study, Daigle examines the background to the 2010 Deepwater Horizon Blowout (also known as the BP oil spill in the Gulf of Mexico) as the largest oil disaster in U.S. history. Considering 10 years of corruption in the watchdog federal agency, he notes that the continuing denial of peak oil has added to technological risks that are becoming systemic in order to continue to feed the world's addiction to oil rather to move toward a green economy and the development of renewable sources of energy. Instead of sewing conflict and confusion, recognizing the reality of peak oil and redesigning our energy system would be a more responsible approach to take if a path to a sustainable future is to be reached.

All forms of media provide a range of powerful tools for shaping culture and influencing public opinion. In public control, these tools can serve the public interest and thereby engender greater public good. Because any tool can be used for good or evil, great vigilance is needed.

The determination of individual and collective values and culture, locally and globally, is foundational to the choices that people make and their aspirations. These values affect everything in this world, from the restoration and maintenance of ecological integrity, to how we adapt to the challenges from the declines in all that support the sustainability of life for both present and future generations. All forms of media that inform and drive that which is foundational to the sustainability of life cannot be left in the hands of those with only self-interested, short-term visions of the world. All interests need to be engaged in such determinations.

CHAPTER TWENTY-FIVE

MEDIA USE AND MISUSE: AT ODDS WITH A SUSTAINABLE FUTURE

ROSE A. DYSON

"Awareness has to be raised in both the East and the West to deglamorize unsurvivable consumerism"
—Julia Whitly, 2010

Introduction

The paradox is that although alleviating poverty may be the fastest way to reduce poverty and family size, it is also the fastest way to increase individual ecological footprints. Clearly, a delicate balance is required with a compromise between fewer people and more people with fewer needs, desires and wants. This means that changing behaviour and common practices to ensure a sustainable future is the primary role for educators, policy makers and researchers today. It also means bridging the gap between theory and practice. No where does the disconnect between the two persist more than in the field of media content, communications technologies and cultural policy - where they are leading us, how to harness their enormous potential, and where the balance between regulation and cherished freedoms ought to lie.

But time is running out. Building safe, liveable communities now includes the added overlay of unfolding complications due to climate change, shrinking resources and financial instability. The need for a new economic order that encompasses modes of social and economic organization based on co-operation and social responsibility designed to serve life rather than accumulation requires an examination of how profit-driven, corporate media block such objectives with emphasis on materialistic definitions of desirability and success. Countless studies have been done and books written on the subject. The task ahead is to translate this accumulated knowledge into effective policy before it is too late.

The Modern Globalised Context for Concern

Peter Nicholson, President of the Council of Canadian Academies, tells us that as we become information-rich, we are becoming attention-poor, an inevitable side effect of the digital revolution. "Economics teaches us", he says, "that the counterpart of every new abundance is a new scarcity - in this case, the scarcity of human time and attention." (Nicholson Sept 12, 2009). Additional observations and reports warn us of radiation overload from too much cell phone use and screen time, ways in which ever intensifying stimulants first initiated by television diminish our capacity for imagination and creativity, and how these accumulating side effects still receive little attention in policy-making circles (Robbins 2010).

American journalist and Pulitzer Prize winning author, Chris Hedges, warns us of the diminishing distinction between reality and illusion (Hedges 2009). Such trends lead to death and we are at a crossroads. Either we will wake up from our state of induced childishness, where trivia, gossip and celebrity worship pass for news and information in our search for an elusive and unattainable happiness, and confront the stark limitations before us, or we will continue our headlong retreat into fantasy and ultimate demise.

As the propaganda machines grind on, too many reporters function as stenographers for corporate elites rather than hard working investigative journalists. Yet, the extent to which our cultural environment is controlled by large conglomerates, dominated by marketing and advertising interests, is poorly understood by the public at large. The irony is that this era is rapidly coming to an end. Journalists and policy makers around the world are watching what is happening in the United States where century old newspaper dailies are folding or declaring bankruptcy. It is the nation that has exported both the ideal of a free press and the pathologies of a corporate mass media system to countless other countries (McChesney & Nichols 2010). Although the Internet is the undisputed future of journalism, the profit driven model of support from advertising dollars is not materializing in sufficient numbers to sustain private newspaper ownership in a way that can guarantee decent wages for essentials such as scientific and investigative journalism.

Educators, activists, and media scholars have for years, chronicled from the margins, the systemic deterioration of journalism as well as entertainment media, increasingly dominated by violent content because it sells well in a global market and translates easily into any language. A deep-seated and long-term crisis has resulted because media owners have made the commercial and entertainment values of the market dramatically higher priorities than the civic and democratic values essential to good journalism and, in turn, a good society. An erosion of standards has led to a rise in stories about sex scandals and celebrities, giving the illusion of controversy. This has led to an increasing displacement of good journalism with sophisticated propaganda which tells people what they need to know to consume products and support spurious wars, but little of what they need to know to be voters and responsible citizens. According to David Suzuki, over $ US 500 billion is spent annually by the advertising industry to get any of us on the Planet to buy things (The11thHour.com). Record profits are frequently announced by Microsoft and other gaming manufacturers who socialize our youth to amuse themselves with endless sedentary, interactive screen time with video and computer games such as Killzone Liberation and the World of Warcraft. Many, such as Manhunt 2, banned in the U.K. are produced in Canada and subsidized by public taxation.

There is an obviously strong link between public education and a free press. The downgrading of journalism leads to ignorance, corruption and misery, the very symptoms that tend to accumulate both within government and the private sector when transparency and vigilance from a free press are reduced. Although the blogosphere and citizen journalism are encouraging symptoms of positive transformative change in the way news is now gathered and reported, close observers stress that most of these developments are still driven by and attached to old media operations. And while non-profit organizations can provide valuable information for public education purposes in their respective areas of expertise, the general trend toward Internet provision of news which results in digital sweatshops is to be avoided.

One of the biggest challenges for educators in this digital age is to dissect the myth that has sprung up in the last century and a half about the evils of government involvement in journalism. The notion that only the corporate sector is the true guardian against censorship has developed into a woeful under-examination of the privileges that media moguls enjoy. Public interest obligations originally promised in exchange for these privileges have virtually disappeared in recent decades. An early commitment to provide professional media services has gradually evolved into a euphemism for "self-regulation". Freedom of the press rests with those who own one, or control the broadcast microphones, or own the Internet servers.

Responsible journalism ought to be regarded as a public good like health care, national parks, and emergency response to natural disasters. Historians have tended to avoid attempts to reconcile the fact that the same enlightened sages of the past who laid the foundations for democratic governance, also created a partisan press system, subsidized by political parties and government contracts which included postal giveaways. Furthermore, these have never been "free" markets. News, advertising and entertainment, either in print or electronic form, have been recipients of huge direct and indirect subsidies, tax breaks and exemptions. Few people understand this.

On the eve of China's 60th anniversary celebrations, a front page article ran in *The Toronto Star* on Beijing's vast and effective propaganda apparatus (Schiller Sept. 27, 2009). But, the reporter's description of how inhibited Chinese journalists are compared to those in the West was overstated. Granted, the firing of Chinese journalist Liu Yuan for breaking a story on the death of a 15-year-old boy at a camp to cure Internet addiction was inexcusable. Such a story in Canada might well have resulted in an award for the journalist. However, that is as far as it would be allowed to go. Corporate media interests would very quickly block any progress toward serious policy development to address the problem.

For example, in 2008, a tsunami of opposition orchestrated by corporate media interests broke out in the mainstream media over a bill proposed in the House of Commons which would have eliminated tax credits for audio-visual productions deemed to be harmful to the public interest. Industry lobbyists rejected the argument that such discretionary funding is expected of any democratically elected government entrusted to set policy on how public money is spent. So the upside-down world of journalism is hardly unique to China. The difference in Canada is that it is the corporate media rather than party officials who call the shots. Indeed, in China, the problem of Internet addiction among youth in particular, is at least acknowledged. In Canada it is ignored. Within the mental health community, growing concerns from parents, educators and family therapists about increasing evidence of Internet addiction among youth have also been ignored. Denial is the order of the day. The problem is considered too messy and difficult to address despite news reports that Canadian children are engaged in 3 times the medically recommended amount of screen time (Ogilvie 2008). In Japan the government has warned parents and teachers of new legislation restricting screen time for youth as a result of evidence of widespread addiction to computer use. In recent months South Korea and Switzerland have both banned violent video and computer games (Grossman 2010).

Parents, on the other hand, are now up against enormous odds in rearing their children. Ways in which the advertising industry knowingly undermines family cohesion through the use of marketing tactics such as "the nag factor" and "pester power" need to be better understood and resisted. Media literacy courses for both adults and children should emphasize ways in which media detract from eco-literacy.

Harmful media content also fuels climate change and social upheaval in ways beyond obvious concerns about fair and accurate reporting. Despite the popularity of Al Gore's celebrated documentary, *An Inconvenient Truth,* vast amounts of fossil fuel continue to light up billboards, and power vehicles for a single film, television or commercial shoot. In 2008, it was pointed out to the Canadian Senate Committee on Banking, Trade and Commerce that, in the previous 12 years, taxpayers contributed over $22 billion to the audio-visual industry.

In 2007, Canada's national newspaper, *The Globe and Mail,* ran an article on a television recycling plant near Toronto's Pearson International Airport where overworked technicians scramble to keep up with the demand for the removal of lead and other harmful metals from old-fashioned tube-style sets as viewers switch to flat screens. But, for every old set shredded, it was reported, 600 new ones are manufactured for distribution in China. Similar stories abound about the life-cycles of computers, cell-phones and other forms of communications technologies. Ironically, at the 2008 Conference of the Canadian International Council, held in Toronto, Wenran Jiang from the University of Alberta spoke of how China was starting to wake up to the ills of environmental devastation and climate change and is especially concerned about becoming the dumping ground for 90 percent of the world's electronic garbage.

Current demands for net neutrality and the right to equal, unlimited access to the Internet, regardless of energy costs or the purposes of its use are shortsighted. What we need, instead, are new lifestyles and changes in basic modes of production and consumption with restriction of advertisements to essential information and a compulsory rationing of products. Pornography, now estimated to involve 40 percent of all Internet use, fuels violence against women and children. The worldwide sex slave trade is identified by the United Nations as the largest illegitimate form of business in the world today. Meanwhile, alternative sources of livelihood, in poverty-stricken developing countries, in particular, continue to shrink in the aftermath of globalization, climate change and growing food and fuel shortages. The pornography industry has actually won the culture war fought with feminists to free women from sexual tyranny. As Hedges points out, "Stripping, promiscuity, S&M, exhibitionism, and porn are now mainstream chic" (Hedges 2009: 86). Pornography is the same disease involving violence and domination that glorifies the cruelty of war and celebrates it in other forms of 'action filled' popular culture commodities. These, in turn, are metaphors for the disease of corporate and imperial power.

Yet the dots remain stubbornly unconnected. In both the 1,000-page Falconer Report on Violence in Toronto Schools released in January, 2008, and the yards of ink which followed, the focus was on the code of silence and fear among teachers and superintendents, with none on the culpability of the media industries themselves. The result is a gradual erosion of our right to feel safe and our responsibility to make others feel safe as the boundaries between entertainment and victimization fade.

What about the Research?

Findings on the harmful effects of violent entertainment from thousands of studies, released into the public domain in recent decades, are usually quickly neutralized to ensure that the debate never gets beyond proof of harmful effects and onto policy (Dyson 1995). In 2001, a leading Japanese brain specialist found that playing Nintendo video games renders parts of the brain inert. The corporate giant approached him and quickly became his number one research donor. Since then, Dr. Rutya Kawashima has reciprocated by calling for more research and for the gaming industry, it has been back to business as usual (Cameron 2006). Findings involving the use of MRI techniques demonstrate that brain cells which normally counsel empathy are shut down in teens who play violent video games (Linn 2004).

Aggressive marketing yields $1 billion U.S. in annual sales for Brainy Baby and Baby Einstein videos despite protests from the Campaign for a Commercial Free Childhood, based at the Harvard University Medical School. The American Pediatrics Academy recommends no screen time at all for children under the age of two, yet only 6 percent of parents are aware of this. Recent evidence demonstrates that children who watch television or video games before the age of two actually show slower vocabulary development than those who experience no screen time (CCFC 2007a). In this case a rare, but qualified win should be noted. Last fall,

Walt Disney was persuaded to refund all purchases of these videos in NA within the last 5 years on the basis of false claims. The offer in Canada and the U.S. was good until March, 2010. This success did, however, result in the CCFC losing their lease at the Judge Baker Center for Children because of corporate donors upset over this development.

What about Censorship?

Demands for the U.S.A. First Amendment to be updated are growing. This is a central theme in McChesney and Nichol's book on The Death and Life of American Journalism. Scholars from various disciplines, particularly in the humanities, are calling for a reinterpretation of both the Amendment and the definition of healthy economic activity in response to mental and physical health issues, looming environmental disasters, and the crisis in American journalism itself (The11thHour.com). While global warming and the coming energy crisis are not major problems, they are symptoms of deeper, entrenched cultural and societal problems. Massive changes to our conceptions of truth, freedom, individual and human rights, and how we see ourselves in relation to the Earth as a whole must change dramatically in the next few years if we are to survive as a species.

Indeed, the threat to our survival goes well beyond the issue of climate change. Violent video and computer games are also helping to fuel terrorism. It is estimated there are now over 8,000 websites associated with extremist groups, many used for recruiting young converts (Dyson 2007). Almost all terrorist activities in recent years have been executed by members of diaspora communities, whether in Spain, England, Holland or Canada. Often they are well educated, technologically-savvy and look up to al-Qaeda (Blatchford 2008). As Thomas Friedman pointed out in a PBS television interview with Charlie Rose recently on the Afghanistan war, the war on terror is now virtual. The threats are no longer contained in that country or along the Pakistan border (May 14, 2010).

Some see involvement in terrorism as a graduation from gangstra rap for young testosterone propelled male teens to a kind of status symbol. Many bring with them skills and lifestyles associated with urban youth gangs where guns, violence and extremism are the norm. At the 6th Annual Summit on Emergency and Disaster Planning for colleges, universities and K-12 schools held in Toronto in 2009, Bill Byrd, Safe Schools Inclusion Administrator for the Toronto District School Board reported on trends toward more gangs and youth violence while family control is deteriorating (Strategy Institute, Oct. 6-7, 2009). Speaking at the same summit, Craig

Peddle, who studies and investigates youth gangs, said he has no doubt that popular culture is a causal factor and reported that for every single website that addresses the problem, there are 100 promoting it, not only on the Internet itself, but through periodicals, rap music and other forms of popular culture. The gangs are growing in number, and many are now inducting girls with the average age-level being pushed up as high as 22 years of age.

One of the central themes in popular culture for young people for decades has been a tendency to undermine all authority figures, parents and teachers included. The subsequent rise in government promises for tougher law and order measures, and the increasing focus on national security issues was predictable. It resonates with the late George Gerbner's definition of "the mean world syndrome" as one of the harmful effects from entertainment violence to the community at large (Morgan 2002).

The Way Forward

Having tolerated the deterioration of our cultural environment despite countless studies warning us of harmful effects, how do we turn things around? First, we must break the international code of silence and fear on the subject. At the 2007 UNESCO annual meeting in Ottawa, Ahmed Djoghlaf, Executive Secretary of the Convention on Biodiversity and Culture, emphasized the importance of transformative change in large urban centers where reverence for nature and biodiversity must be cultivated. But, he somehow managed to avoid entirely the issue of consumer driven lifestyles fueled by billions of advertising dollars and government subsidies for harmful content and with universities collaborating through skills training.

Plans for safe schools in large urban centres, such as Toronto, admittedly hinge on funding. But, surely it is a no brainer that they must involve more than the millions earmarked to hire more social workers, policemen and psychologists to deal with cyber-bullying, sexual predators and expelled students? One wonders if officials from Ministries of Culture ever talk to those in Ministries of Education. All applications for government funding - anywhere - should be carefully monitored for their impact, on both the cultural and natural environment. A lesson can be learned from Leonardo Di Caprio who concludes his film, *The 11th Hour*, with the statement that it was produced with the smallest possible impact on the environment.

Health care providers, on the whole, must become better advocates for change. Given the evidence of the potential harm of commercialized culture, it is essential for them to take on more responsibility for the education of parents about the negative effects of media and marketing on children and to work with parents to limit children's access to screen media. But, that is not enough. Limits need to be set on the access marketers have to children as well. As pointed out by educational psychologist, Susan Linn, Director of the Boston, Mass. based Campaign for a Commercial-free Childhood, "Toward that end, health care providers need to move beyond their offices and become public advocates for policies that restrict and/or prohibit advertising and marketing to children" (April, 2010).

Canadian organizations such as Edupax, the Canadian Centre for Policy Alternatives, and Canadians Concerned About Violence in Entertainment have taken public stands against media violence marketed to children, in particular, to in-school commercialism and to advertising directed to children. The Canadian Paediatric Society and its individual members should also become more proactive and join with municipal bodies such as the Toronto Board of Health, organizations such as the Ontario Public Health Association, the Toronto Elementary Teachers Association, and the Ottawa based Centre for Science in the Public Interest and become part of a growing movement to extend Quebec's ban on advertising directed to children 13 years and under to the rest of the country. Analogous strategies could be implemented in other countries.

Policy-making everywhere in the developed world must measure up to standards in the province of Quebec and other parts of the world where advertising to children is banned. Such rules apply to violence as entertainment, the marketing of junk food and the sexual exploitation of children. The Scandinavian countries, Malta, Greece, Turkey and New Zealand adopted such standards years ago, with the U.K., Switzerland and Italy joining the list last year. The Canada-wide ban called for by the Toronto Board of Health which has led to the introduction of private members bills in the Ontario Legislature and at the federal level in Canada should, with urgency, be passed and implemented (Goar, Feb. 28, 2010; Jeffery Fall, 2007).

Regulatory bodies such as the Canadian Radio, Television and Telecommunications Commission must be discouraged from corporate friendly plans toward further deregulation while ignoring findings of growing problems such as obesity from the sedentary nature of computer use and TV viewed, and the relentless marketing of junk food. In 2004, researchers at Laval University found that, in 10 years, acts of violence on

Canadian TV rose 286 percent with 81 percent of it before 9:00 p.m., the watershed hour established for the protection of children (Hansard 2007). Clearly, industry self-regulation does not work without a little help from government and the rule of law. Cultural policy must become more topical during all election campaigns and better connected with global warming and other looming disasters. Only then will we begin to move toward real systemic change.

References

Cameron, D. (2006, May 13). Moving to the dark side of the screen. *Sydney Morning Herald.*

Campaign for a Commercial-Free Childhood. (2007). www.commercialfreechildhood.org/index.html.

Colbourne, S. (2006, May 10). Gamers get set to break a sweat. *The Globe and Mail.*

Dyson, R. (2007, Jan.) Hate on the Internet. *The Learning Edge.* www.oise.utoronto.ca/CASAE.

—. (2000) *Mind Abuse: Media Violence in an Information Age.* Montreal: Black Rose Books.

—. (1995) *The Treatment of media Violence in Canada since publication of the LaMarsh Commission Report in 1977.* Doctoral Thesis. OISE/UT.

Dude, R. (2008, July 14) Beyond Blackberry Thumb. *The Globe and Mail.*

Goar, C. (2010, Feb. 26) Shielding Youth from Pressure to Buy. *The Toronto Star.*

Grossman, D. (Mar. 21, 2010) *Trained to kill.* LtColDaveG@aol.com.

Houpt, S. (2008, July 14) Summer camp for mystery hopefuls. *The Globe and Mail.*

Hedges, C. (2009) *Empire of Illusion: The End of Literacy and the Triumph of Spectacle.* Knopf, Canada.

Jeffery, B. (Fall, 2007) Hitting the easy mark: The law on marketing to children is ripe for reform. *Our Schools/ Our Selves: Media Education and Educating the media.* The Canadian Centre for Policy Alternatives. V. 17 N.1 (#89) Fall, Ottawa, 2007.

Jegatheesan, R. (2008, July 14) As campaigns heat up, U.S. braces for 'swiftboat' season. *The Globe and Mail.*

Linn, S. (April, 2010) The commercialization of childhood and children's well-being: What is the role of health care providers? *Paediatrics & Child Health.* V. 15 N. 4 April, Ottawa, 2010.

—. (2004). *Consuming Kids: The Hostile Takeover of Childhood.* New York: The New Press.

McChesney R. W. and J. Nichols (2010) *The Death and Life of American Journalism.* Nation Books, Philadelphia.

Morgan, M. (2002) *Against the Mainstream.* New York: Peter Lang Publishing.

Mortished, C. (2008, July 10). Brown stumps while economy slumps. *The Globe and Mail.*

Nicholson, P. (2009, Sept. 12) Information-rich and attention-poor. *The Globe and Mail.*

Oglivie, M. (2008, May 28) Kid's couch-surfing hits new high. *The Toronto Star.*

Ontario Media Development Corporation. (2008, Feb. 12) OMDC Invests $1 million in Video Game Prototype initiative. *News Release.* Ontario. Author.

Ontario. (2008). *The Review Of The Roots of Youth Violence.* Queen's Printer for Ontario.

Robbins, J. (Spring, 2010) Missing the Big Picture: Studies of TV's Effects Should Consider How HDTV is Different. *The New Atlantis.* No. 27, Washington, D.C.

Saltzman, M. (2007, Oct. 13). British still unable to stomach bloody Manhunt 2. *The Toronto Star.*

Schiller, B. (2009, Sept. 27). 'The aim is to make people docile'. *The Toronto Star.*

Schuker, L. (2008, May 27). For studios, a summer blockbuster isn't what it used to be. *The Globe and Mail.*

Stewart, S. (2008, July 9) In media skirmish, Fox aims for the head. *The Globe and Mail.*

Strategy Institute Inc. (2009, Oct. 6-7) 6th Annual Summit on Emergency & Disaster Planning. www.strategyinstitute.com.

Whitly, J. (May, 2010) Population: The Last Taboo. *Mother Jones* <motherjones.com/environment/2010/05population-growth-india-vatican>.

CHAPTER TWENTY-SIX

COMPUTING OUR WAY TO PARADISE

ROBERT RATTLE

Introduction

Globalization, quite likely a process that has been occurring for centuries, has bounced onto the world stage - indeed *has become* the world stage - in the last few decades. It is characterised by the shifting of time and space, enabling peoples of different cultures and geographies to understand, learn from and exploit the opportunities of distant practices. The Internet and communication technologies (ICTs) have performed and continue to perform, important roles in modern globalization.

Internet and communication technologies are quite simply any product or system that communicates, stores, and/or processes information (O'Meara 2000). ICTs have unique characteristics and were originally conceived as a virtual service. Partly attributed to this 'weightless' sector assumption, ICTs became very popular, especially in environmental circles. Their popularity has served to propel human activity around the globe, often with the vague assumption that perfect substitutability of less efficient processes with more efficient ones would enable a diminished ecological footprint of the service provided.

Evidence now indicates that the ideal of perfect substitutability is largely an illusion as many factors appear to militate against pure substitutability. Moreover, efficiency, as a primary method to reduce the ecological impact of an activity, has proven to be a double-edged sword. In fact, ICTs have often been applied as yet another layer of social innovation to advance the physical growth of economies and businesses, and their concomitant physical impacts.

ICTs are far from the benign service originally conceived. The buildings that house servers - the machines that store our text, videos, emails and other data - are very much physical in nature, and the energy demands of these servers have become a significant strain on energy grids around the world. The manufacture of ICT devices and components draw physical resources that place increasing demands on - even create conflicts over - water and non-renewable resources. The lifespans of ICTs are incredibly short, and once retired, old ICTs often find they serve little more purpose than toxic pollution, despite a proliferation of international trade agreements and local initiatives to prevent their inappropriate disposal, and to encourage more sustainable options.

Yet the impact of ICTs on ecological integrity may prove more pervasive than the above less than appealing scenarios might suggest. The emerging global information society may be establishing practices and lifestyles that entrench, rather than liberate, citizens in their role as global consumers. Perhaps more profound are the roles of certain social actors in the substantiation and advance of technological progress and the consumer ethic. These actors reflect a profoundly unequal distribution of social wealth and control vast resources, including many media platforms. Following rules laid down by global markets, these actors remain alert to opportunities that seek to exploit globalization for their purposes of wealth accumulation. ICTs are consistently applied to manipulate global social thought, normalised along global cultural trajectories that predominantly covet economic growth and the consumer as a central actor in market economies.

Perhaps even more perplexing and confounding are the effects ICTs exert through global financial mechanisms, and the subsequent implications for global ecological integrity. As consumers enter the formal economy, they become indebted to modern global financial activity, and in so doing contribute to the mechanics of abstract wealth generation. If anything, the global financial meltdowns of recent years have taught us that there is an increasingly tight coupling of global financial activity with the physical world we inhabit. As these connections strengthen, peoples and cultures around the world become increasingly beholden to the dictates of modern global market mechanisms manipulated by ICTs. Investment practices are exclusively directed to the growth of abstract wealth, once perceived as the invisible hand that would create social wealth. Through ICTs, rapid information flows direct modern investment practices. One need only look at trading volumes and market activity since the emergence of the computer as a central technology in markets in the mid-1990s to observe the rapidly increasing growth in market activity. As people receive, interpret, and transmit vast amounts of data and information in all manner of formats from text to multimedia, they contribute to the opinions, conditions and trends of markets, shifting wealth at quite literally the speed of light, and directing physical changes in the environments we inhabit. Actors have, through increased access to ICTs, greatly improved access to modern market mechanisms, contributing to the speed and swiftness of global market transactions well in excess of historical assumptions laid down by Adam Smith and his colleagues. In so doing, ICTs have generated a level of conscious global activity that feeds off itself - information - and, as we have seen, this can have tremendous implications for people and societies around the planet.

ICTs are neither beneficial nor harmful; they merely develop in a coevolving manner with social values, and the institutions, practices and expectations society has developed to express those values. Yet in today's globalizing world, those values seem to be growth, competition, wealth accumulation, and hierarchical control. This does not bode well for ecological integrity. Despite this rather somber outlook, ICTs hold tremendous potential to enhance ecological integrity, and may in fact be doing just that. There is increasing evidence that ICTs are contributing to a self-emerging phenomenon, but one that, despite all appearances, does not covet growth, consumerism, and wealth inequalities.

Many adoptions and adaptations of ICTs deeply integrate the concepts of distributive sharing, cooperation, and egalitarianism, qualities evident in many online platforms such as social media, micro-volunteering, and even e-health.

Our evolutionary past and many traditional societies reveal evidence of non-hierarchical, cooperative, sharing and egalitarian human qualities. Amongst those qualities was an inherent brake on the 'getting ahead of, or keeping up with, the Joneses' objective through this non-hierarchical, or distributive, manner of sharing power and control. In other words, humans demonstrated effective practices against the emergence of permanent leadership over the majority of our existence. The goal would be "not so much keeping up with the Joneses, as making sure that none of the Joneses gets ahead in the first place" (Erdal & Whiten 1996: 148). The potential for these structures to transform the concept of modern 'leaders' who now manipulate the balance of global power structures - effectively wielded through control and ownership of the media and communications infrastructure that supports ICTs - suggests a return to the distributive decision-making of our evolutionary past.

In effect, it is the attachment to conventional global economic ways of operating - global market ideologies characterised by growth and competition - that is militating against the emergence of a sustainable society. Beyond that barrier exists tremendous hope for ecological integrity.

We are now at a period in human history where we no longer need rigidly controlled hierarchical structures to distribute the wealth of nations. Global society can - indeed already may be - escaping those shackles. If we can nurture the distributive nature of ICTs for expanding cooperation, egalitarianism, and sharing, humans may embrace social and ecological sustainability on a global scale.

In this Chapter, I discuss the potential ecological consequences of ICTs and their role in a more sustainable world.

On the Surface

To the casual user, ICTs appear ecologically benign. Most devices are small, contained, and clean. Their computing power, while improving at an astounding rate, is already incredible. New devices with updated and expanded functions enter the market regularly, allowing faster and more effective processing and communications. They enhance process efficiencies, cutting costs and time, improving economic productivity. Growth in the ICT sector has been phenomenal, notwithstanding the tech bubble of the late 1990s. Modern life has become utterly dependent on ICTs, and their technological ubiquity is unsurpassed.

These features have made ICTs indispensable in modern life. We all know of their benefits, and most of us could not function without them. However, ICTs also have drawbacks, the least of which is a direct result of their popularity: the sheer volume of electronic waste.

Electronic waste - or e-waste - is the discarded electrical and electronic devices and components of our modern era, including such devices as televisions, computers, monitors, mp3 players, cell phones and other mobile communication devices. These devices are manufactured with and contain many heavy metals and toxic compounds, such as lead, mercury, arsenic, cadmium, brominated flame retardants and polyvinyl chloride.

The market for e-wastes is about $11 billion dollars annually. Each year, some 50 million tons of e-waste are generated, adding to existing accumulations. Despite the establishment of Waste Electrical and Electronic Equipment (WEEE) and Restriction of Hazardous Substances (RoHS) Directives, e-waste continues to grow, and the illegal, dangerous and unsustainable deconstruction of e-waste is a growing business worldwide. Only 15% - 20% of e-waste is recycled in the United States, while the vast majority continues to be exported.

At a basic level, all ICTs contain microchips that are used to power their computational properties. A basic microchip used in pentium central processing units requires 11.44 cubic metres of deionised water, 120.8 cubic metres of bulk gases, produces 0.83 cubic metres of hazardous gases, 14 cubic metres of waste water and 4 kilograms of hazardous wastes (SVTC 2000). So substantial are the refinement needs for device construction, the average input-output ratio for a single DRAM chip is 800:1. For comparison, the same input-output ratio for an automobile is 2:1 (Williams et al. 2002). Nitrogen trifluoride, a gas used in the cleaning of circuit boards, has a global warming potential 17,000 times that of carbon dioxide. Ironically, nitrogen trifluoride was originally adopted for use as a replacement for perfluorocarbons - a powerful ozone depleting substance.

Beyond the rising global problem of e-wastes are the energy demands and carbon emissions of ICTs. This is attributed to the higher energy demands of newer applications and devices to accommodate their processing power and speeds, as well as the strict tolerances these devices demand. As consumers seek higher bandwidth, so too do they demand higher energy requirements. As the world shifts towards broadband as a universal right (United Nations 2010), the energy demands of the Internet are poised to soar.

Data server farms - the location where thousands of servers store and route our requests for text, web pages, multimedia experiences and emails - are constructed to rigid tolerances. These farms must operate without failure, and they contain very high chip densities. As a result, most server farms are over-constructed, requiring substantial dependencies on primary and backup power, and subject to considerable cooling needs.

Kawamoto et al. (2001) and Koomey (2007) assessed server farm energy requirements. Estimating only the server power and cooling and auxiliary energy demands directly associated with the servers, these studies suggest the energy demands for servers in the United States rose from about 20 billion kilowatt hours per year to 40 billion kilowatt hours per year between 2000 and 2005. Globally, these values translate to 60 billion kilowatt hours per year in 2000 rising to 120 billion kilowatt hours per year in 2005. Despite these high values, data storage power, network equipment power, and cooling and auxiliary power for data storage and for networking were not included (Koomey 2007). Furthermore, these studies were focused on data farms, a small segment of the ICT sector.

Resulting from these high energy demands, the equivalent carbon greenhouse gas emissions from server farms in the United States alone exceed the total national greenhouse gas emissions from Argentina or The Netherlands and approximate those of Malaysia (Lucente 2010).

Many approaches have been taken to resolve this problem. The most common is to employ the efficiencies embodied by the process technologies and practices ICTs enable. The SMART 2020 report reflects this approach: "While the [ICT] sector plans to significantly step up the energy efficiency of its products and services, ICT's largest influence will be by enabling energy efficiencies in other sectors, an opportunity that could deliver carbon savings five times larger than the total emissions from the entire ICT sector in 2020" (SMART 2020).

As processes and devices have become more efficient, businesses have benefitted due to the decreased costs of operations. This has led to growth opportunities not previously realized. As Stanley Jevons (1866) noted in the 19th century, the efficient use of resources is not equivalent to the diminished consumption of resources. Indeed, he noted the converse is true. Over a century later, Harry Saunders similarly concluded that "energy efficiency gains can increase energy consumption by two means: by making energy appear effectively cheaper than other inputs; and by increasing economic growth, which pulls up energy use" (Saunders 1992: 139).

It seems that the impact of ICTs depends more on human choice and behaviour than on the predictable theory of technological energy characteristics. It is to these factors that I now turn my attention.

Reality Check

While ICTs hold immense potential to reduce the ecological consequences of human activity, this potential ultimately depends on the way ICTs are applied, which, in turn, depends on human behaviour and choice. Human behaviour is far more complicated than the behaviour of specific technologies, and in every situation where social and ecological benefits may be envisioned from a technology, less than ideal applications can also be envisioned. Indeed, the fact that ICTs have evolved a new world of challenges - security, health, social, ecological, etc. - should in itself be pause for concern.

Understanding interactions between ICTs and human behaviour has been a challenge, as they require understanding complex social environments.

Policy- and decision-makers generally simplify, misunderstand or ignore these complex processes. However, marketers and advertisers seem to have thoroughly grasped at least some of these complex interactions.

Research is beginning to tease apart the intricate, complex pathways of consumption that can be established. For example, Ropke (2001) discusses several social relationships - identity formation and domain conflicts - that affect the adoptions and adaptations of ICTs. Identity formation, closely affiliated with materialism in western economies, is an important social process that helps define who we are. This process can fuel the advertising and marketing promotions as effective tools to elicit product and company loyalty. Similarly, domain conflicts exist as contradictions between the various roles people perform in society. For example, family, work and leisure domains often generate conflicts in their use of products, services and space. Multiple devices that serve the same function, such as a computer that allows portability, one for the kitchen, one for the family room and one for personal uses help to resolve those domain conflicts (Ropke 2001).

Similarly, families often had a single telephone when land lines were prominent. Today, each member of the family will have their own telephone to resolve domain conflicts as well as social issues, such as safety and security concerns, and each of these telephones last on average a few months before being upgraded and updated to satisfy identity formation roles. Notwithstanding the expansion of cellular telephone usage, land lines also continue to expand, both in terms of number of lines, average number of calls per line and per capita, and the frequency of calls and length of calls (Statistics Canada 2006). The replacement of older technology variants, as can be seen, does not necessarily result in their substitution, and, in fact, can generate substantial increases in the use of the older variant. However, in some cases, variants do become obsolete.

As products become obsolete, they are replaced with newer (often more efficient, flexible and versatile) variants, offering similar experiences or services. Despite the challenges often accompanying the adoption of a new technology into the family lifestyle, they permit the expansion and deepening of the consumer process. In general, Ropke's research suggests these relationships tend to accelerate consumption.

The cellular telephone is an interesting case study. As cell phones advanced, newer varieties provided expanded features and versatility, motivating the constant renewal of older products. Owning the latest and newest variety with the latest functions symbolizing the latest trends (which has important implications for access and universal equalities) conveys important social signals to peers. As conventions and institutions change, so too must their tools. In some cases, older variants of a technology simply become dysfunctional, such as analogue cellular telephones, analogue televisions, and the 3 ½ inch floppy disk and tape for data recording. The cellular telephone is also headed in this direction with the advent of the iPhone and similar technology variants.

Lifestyle and behavioural choices make many ICT devices outdated on incredibly short cycles. Interest in new devices and applications, and an expanding arsenal of opportunities to which ICTs are well suited drive the renewal process. In the electronics world, devices become antiquated after as little as 18 months, and many are discarded long before this time to make way for the most updated, popular, interoperable, fashionable, and advanced versions. Similarly, meta-level analyses have demonstrated such consumption effects at the industry level (Sonntag 2000; Traxler and Luger 2000).

Moreover, many of these devices establish lifestyles that can become incredibly energy intensive. Many products, especially geared towards portability and mobility require battery power and charging, which, despite a full charge, are often left plugged in, drawing energy well past their full charge. Their mobility features actually encourage greater mobility, such as a laptop or cell phone that can permit a mobile office, contributing to greater travel, precisely the opposite of which ICTs were supposed to generate for issues such as telecommuting. In fact, video conferencing has increased, rather than reduced physical travel by expanding contacts and opportunities. Modern digital televisions and other devices that operate on an 'always on' mode draw power even when turned off. Despite simple solutions to dramatically reduce domestic energy demand (estimates suggest 30% of domestic energy demand (Aebischer and Huser 2000)), various social and infrastructural reasons limit these solutions. For example, placing a television (and it's associated cable box and set top converters) on a power bar that can be easily switched off is generally unacceptable as this would require the re-programming of several devices each time the power is switched back on.

In addition to identity formation and domain conflicts, ICTs contribute to complex socio-technical processes that include habituation, planned obsolescence and creative destruction, group association, symbolism, flexibility and mobility, a perceived level of time savings, multi-tasking and efficiencies, and multiple varieties of the same devices to serve different time and place settings.

Ultimately, the way people, institutions and nations respond to and interact with ICTs is greatly affected by the social environment and the cultural ideals that direct those environments (Green 1992; Rip 1995; Ropke 2001). While there is a distinct duality of structuration, the establishment of values and belief structures laid down over centuries greatly influence which actions people may, and are encouraged to take, and even more telling, which actions are discouraged. Despite the perception of control and choice that individuals feel they have when it comes to their consumer decision-making, individuals actually have remarkably little control over the structures that affect those consumer decisions. At the level of everyday consumer choice, "the ordinary consumer will have little or no control over most of this decision architecture" (Jackson 2005: 29). Our free choice is not as free, rational or objective as we might be led to believe. Our choices are largely influenced by cultural factors laid down over many years, and cultural change occurs on a far slower scale than do technological changes. It is to this cultural layer that I now turn my attention.

Below the Surface: Virtual Morality

Power and wealth have gradually shifted throughout human history away from the egalitarian, cooperative, sharing structures evident in small group societies that survived sustainably within their local environments for the majority of human history (Knauft 1991; Erdal and Whiten 1996). The modern shift has been towards the increasingly rigid, controlled, unequal, hierarchical, individualistic and competitive social structures manifest today (Naiman 1997).

Almost entirely erased today are the social obligations of sharing and egalitarianism evident in early social organisation. Similarly, the cooperative aspects are also much less prevalent, replaced by a competitive global market standard. So too eliminated are most small social group controls on dominance hierarchies, and similarly any restraints on wealth accumulation. Put another way, gone are any constraints on the Joneses getting ahead or, indeed, constraints for anyone trying to get ahead of the Joneses. These constraints have been replaced by a collection of beliefs, values and sophisticated techniques that instead encourage wealth accumulation and growth.

Lower order effects of ICTs have been considered during the course of the last 15 years. In this section, I briefly consider how ICTs are affecting our global cultural fabric, entrenching social rules that guide how we interact with the ecologies we inhabit.

Where these social rules are most profound is in the nature of financial transactions. Across the western world, money creation has become primarily the prerogative of private institutions (Daly and Cobb 1994). Its creation is accomplished through a creative lending process. Low reserve requirements stimulate the creation of money through the practice of lending. Once debts are repaid, money is created through the interest charged.

As this new money is created faster than actual goods and services circulating in the economy, the value of money declines - it requires more fiat currency to equal the same value of a good or service (Hoogendijk 1993; Schnaiberg 1980). The inevitable result is the relentless pressure for people, businesses and nations to grow their incomes. This growth compels economic expansion, and collectively generates a remarkably powerful normalising growth force (United Nations 1998).

Anyone with the ability to invest also contributes to this process. The remarkable ability of ICTs allows people to connect directly to global financial trading mechanisms. The resulting millions of individual decisions slosh around in the electronic financial Ethernet contributing to the ebb and flow of global financial markets.

Recent global market periods, characterised by irrational exuberance and rapid capital withdrawal, demonstrate the synchronous nature of global financial decision-making. Governments are hard-pressed to affect these movements, and even leviathan efforts have been met with trivial effect, so large and swift are the global capital movements. During less turbulent times, markets also shift in a remarkably fluid fashion, ebbing and flowing with the mood of the day.

One might compare these movements to species' groupings, such as a flock of birds or a school of fish. John Livingston (1994) observes how such species have a seemingly instinctive ability to communicate effortlessly and instantaneously - a form of group consciousness. "The flashing angles, turns, slants, starts, and stops are too swift, too fine-tuned - indeed, too erratic - to be executed by the synchronized separate movements of hundreds of individuals" (Livingston 1994: 106).

By comparison, through global social structures, humans have been furnished with tools that accentuate individualism and mechanistic reductionism. Conversely, group consciousness, argues Livingston, remains imperceptible to the human mind, so engrossed we are in the individual. Humans are, as a consequence of our technological prosthesis, unable to comprehend this higher level of connectedness.

Now, however, through the power of our connected and wired Internet and communication technologies, humans may have (re-)gained that instinctive ability to communicate across society and, more than merely communicate, respond to and act as a group in unified form.

Until recently, human ecologists have been able only to marvel at the synchronous movements, speed, eloquence and the extraordinary maneuverability of such species' groupings - their "oneness." Could it be that we have re-created this "group-consciousness" through our technological prostheses in servitude of capital and profit growth? "Have the institutions of market-driven globalization and the role of ICTs given birth to an emergent global super organism capable of functioning, not as a mere aggregation of individuals, but as a self-aware, coherently functioning, synchronous growth entity?" (Rattle, 2010)

The expanding global arsenal of financial mechanisms have been made possible through the computing prowess, communications and connections of ICTs. Programs generate and automate the entire process in most situations, often triggering rapid cascading effects that can radiate globally in mere seconds, with very real human and ecological consequences.

ICTs afford access to this computing power for untold numbers of individuals in ways unimaginable barely a decade ago (Scholte 1997). The global power structures of wealth - still securely established today - help maintain certain ideological beliefs and values. The social structures we have developed over the last few centuries support these power structures.

ICTs have contributed to the democratisation of trade and financial planning. Indeed, ICTs have contributed in numerous complex ways to the indebtedness of consumers to the financial process. Their global communications and processing capabilities have expanded access to information - the social, political, economic, health and even environmental events that affect the (financial) decision-making processes. The swiftness afforded by ICTs can induce rapid rippling of financial activity - the greed and fear that drives markets. This effect can direct abrupt capital withdrawal or entrance, affecting physical realities in far off lands, impacting the ecological health and population health of diverse geographic areas. Opinions as much as gossip, rumour and competent analyses can inform financial decision-making. Media in particular, and control of that media, can effect enormous social influence and control, despite the distributive and more democratic openness that ICTs are believed to engender. Disinformation, information overload and confusion, not to mention contradictory messages, can rapidly spread contagion around the globe, affecting financial activity in utterly unpredictable ways in the blink of an eye.

Today, these structures appear to be contributing to the rise of a global growth consciousness, securely anchored in the global psyche by historic belief structures and the infrastructures and processes that reinforce them.

But, are they securely anchored, or might they be in a state of flux? The global ideologies that have given rise to the consumer ethic and growth economics germinated and thrived under hierarchical structures. ICTs invert those structures. By giving rise to ICTs, hierarchical growth structures of private wealth accumulation may have thus sown the seeds of their own demise. It is worth noting that a similar shift in power previously occurred when the landlord class, sensing the potential for better profits from resources rather than rent, evicted peasant workers and began using the monied economy, giving rise to the shift in power from land to capital as evicted peasants moved to urban areas and contributed to the expansion of the capitalist movement. Might a similar shift in power be occurring today as ICTs give rise to egalitarian, distributive, cooperative networks?

Globalisation and Ecological Integrity: The Role of ICTs

While earlier sections of this Chapter paint a rather bleak picture, the key lies in the fact that ICTs do not define how they will be applied - that decision lies with social actors, and the variety of institutional structures to express our values. As such, where ICTs may be applied to create a coherent growth consciousness, they could just as readily be applied to generate a coherent sustainable de-growth consciousness.

To overcome this barrier to sustainable de-growth will require a transformation of values, which, beneath the turbulent surface, may already be underway.

If ICTs can contribute to the global growth of consumerist messages, why could they not also contribute to the global expansion of conserver messages? In effect, this may be occurring, as micro-volunteerism, non-profit platforms, new definitions of wealth, and social networking expand distributive frameworks of power and wealth. In so doing, there is some expectation that these transitions may bring with them a better understanding of our place and role in the ecosystem - the complex linkages, not only between individual people and separate cultures, but also of the human and ecological systems that define and shape this planet.

Hierarchical social structures sanctioned the message of getting ahead of the Joneses. This served as a central factor in the growth ethic of the 20th century. That message is now being challenged. With a distributed, egalitarian, cooperative global network, the emerging message may be one of preventing the Joneses from getting ahead in the first place.

Conclusions

Globalisation has altered relations across space and time, transforming cultures and geographies. In our global world, ICTs perform indispensable roles, and perhaps ones which may prove transformational. Human civilisation may be returning to our socially egalitarian and cooperative past. Since the agricultural period began, humans required complex hierarchical structures to maintain the distribution of social wealth. This led to complex power relations that remain to this day. Those complex power relations established unequal structures characterised by competition, growth and the accumulation of material wealth.

ICTs have now created a global environment in which sharing, cooperation and egalitarianism have once again become prominent features of human society.

The distributive powers of ICTs reflect a return to historic ways of social provisioning, only this time on a global scale. The culturally evolutionary implications of this phenomenon suggest not only a redistribution of wealth, but a potentially sustainable de-growth of the human economy to accommodate and be respectful of ecological integrity. The direct and indirect ecological impacts of ICTs - e-waste, rebound effects, lifestyle changes - may prove trivial by comparison.

References

Aebischer, B., & Huser, A. (2000, November). Networking in private households: Impacts on electricity consumption. Ittigen, Switzerland: Swiss Federal Office of Energy. Retrieved July 6, 2005, from www.electricity-research. ch/SB/haushaltsvernetzung-00-english.pdf.

Daly, H. E., & Cobb, J. B. Jr. (1994). For the common good: Redirecting the economy toward community, the environment, and a sustainable future. Boston: Beacon Press.

Erdal, D., & Whiten, A. (1996). Egalitarianism and Machiavellian intelligence in human evolution. In P. Mellars & K. Gibson (Eds.), Modelling the early human mind (139–50). Cambridge: McDonald Institute Monographs.

Green, K. (1992). Creating demand for biotechnology: Shaping technologies and markets. In R. Coombs, P. Saviotti, V. Walsh (Eds.), Technological change and company strategies: Economic and sociological perspectives (164–84). San Diego: Academic Press Limited.

Hoogendijk, W. (1993). The economic revolution: Towards a sustainable future by freeing the economy from money-making. Utrecht, Netherlands: Jan van Arkel.

Jackson, T. (2005). Live better by consuming less: Is there a "double dividend" in sustainable consumption? Journal of Industrial Ecology, 9(1–2), 19–36.

Jevons, W. S. (1866). The coal question—An inquiry concerning the progress of the nation and the probable exhaustion of our coal mines (2nd ed.). London: Macmillan.

Kawamoto, K., Koomey, J. G., Nordman, B., Brown, R. E., Piette, M. A., Ting, M., et al. (2001, February). Electricity used by office equipment and network equipment in the U.S.: Detailed report and appendices. Washington, DC: U.S. Department of Energy.

Knauft, B. M. (1991). Violence and sociality in human evolution. Current

Anthropology, 32, 391–428.

Koomey, J. G. (2007, February). Estimating total power consumption by servers in the U.S. and the world. Oakland, CA: Analytics Press.

Leake, J. and Woods, R. (2009). Revealed: the environmental impact of Google searches.

Livingston, J. (1994). Rogue primate. Toronto: Key Porter Books.

Lucente, Edward J. (2010). The Coming 'C' Change in Datacenters. HPC Wire, June 15, 2010.

Naiman, J. (1997). How societies work: Class, power and change in a Canadian context. Scarborough, ON: Irwin Publishing.

O'Meara, M. (2000). Harnessing information technologies for the environment. In L. Stark (Ed.), State of the world 2000: A Worldwatch Institute report on progress toward a sustainable society. New York: W. W. Norton & Company.

Rattle, R. (2010). Computing Our Way to Paradise? The Role of Internet and Communication Technologies in Sustainable Consumption ad Globalization. New York: Altamira Press.

Rip, A. (1995). Introduction of new technology: Making use of recent insights from sociology and economics of technology. Technology Analysis and Strategic Management 7, 417–31.

Ropke, I. (2001, September). New technology in everyday life—Social processes and environmental impact. Ecological Economics, 38(3), 403–22.

Saunders, H. (1992). The Khazzoom-Brookes postulate and neoclassical growth. Energy Journal, 13(4), 131–48.

Schnaiberg, A. (1980). The environment: From surplus to scarcity. Oxford: Oxford University Press.

Scholte, J. A. (1997, July). Global capitalism and the state. International Affairs, 77(3), 427–52.

Silicon Valley Toxics Coalition. (2000). Water use and other materials and wastes associated with semiconductor production. San Jose, CA: Silicon Valley Toxics Coalition.

Sonntag, V. (2000, July). Sustainability—In light of competitiveness. Ecological Economics 34(1), 101–13.

Statistics Canada. (2006, November). Our lives in digital times: Connectedness series (Research Paper, Catalogue no. 56F0004MIE–no. 014). Ottawa: Statistics Canada Science, Innovation and Electronic Information Division.

Traxler, J., & Luger, M. I. (2000, December). Businesses and the Internet: Implications for firm location and clustering. Journal of Comparative Policy Analysis, 2(3), 279–300.

United Nations, Department of Economic and Social Affairs, Population Division. (1998). Measuring changes in consumption and productions patterns. New York: Author.

United Nations, Broadband Commission for Digital Development (2010). Broadband: A Platform for Progress. September 2010.

Williams, E. D., Ayres, R. U., & Heller, M. (2002, December 15). The 1.7 kilogram microchip: Energy and material use in the production of semiconductor devices. Environmental Science & Technology, 36(24), 5504–10.

CHAPTER TWENTY-SEVEN

REVERSAL OF FORTUNE: "CLIMATEGATE" AND THE DEEPWATER HORIZON BLOWOUT

DOUG DAIGLE

For those who accept the premises that global warming and "peak oil" are major problems facing humanity, we appear to be in a bellwether period. Unfortunately, this is true in a negative sense. Events through 2010 indicate that American society has made an implicit decision not to address those issues, and U.S. policy obviously has a major effect on international agreements. This paper examines the impacts of two of those events – the "Climategate" controversy and the Deepwater Horizon oil leak in the Gulf of Mexico – and their implications for the future state of energy policy and the environment.

Anatomy of a "Scandal"

The year 2010 may be remembered as the year that climate policy came to a standstill. The failure of the Copenhagen Summit in 2009 and the collapse of a climate bill in the U.S. Senate in 2010 have left advocates of action on the problem of global warming at an impasse.

The impact of the controversy known as "Climategate" on this state of affairs should not be underestimated. Far from being merely a spat between warring scientists and advocates, or just another media circus, "Climategate" was part of a calculated strategy to undermine public confidence in science and its institutions. Its immediate effect in weakening support for both federal legislation and a global accord is only a short-term impact. The longer term implications point to an end of public policy not directly controlled by private interests.

The release of e-mails, documents and computer code from the Climatic Research Unit (CRU) at the University of East Anglia (UEA) in

November 2008 was described initially as a "hack" and subsequently as a "leak." The source of the leak remains unknown as of this writing. The sequence of events by which "Climategate" was brought into the world, however, is fairly clear.

During the weekend of November 13, 2009, files from a backup server at the CRU were copied by an unknown party. The 4,660 files included 3,587 that were documents, raw data, and computer code, and 1,073 e-mails. Reporters from the *Guardian* newspaper concluded after a digital forensic examination that this archive was created around September 30.[1] Additions were made during October and November until shortly before the leak. One suggestive episode involved a weatherman for the BBC who was sent a sample of the emails a month before the leak, after he had expressed some skepticism about global warming in a broadcast.[2]

The documented hack of the event occurred early on November 17 when a copy of the archive was uploaded onto the server of the website RealClimate. The RealClimate site director quickly detected the upload (labeled "FOIA.zip"), deleted it, and contacted the CRU. In the meantime, the hacker had posted a link to the files on RealClimate to the skeptic site Climate Audit. Links to the stolen archive were subsequently sent to the websites Watts Up With That (WUWT), the Air Vent, and Coyote Blog.[3] From there the e-mails eventually went "viral" to print and broadcast media.

An interesting step in that process occurred when the hacker posted a message to WUWT to complain nothing was apparently happening with the leak, only to be assured that the site managers were coordinating with some members of the media and other players to plan a wider release.[4]

On November 19, a contact within the UEA forwarded Climate Audit's director a departmental alert about the theft of the documents, which confirmed for the bloggers that the files were likely authentic, although their sheer number meant that no complete assessment had been made.

The news that a major scandal had been uncovered spread quickly through denialist[5] and partisan websites and media outlets. A story appeared in the New York Times on November 21, as the controversy entered the mainstream media.[6]

Initial speculation that foreign governments might have been involved was fueled by the sources of some of the posts and uploads[7] – the RealClimate site was hacked from an IP address in Turkey, the links posted on Climate Audit and WUWT were from Russian servers, while the post to the Air Vent had a Saudi origin. But, subsequent investigation by the *Guardian* found that the stolen material was "simply sitting in an archive on a single backup CRU server, available to be copied," and that even the hack was a fairly straightforward operation.[8]

What do the E-mails Show?

One of the striking things about the "Climategate" controversy is that it rests largely on five e-mails between a few prominent climate scientists and others which contained incriminating sounding phrases and discussions. These include the infamous expressions "trick" and "hide the decline" that were seized on by a number of politicians and pundits as evidence of scientific fraud:

> "I've just completed Mike's Nature trick of adding in the real temps to each series for the last 20 years (i.e., from 1981 onwards) [and] from 1961 to Keith's to hide the decline." Phil Jones, November 16, 1999.[9]

This message was widely claimed to have had a serious impact on the state of the science, since the phrase "hide the decline" fit into the counter-narrative of a cooling climate. Knowledgeable observers were easily able to refute the idea that the email referred to a decline in observed global surface temperatures. At the time of its writing, the previous year (1998) had been the warmest on record (in a record warm decade), so the idea of a decline in temperatures was not being discussed, and since the email dated from 1999, it had no connection to more recent instrumental records at the time of the leak (2009), when some denialists were claiming a decline or cooling temperatures .[10]

In-depth analysis of any kind is rare in the media, which is where most of the public got their information about the controversy. It may be useful to divide "media" into "partisan" and "mainstream" components. The partisan media would include "denialist" and "skeptic" blogs, right-wing in political orientation, and their role in the controversy was a familiar one, illustrated by two of the more egregious headlines: "Science scandal of the century" and "Climategate: Final Nail in the Coffin of 'Anthropogenic Global Warming?'"[11]

The partisan media clearly served as a catalyst for the story to make it into the mainstream media. Their influence reached further than that, however, since most of the mainstream media accepted the denialist/skeptic account of the emails' significance, and never looked far past the infamous five messages. Had they done so, even just for messages in the few months before the leak, they could have discovered other points of interest:

1) The emails contain a number of trenchant discussions of the flaws in the science skeptics' work, and in the denialists' allegations, which could have formed the basis for news stories.[12]
2) All of the issues of contention raised by the emails – the Yamal tree ring records, accusations of data destruction, etc. – were already being furiously hurled about in the partisan media in the months leading up to the leak, and the climate scientists were being subjected to an accelerating level of attacks and demands from these sources.[13]
3) The scientists discerned that these attacks (and related media pieces) were often timed in relation to external events: the "Greenhouse Gas Endangerment" Finding of the U.S. Environmental Protection Agency, hearings on the Senate climate bill, and the Copenhagen Summit (the timing of the emails' leak so close to the latter did get some notice in the media.)[14]

Overall, it is clear that there was little if anything that was new in the emails – all of the issues, allegations, etc. that they involved had been brewing in the denialist media (with ripostes from sites like RealClimate) for months or longer. Since the mainstream media generally does not follow these sources, they weren't likely to have known this – but they could have learned it by reading a representative sample of the emails.

They would have been reminded as well that "Climategate" is only one part of a connected series of events.[15] Following the email leak, there were, in short order, attacks on the Intergovernmental Panel on Climate Change (IPCC) and NASA for alleged flaws and fraud in their collection and treatment of data. While two of the errors found in the IPCC's 2007 Third Assessment Report (TAR) were genuine (and identified by scientists), a series of articles by London *Times* reporter Jonathan Leake erroneously gave the impression that the TAR was riddled with more serious mistakes, a theme that partisan media and some mainstream outlets took up and disseminated.[16]

"Climategate" is clearly also part of a strategy. The Competitive Enterprise Institute (CEI) has been a major actor in the denialist campaign against climate science, and perhaps in whatever plan lay behind the leak. Following the initial media eruption over Climategate, CEI announced its intention to sue NASA for failure to respond to FOIA requests, and alleged that the agency suffered from data problems similar to those of which CRU was accused. There was obvious collusion with sympathetic media outlets. Fox News.com carried a story with the headline "NASA Data Worse Than Climate-Gate Data, Space Agency Admits," (3/30/10). A reading of the article shows that the agency did not actually "admit" this, and that the headline is, in fact, a quote from the CEI "senior fellow" who had filed the notice to sue.[17]

The attacks on the IPCC and NASA were obviously timed to take advantage of media and public attention, so that an impression of a widening scandal could be created as part of an effort to discredit scientific institutions that serve as authorities on the issue of global warming. Once discredited, none of these institutions, their research, or their scientists would be accepted as sources of unbiased information by the denialist sector and their audience, nor by a potentially growing segment of the public.

The strategy utilized in "Climategate" is a variant of the "Swiftboat Attacks" carried out in recent political campaigns. The Swiftboat strategy aims at hitting the target at his/her perceived area of strength, to discredit and degrade what would have been a political advantage. For the climate issue, this is its grounding in established science. But the Swiftboat strategy also involves a keen understanding of how the media works. Since many outlets can be counted on to merely repeat allegations, rather than doing even a rudimentary investigation, they aid in the process – the allegations themselves become the story, and each repetition serves to further implant them and their plausibility in the public mind.

The full degree to which the controversy has impacted public perception in the U.S. and elsewhere will only become known over time. A working project by George Mason University and the Yale University School of Forestry polled a sample of Americans from December 2009 to January 2010 and found a significant decline in public acceptance of global warming. The loss of trust in climate science and scientists was greatest among individuals "with a politically conservative worldview."[18] This is hardly surprising, since the denialist narrative has been most strongly promoted over conservative blogs and media outlets.

That narrative now involves complete rejection of each of the official Climategate inquiries conducted in the U.K., along with the University of Pennsylvania's investigation of Dr. Michael Mann, as "whitewashes."[19] It also promotes total rejection of any scientific validation of AGW and any institution whose work supports it. Each of the U.K. inquiries concluded that Climategate had not affected the basis of the science of AGW, yet they did direct criticism at some aspects of the scientists' conduct (as well as at UEA policies on disseminating information.)[20] But, with the theme of fraud firmly planted in a constantly reinforced narrative, anything short of complete rejection is inadequate for the denialist /partisan media and its audience.

This situation in itself represents a victory for the denialists. The overall strategy of global warming denial always prioritized confusing the public over converting them, so that they disengage from the climate issue. Having the "debate" become primarily one of ideology and politics, in which science figures as merely one more vehicle for disagreement, guarantees an argument (as Alisdair MacIntyre put it) without a terminus.[21] That the scientific basis of AGW is still being debated in 2010 – along with the fact that denialist opposition is stronger and more implacable than it was ten years ago – bodes ill for any comprehensive action on a problem that is, as the saying goes, "time sensitive."[22]

The Deepwater Horizon Blowout and the Oil Dilemma

The Deepwater Horizon blowout is the largest oil disaster thus far in U.S. history. Recent estimates of the total amount of oil that flowed into the Gulf of Mexico during the 84 days of the leak stand at 4.9 million barrels.[23] The ultimate effects on the Gulf ecosystem, as well as human health, will not be known for some time.[24] The effects on U.S. energy policy, for which this might have been a defining moment, are also unclear.

It *was* clear early on that the disaster had its roots in a decade of corruption in the federal Minerals Management Service, the agency charged with managing offshore oil and gas on the outer continental shelf, though it directly resulted from flawed operational decisions made at the scene.[25] The size of the disaster would have presented an unprecedented challenge to government agencies even if they had been prepared and operating at optimum effectiveness. Federal agencies did respond on an unprecedented scale, but many of their protocols had been developed in response to the Exxon Valdez spill in 1989, and were less suited to deal with a massive leak occurring at a depth of 5000 feet in an open and dynamic sea.[26]

A state of conflict arose early on between federal agencies and state governments, particularly in Louisiana.[27] The federal response seemed focused primarily on the site of the blowout, 41 miles offshore, while Gulf Coast states were desperate to try to block oil from washing onto beaches or entering wetlands. The sheer scale of the area involved and the mobility of the oil made some proposed remedies ineffective or obsolete, but the federal agencies had difficulty from the beginning explaining the basis for their decisions regarding those options. This was a major liability in a situation that became highly politicized.

Political tensions rose even higher after May 27, 2010, when the administration declared a halt to drilling in waters deeper than 500 feet for a period of six months, along with the cancellation or postponement of several offshore leases, so that regulations and safety procedures could be thoroughly reviewed.[28] Local officials and residents already stressed by the economic effects of the on-going spill – and worst case scenarios that suggested even more pervasive damage was possible[29] - feared greater job losses and reacted with anger, augmented by active legal and media opposition from some segments of the oil industry.

Given the ongoing and unprecedented scope of the leak, and the raw emotions generated, it is understandable that longer-term energy implications didn't receive adequate discussion. At the same time, the failure to grasp those implications would seem to indicate both an incomprehension and an unwillingness to face our oil dilemma. While local and state officials railed against even a temporary halt in drilling, a number of environmental organizations focused their efforts on stopping all offshore drilling in campaigns reminiscent of those from the 1970s.[30] Our "addiction" to oil was widely decried, while our dependence on it as our central energy source was not closely examined. This is unfortunate, since both the addiction and the dependence are results of an era of cheap, easily accessible petroleum, which is now coming to a close.

Richard Heinberg was one of the few commentators to try to put the Deepwater Horizon disaster into this wider context:

> "The International Energy Agency projects that by 2020 deepwater will be providing 40 percent of all oil being extracted. Why the emphasis on deepwater? Because we've already chewed our way down through the higher levels of the oil pyramid: there's very little onshore or shallow-water oil left to find… We're not even close to *running out* of oil, coal, gas, or most minerals. But we face a convergence of entirely predictable but severe consequences from the depletion of the concentrated, high-grade resources at the top of the pyramid."[31]

Despite the fact that some agencies of the federal government have acknowledged the looming challenge that the U.S. (and the world) face, official acknowledgement seems to be another "third rail" in American politics.[32] The Government Accountability Office (GAO) tried to bring the problem mainstream in 2007, but attracted little notice:

> In the United States, alternative fuels and transportation technologies face challenges that could impede their ability to mitigate the consequences of a peak and decline in oil production, unless sufficient time and effort are brought to bear. For example, although corn ethanol production is

> technically feasible, it is more expensive to produce than gasoline and will require costly investments in infrastructure, such as pipelines and storage tanks, before it can become widely available as a primary fuel. Key alternative technologies currently supply the equivalent of only about 1 percent of U.S. consumption of petroleum products, and the Department of Energy (DOE) projects that even by 2015, they could displace only the equivalent of 4 percent of projected U.S. annual consumption.[33]

Hence, the silver lining that some hoped would result from the BP disaster – acknowledgement of our oil and energy dilemmas in their actual context – has as yet not been grasped, and instead the event has largely resulted in further conflict and confusion.

Danger and Opportunity

One of the most alarming things about the trajectory of failure that American society seems to be on - the failure to acknowledge and address the climate and energy problems - is that time is being lost. Those grasping for the right metaphors tell us that the window is closing on actions that can be taken fairly comfortably, and that we risk hitting the wall, or alternatively, a hard landing. Are there any grounds for guarded optimism?

It is clear that both problems need to be addressed together, and that necessity offers opportunity. The effects of the energy dilemma will likely be felt first, since most of our economic activity and infrastructure are designed around the assumption that cheap, accessible, and seemingly limitless oil will be available. The most difficult aspect of that problem is that there are no ready solutions – no alternative fuels that can be produced at the same scale as oil, even if they could be expanded in a timely enough manner, and there are numerous wrong turns to make. Obsessing over drilling in every last oil patch (including wilderness areas), throwing large sums at biofuels that cannot close the gap, and failing to invest quickly enough in mass transit are three such mistakes that come to mind. These are, in fact, likely to happen because of the inertia caused by patterns of previous investment.

At the same time, the redesign of energy and transportation infrastructure needed to reduce our carbon footprint is largely congruent with that needed to reduce our dependence on oil. The greatest obstacle to acting on those needs is the incompatibility of political and ideological worldviews that Climategate and the Deepwater Horizon blowout both reflect in their own way. Shocks such as the Gulf disaster (or Hurricane Katrina) do not necessarily lead to greater clarity, especially without a basis of commonly accepted fact. There is a pervasive media network (cyber, broadcast, and print) that is working to politicize science to the point where it is degraded into conservative vs. liberal versions, and thus loses its power as a central measure of reality. Preventing that state of affairs seems a necessary though not sufficient condition for grasping whatever opportunities we have left for building a sustainable future.

Notes

[1] "Climate email hackers had access for more than a month," *The Guardian*, Nov. 27, 2009; www.guardian.co.uk/environment/2009/nov/27/climate-email-hackers-access-month. The Guardian found that the e-mails seemed to show targeting of backup files of a few key CRU staff, such as CRU Director Phil Jones, and that many appear to have been filtered for key words that appeared repeatedly, such as "tree ring data." (Lead Guardian reporter Fred Pearce expanded his articles into an early book about "Climategate," *The Climate Files* (2010), Guardian Books/Random House).

[2] Leigh, Arthur, Evans, and Pierce, "Climate emails: were they really hacked or just sitting in cyberspace?" *The Guardian*, February 4, 2010; www.guardian.co.uk/environment/2010/feb/04/climate-change-email-hacker-police-investigation.

[3] Leigh, et al, "*The Guardian*, February 4, 2010.

[4] Ibid.

[5] Despite debate over the best application of the terms "denialist", "skeptic," etc., it seems clear that the parties and sites promoting Climategate and attacking climate scientists and institutions are overwhelmingly "denialists."

[6] Revkin, Andrew, "Hacked E-Mail is New Fodder for Climate Dispute," *New York Times*, November 20, 2009; www.nytimes.com/2009/11/21/science/earth/21climate.html?_r=1&scp=6&sq=climategate&st=cse.

[7] Stewart. W., Delgado, M., "Were Russian security services behind the leak of 'Climategate' emails," *Daily Mail* online, December 6, 2009; www.dailymail.co.uk/news/article-1233562/Emails-rocked-climate-change-campaign-leaked-Siberian-closed-city-university-built-KGB.html.

[8] Leigh, et al, *The Guardian*, February 4, 2010: "The leaker… knew something about computers, just as they knew something about climate science. But it didn't require the skills of Government Communications Headquarters or a foreign intelligence agency, as has been suggested. The hacking of the RealClimate blog exploited the fact that its wordpress platform has security holes well known to hackers." They also describe how one of the central antagonists of CRU Director Phil Jones, Stephen McIntyre (founder of the Climate Audit blog) had, in fact, managed to lift some data off of the public CRU site in July that he had previously requested through the U.K. Freedom of Information Act (FOI).

[9] There are a number of websites that display the email files – I utilized www.eastangliaemails.com; Jones' message is document number 942777075.txt, www.eastangliaemails.com/emails.php?eid=154&filename=.txt.

[10] Cf. Pearce, "How the 'climategate' scandal is bogus and based on climate sceptics' lies," *The Guardian*, February 1, 2010, www.guardian.co.uk/environment/ 2010/feb/01/climate-emails-sceptics; Schmidt, G., "The CRU Hack: Context," RealClimate, November 23, 2009,
www.realclimate.org/index.php/ archives/2009/11/the-cru-hack-context/.
[11]"Science Scandal of the Century," http://rockcreekfreepress.tumblr.com/post/ 283942989/climategate; http://blogs.telegraph.co.uk/news/jamesdelingpole/10001 7393/climategate-the-final-nail-in-the-coffin-of-anthropogenic-global-warming/.
[12] Email from P. Jones to M.Mann, et al, July 28, 2009, and email from M. Mann to Trenberth, July 23, 2009 (1248790545.txt); email from Mann to G. Foster, July 29, 2009 (1248877389.txt); email from B. Santer to T. Karl, August 7, 2009 (1249652050.txt); email from Mann to Revkin, September 29, 2009 (1254258663.txt); email from Santer to Plitz, October 8, 2009 (1255095172.txt).
[13] Emails from Jones to Mann, Schmidt, September 30, 2009 (1254323180.txt); email from M. Lutyens to A. Manning, October 5, 2009, and email from Jones to Manning, October 6, 2009 (1254832684.txt).
[14] Email from Jones to T. Osborn, Mann, September 29, 2009 (1254230232.txt), email from Plitz to Santer, et al, October 8, 2009 (1255095172.txt).
[15] The history of organized climate controversy has been documented in a number of recent books and reports, such as Oreskes and Conway, *Merchants of Doubt* (2010), Bloomsbury Press, and J. Hogan, R. Littlemore, *The Climate Cover-Up*, Greystone Books (2010).
[16] "IPCC errors: facts and spin," RealClimate, February 14, 2010,
www.realclimate.org/index.php/archives/2010/02/ipcc-errors-facts-and-spin/.
[17] Snow, B., FoxNews.com, March 30, 2010; www.foxnews.com/scitech/ 2010/03/30/nasa-data-worse-than-climategate-data/.
[18] Leiserowitz, et al, "Climategate, Public Opinion, and the Loss of Trust," Yale University School of Forestry, George Mason University,
www.climatechangecommunication.org/resources_reports.cfm.
[19] S. Fred Singer, "Climategate Whitewash," American Thinker, April 14, 2010, www.americanthinker.com/2010/04/climategate_whitewash.html; E. Barnes, "Penn State Probe into Mann's Wrongdoing a 'Total Whitewash,'",
FoxNews.com, February 5, 2010.
[20] Science and Technology Committee, "The disclosure of the climate data from the Climatic Research Unit at the University of East Anglia", House of Commons, Eight Report of Session 2009-10, March 24, 2010, www.parliament.uk/business/ committees/committees-archive/science-technology/s-t-cru-inquiry/; University of East Anglia, "Report of the International Panel set up by the University of East Anglia to examine the research of the Climatic Research Unit", April 12, 2010, www.uea.ac.uk/mac/comm/media /press/CRUstatements/SAP; "The Independent Climate Change Emails Review," July 7, 2010, www.scribd.com/doc/3400 3747/Muir-Russell-Final.
[21] A. MacIntyre, *After Virtue: A Study in Moral Theory*, 1984, University of Notre Dame Press, p.6.

---

[22] Recent articles and reports have shown that the denialist effort has essentially unlimited funding as well as strong political support in Congress (J. Mayer, "Covert Operations," *The New Yorker*, August 30, 2010;www.newyorker.com/reporting/2010/08/30/100830fa_fact_mayer?currentPage=all). One of the two major political parties in the U.S. seems to have adopted "skepticism" or denialism as a central tenet. Congressional GOP members have pledged investigations of the scientists named in Climategate, while the Attorney General of Virginia has been attempting to prosecute one of them; R. Helderman, "Judge rejects Cuccinelli's probe...", Washington Post, August 31, 2010, www.washingtonpost.com/wp-dyn/content/article/2010/08/30/AR2010083005004.html.

[23] Crone and Tolstoy, "Magnitude of the 2010 Gulf of Mexico Oil Leak," *Science*, October 2010, Vol. 330, www.sciencemag.org/cgi/content/abstract/science.1195840.

[24] Numerous sources are assessing and compiling data and other information; "Fish and Wildlife," Restore the Gulf, www.restorethegulf.gov/fish-wildlife; Institute of Medicine of the National Academies, "Assessing the Effects of the Gulf of Mexico Oil Spill on Human Health," A Workshop, August 10, 2010; www.iom.edu/Reports/2010/Assessing-the-Effects-of-the-Gulf-of-Mexico-Oil-Spill-on-Human-Health.aspx.

[25] J. Eilperin, "U.S. exempted BP's Gulf of Mexico drilling from environmental impact study," *Washington Post*, May 5, 2010, www.washingtonpost.com/wp-dyn/content/article/2010/05/04/AR2010050404118.html; D. Hammer, "Hearings: BP representative overruled drillers, insisted on replacing mud with seawater," *Times Picayune*, May 26, 2010; www.nola.com/news/gulf-oil-spill/index.ssf/2010/05/hearings_bp_representative_ove.html.

[26] C. Cleveland, "Deepwater Horizon Oil Spill," The Encyclopedia of Earth, October 9, 2010 (rev.); www.eoearth.org/article/Deepwater_Horizon_oil_spill#gen38; "Macondo Prospect, Gulf of Mexico, USA," www.offshore-technology.com/projects/macondoprospect/.

[27] The state of Louisiana initiated a controversial sand berm project that was supposed to act as a barrier to block oil from entering bays and wetlands, but it received opposition from federal agencies and the scientific community at the national and state level; see J. Rudolf, "Louisiana builds barriers even as oil disperses," New York Times, October 21, 2010, www.nytimes.com/2010/10/22/us/22berms.html?emc=eta1.

[28] U.S. Department of the Interior, "Salazar Calls for New Safety Measures for Offshore Oil and Gas Operations," May 27, 2010; www.doi.gov/news/pressreleases/Salazar-Calls-for-New-Safety-Measures-for-Offshore-Oil-and-Gas-Operations-Orders-Six-Month-Moratorium-on-Deepwater-Drilling.cfm.

[29] "The Oil Spill's Worst Case Scenario?" Newsweek, June 23, 2010; www.newsweek.com/2010/06/23/the-oil-spill-s-worst-case-scenario.html.

[30] Sierra Club, "Rallies Against Drilling, For Clean Energy," http://sierraclub.typepad.com/compass/2010/05/rallies-against-oil-drilling-for-clean-energy.html.

[31] R. Heinberg, "Deepwater Horizon and the Technology, Economics, and Environmental Impacts of Resource Depletion – The End is Nigh," The Oil Drum, June 2, 2010; www.theoildrum.com/node/6544.

[32] U.S. Government Accountability Office, "Crude Oil: Uncertainty About Future Oil Supply Makes it Important to Develop a Strategy for Addressing a Peak and Decline in Oil Production," GAO-07-283 February 28, 2007, www.gao.gov/products/GAO-07-283; "Heads in the Sand: Why Governments Don't Talk About Peak Oil," The Oil Drum, January 5, 2010; www.theoildrum.com/node/6100.

[33] U.S. GAO, GAO-07-283 (2007).

CONCLUSION

THE PROPHETIC VOCATION OF THE GLOBAL ECOLOGICAL INTEGRITY GROUP

J. RONALD ENGEL

At a recent meeting of conservationists in Chicago, Ashok Khosla, founder of Development Alternatives, a not-for-profit organization creating large scale sustainable livelihoods in India, now President of the World Conservation Union (IUCN), spoke about the series of "cascading crises" that are enveloping the world–economic recession, resource scarcity, growing divisions between rich and poor in both developed and developing worlds, over-population, military conflict and the nuclear arms race, lack of just and sustainable food systems, infringements of human rights, and last but not least, climate change, loss of biodiversity, and the unraveling of the integrity of the biosphere. He noted that the crisis that must be addressed as a precondition for successfully addressing any of the others was the crisis occasioned by the loss of biodiversity and ecosystem integrity, and this was the crisis that IUCN had primary responsibility to address. Ashok spoke with passion, and climaxed his comments by sharing his conviction that the global environmental crisis could not be met short of new moral understanding and action.

The Call for Prophecy

In my subsequent response to Ashok, I argued that if IUCN and the world conservation movement generally is to succeed in its mission, it needed to make this "big picture" of our global situation clear to the public. I suggested that when someone opened the IUCN website (www.iucn.org) they should be met immediately with the message that if we are to make headway on the "great issues" of the time we must sustain the natural foundations of human civilisation, and that this imperative was ethically as well as prudentially mandated not only because of our responsibilities to the human community, including future generations, but to life itself. My remarks were premised on the fact that IUCN is the world's most politically inclusive forum for discussion of global environmental science, policies and issues. It therefore has the obligation to direct the debates in that forum toward the ultimate ethical question: Do we not have an over-arching responsibility as global citizens to the continuation of life on planet Earth and a healthy life and environment for every person?

The prophetic vocation that I was urging on IUCN was no doubt influenced by my participation in the Global Ecological Integrity Group (GEIG). In reviewing the papers for this volume and reflecting upon what transpired at the Vancouver conference in 2010, I realised that one of the defining characteristics of GEIG was its "prophetic stance" toward the world. While this stance is chiefly rooted in the prophetic heritage of the Western religious traditions, it also has roots in other cultures such as those of indigenous peoples, as the papers by Jack Manno on the Haudenosaunee and Linda Te Aho on the Maori in this volume attest. I doubt if many members of GEIG are themselves aware of just how deep this common prophetic theme runs but I am convinced it is one of the underlying motivations for participation in the Group and why, as I will show in these concluding remarks, this particular conference took the direction it did and ended on such a powerful questioning note.

Klaus Bosselmann writes in his introduction to Part II, "Human beliefs and perceptions are fundamental to how we see the world and how we rate the potential for change." As the papers in this volume well demonstrate, it is with prophetic eyes that GEIG sees the world and assesses what changes are needed and possible.

Prophetic Argument

What is the prophetic vocation? In briefest answer, it is to make an argument, a public argument aimed at the ears of those who occupy the centers of power and influence in society as well as those citizens of every nation who are receptive to placing the lively experiment we call human co-evolutionary history within the context of ultimate meaning, *sub species aeternitatis*. The vocation of the prophet is to mobilise by persuasion the political will necessary to achieve justified ethical purposes. The following thumbnail definition lifts up seven elements in what we might call a prototypical prophetic argument which often unfolds as seven successive moments in a single narrative event.

First in the order of progression is the presupposition that history is morally meaningful, and however human beings may differ in their personal and cultural actualisations of the moral life, they all have by nature the capacity and obligation to live by that meaning. Second is the conviction that the moral meaning of history is to be found in "the higher law," or what is now commonly referred to as a "comprehensive vision of the good." While the higher law requires human reason and experience for its articulation, which is never perfect, it is not a human invention. It is rooted in the nature of reality and stands in judgment of human history. It is the higher law, however characterized or derived, that provides the source of authority for the individual or committed association of individuals to challenge the reigning principalities and powers.

The third element is a deep sensitivity to the evil our species perpetuates upon the world, the violence we do to nature and others and ultimately to ourselves, in defiance of the higher law. No wonder, then, that those who follow the prophetic vocation are constantly engaged in a fourth step, "reading the signs of the times" to discern whither we are tending in light of the requirements of the higher law, and to gain a realistic grasp of our human shortcomings and our ever renewed hopes for worldwide earthly redemption. This fourth moment leads inevitably to a fifth element, the pronouncement of an epochal crisis which demands profound personal and societal decisions to be made if doom is not to fall upon us. At this point the prophetic vocation requires the courage to speak truth to power and to issue a resounding call for *metanoia* or conversion on the part of the people and their leaders. This then is the sixth moment. The mandated moral transformation goes beyond a change of heart or intention, however; it requires a seventh element, practical actions and policies, including sacrifices, which those who assume the prophetic mantle are not hesitant to specify.

GEIG's Prophetic Argument

The Global Ecological Integrity Group is an island of prophetic truth-telling for those who come together each year for a week of rigorous interdisciplinary exchange and mutually supportive interpersonal relationships and for those who turn to its published proceedings for greater understanding of our global problematique. GEIG is a place where we can pursue the prophetic vocation of making the arguments which alone can give us genuine hope for our future — the arguments which will articulate the major contemporary moral issues crying for attention, provide greater scientific, ethical and spiritual insight into the higher law, allow us to celebrate and draw hope from the ways in which the universal moral law is instantiated in our legal and cultural systems, cast judgment on the ways it is neglected or contradicted by these same systems, analyse the power relationships and other dynamics of our society in order to discern portents of the future and the full size of the decisions required at this epochal moment in history, and prescribe what actions and policies are most urgently needed to avoid catastrophe and realize what is most promising for the well-being of humanity and the planet. GEIG is a prophetic reality check and a sanctuary of hope.

Vancouver was no exception.

GEIG's distinctive forms of prophetic argumentation are evident throughout this volume, beginning with the lead chapter by James Karr. Karr, who keynoted the Vancouver conference, and is one of the major theorists of the concept of ecological integrity, assumes the prophetic voice early in his essay. His fourth paragraph casts judgment on the reigning technological optimism of modern societies and predicts doom for the planet if present trends continue:

> Touting the uncanny ability of humans to be innovative and thus improve their lot, optimists—including economists, technologists, and futurists—see progressive improvement in the human condition as an inevitable outcome of human ingenuity. But the world we have created may not be the ideal world we intended to create. Nature continues to challenge us through the very by-products of our own ingenuity: human ingenuity has had serious unintended consequences, which can no longer be ignored. The most serious unintended consequence of this attitude is growing biotic impoverishment, the systematic reduction in Earth's ability to support living systems.

Karr proceeds to define the higher law of ecological integrity which has functioned throughout the history of GEIG as a primary standard of prophetic critique:

> [Biological integrity] refers to the capacity to support and maintain a balanced, integrated, adaptive biological system having the full range of elements (genes, species, assemblages) and processes (mutation, demography, biotic interactions, nutrient and energy dynamics, and metapopulation processes) expected in the natural habitat of a region.

The scientific natural law standard of ecological integrity is here stated in summary fashion. In his oral presentation Karr emphasized the site-specific character of ecological integrity. Ecological integrity describes the relationships between processes and elements that have successfully evolved in a particular locality or region with minimal human disturbance, and thus are geographically and biologically specific to that place. Karr here assumes the mantle of Aldo Leopold whose well-known "land ethic" stemmed from his prophetic judgment on the loss of ecological integrity on a particular piece of land in Sand County, Wisconsin in the 1940s.

In his presentation, Karr filled out his understanding of the higher law by quoting Ken Saro-Wiwa's prophetic declaration: "The environment is man's first right. Without a safe environment, man cannot claim other rights." On the basis of these two moral affirmations — the twin imperatives to preserve natural ecological integrity and honour the universal human right to life — Karr offers his normative policy prescriptions:

> If we are to stem biotic impoverishment and reverse environmental degradation, we must: (1) set societal goals based on broad concepts of ecological integrity and ecological health; (2) forge partnerships among scientists, engineers, policymakers, resource managers, and citizens to develop approaches for attaining those goals; (3) revise the legal framework guiding environmental policy to ensure that both ecological risks and threats to human health are minimised; (4) protect existing resources; (5) restore resources that are degraded; and (6) reduce resource consumption.

Expanding the Argument

Each step in the prophetic argument that Karr sets in motion in Chapter One is repeated and developed in a fresh way by each of the subsequent authors of this volume.

Like Karr, members of GEIG are morally engaged in history. To read their resumés is to read the profiles of active global and local citizens. This book contains a representative sampling of the issues in which they are passionately involved: sustainable regional and global ecosystems (Karr; Cudlín, Seják, and Pokorný; Kissinger, Timmer, and Rees), climate change (Goodland and Ahnang, Don Brown), food systems (McLeod-Kilmurray), global economics (Peter Brown), international law (Dellapenna, Seck, Bosselmann, Kintzele), water resources (McIntyre, Karageorgu), human security (Sabina and Alex Lautensach), effects of globalisation on developing nations (Crabbé), indigenous peoples and their environments (Aho, Lam and Borch, Manno), human health (Spady, Soskolne and Kramer, Bencko), and media and communication technologies (Dyson, Rattle, Daigle). In each case the particular issue addressed is analyzed in relation to other leading moral issues, for example, the relation between human and environmental health. Indeed, the precondition of ecological integrity — as a scientifically-based form of modern natural law, for all other forms of higher law, especially the human right to health — is perhaps the central insight of GEIG and is one of the most critical linkages between ethics, science and law that it seeks to advance.

Each author of this volume has an allegiance to one or another expression of the higher law — the unwritten, universal principles of fairness, morality, and justice that hold in judgment all instances of political and economic decision-making. Although the authors differ among themselves regarding the ultimate nature and justification for higher law principles, there are at least four expressions of the higher law that are widely shared and rationally defended.

First and foremost, the authors share the conviction that we can speak objectively and truthfully about matters of morality as well as matters of empirical fact. They recognise and affirm the diversity of moral cultures in the world but are not moral relativists — or triumphalists. They are ready to call into question any claims to pristine heritages or the idea of any civilization having a monopoly on virtue or insight. Second, ecological integrity is both an empirically verified fact and a universal normative value; it exists and it ought to be maintained. In the view of Helmut Burkhardt, indeed, "integrity" is broadened to become a universal principle uniting the individual, the community, the ecosystem and the cosmic environment and is the essence of sustainability. Third, the authors affirm the higher law of universal human rights, with special attention to the right to life and health. Fourth, they are all deeply committed to justice conceived as equality and fairness within and between the generations. Fourth, they share a belief in non-violence as the guiding principle for human security and social and ecological reconciliation and peace. There are also a number of "higher order" virtues that GEIG members affirm, such as compassion for our fellow humankind, especially the victims of history, intellectual humility, and mutuality.

GEIG endorsed the Earth Charter shortly after its publication in 2000 precisely because of its clear articulation of the higher laws mandated by our universal responsibility for the flourishing of the commonwealth of life. In his essay for this volume Bosselmann draws on the Earth Charter to construct the legal architecture for an ecologically and ethically-based world constitution.

As the reader of this volume will appreciate, the members of GEIG are never so much in their element as when they are casting judgment on the ways in which our species fails to live by the higher law. They do not hesitate to condemn the pride, selfishness, violence and greed of their governments and fellow citizens, and the false gods of marketplace, nation, war and religion that visit such destruction upon the world. They are forever on the outlook for hypocrisy and their own complicity in what they are condemning.

The burden of the argument in most essays in this volume is devoted to discerning the signs of the times, for example, Colin Soskolne and Shira Kramer, analysing the conflicting interests that led to suppression of research and public understanding of the effects on human health from the Canada's Sydney Tar Ponds, or Meidad Kissinger, Vanessa Timmer and William Rees, analysing the "historically unprecedented extension and thickening of the web of material connections among regions" and the consequent threats to human sustainability and the global ecological system.

The authors understand that the comprehensive moral, social and spiritual question of our epoch (referred to by some as the Anthropocene because of the unprecedented impacts of human activities upon the viability of the biosphere) is whether or not we are able to make the "great transition" to new forms of governance which will enable our global civilization to live within ecological limits and honour universal human rights and the just rule of law. We are now dangerously close to passing the point where Earth's life supporting systems have the capacity for self-renewal. The obligation of the prophet — in the sense of foretelling the future — must be to warn of such impending doom.

Sheila Collins addresses this overarching question in her discussion of alternative scenarios for the epoch we are now entering and the decisive demand for affirming the "commonwealth of life." Peter Brown addresses the question in his discussion of the need for a new political economy in keeping with the worldview and attendant ethics of the ecological sciences. Brown joins Michael Schröter in seeing many of the failures of contemporary public policy as traceable to the western philosophy of liberalism, including but not limited to neoliberal economic theory and its paradigm of unlimited growth fueled by the idea that the common good emerges as an automatic harmony from the unrestrained pursuit of individual interests. The prophetic imperative for the 21st century is an ethically-grounded, macroeconomic system that maintains a mutually enhancing relationship between the economy and Earth.

For many GEIG members, the acid test for whether this kind of fundamental intellectual and moral transformation is taking place is the state of the law and the reigning legal systems of human governance, international as well as domestic, so that the criterion of good or deficient, just or unjust, ecologically-informed or market-driven law and governance becomes itself a matter of primary moral concern. Kathryn Kintzele's presentation citing examples of the positive influence of higher law ethics in shaping national legal constitutions and judicial decisions was therefore most welcome. Laura Westra presented a more sobering picture of the role of law in today's world. This picture will be spelled out in her new book, *Globalisation, Violence and World Governance* (Leiden: Brill, 2011). In more cases than not, in her judgment, current international law should be viewed as a symbol of plunder rather than of the promise of making ecological integrity and human rights fundamental norms for global development.

The Unanswered Question

I am writing this conclusion on the first day of January, 2011, as this volume is on the verge of going to press. It is a time traditionally set aside to review the past year and make resolutions for the future. I feel that in many ways I already passed through the New Year's Day ritual at the GEIG conference in Vancouver, and then again as I read the papers for this volume.

In its search for positive signs of progress in 2010 toward a sustainable global society, the January issue of "Wavefront," the newsletter of the AtKisson Group in Stockholm (www.AtKisson.com), cites the September Nagoya meeting on the Convention on Biodiversity for significantly raising the bar on humanity's commitment to preserve and protect other species, the December Cancun meeting on climate change for putting the negotiating process back on track, the decision by the United Nations to return to Rio de Janeiro twenty years after the original Earth Summit, the global movement launched by writer/activist, Bill McKibben, which is managing to seriously raise the bar on what successful global climate activism looks like, both in substance and in symbol, and a few other encouraging events on the world stage. But as global sustainability consultant Alan Atkisson would be the first to admit, this is far from enough to put the world on a just and sustainable development path and the news of violence against nature and humanity in 2010, from the Gulf of Mexico to Afghanistan and far beyond, was grim indeed.

For me, and I believe many other participants in the 2010 GEIG meeting, the global reality check we received at Vancouver was almost uniformly discouraging. Indeed, perhaps for the first time in my life, I actively contemplated the possibility that the so-called Anthropocene, or what McKibben now refers to as Eaarth, may be the true "end times" of human co-evolutionary history.

As a member of the panel charged to make concluding observations at the end of the Vancouver conference I said that in my view we have a good idea of what should and must be done, both in the large picture, and in specifics; we know to our dismay how the higher law is being violated, and we have good evidence for what and who is responsible for its violation; we know that significant change will only come by a movement that marshals political power with scientific understanding and moral argument, but we do now know how to bring that movement about. In other words, we are long in our analysis of what is going wrong in the world, long on the right ideas for what needs to be going on and why, but come up short on how to close the gap, on how to bring about significant societal transformation.

How, I asked, will the *metanoia*, the change in personal and political will, the moral and social transformation we seek, come about? How will we reach binding collective agreements on the limits that need to be placed on our respective ecological footprints? How do we promote on the part of as many persons and communities as possible what Pavel Cudlín in his paper on ecological citizenship called "non-reciprocal responsibility?" Is it even possible to consider a transformation on the scale that is required?

My fellow panelists, William Rees, Don Brown, and Colin Soskolne, basically agreed with my concluding question. Don Brown argued that markets are simply unable to solve the problems we are facing. William Rees pointed out that civilization is a one-shot affair: we will never have the opportunity to make a new try, and we have no precedents for our current global experiment, if we can call it that. Colin Soskolne reaffirmed his discouragement regarding the prospects for social change with corporate influences undermining democratic processes operating even in the most advanced of technological societies.

Reflecting back upon these concluding exchanges at Vancouver, and on the papers collected in this volume, I am led to think that our conference constituted a significant turning point for GEIG. We might say that at Vancouver GEIG took a "Niebuhrian turn" — so named after the American theologian of international stature, Reinhold Niebuhr. In the 1940s and 50s, as the world tried to recover from the devastation of the Second World War, Niebuhr made an incisive critique of the presuppositions that were driving the predominant liberal internationalism of his time. Niebuhr, as few others, stood squarely in the universalist Christian prophetic tradition, and he argued that liberals were dangerously misguided if they thought fundamental change could be brought about by more rational persuasion, scientific facts, education, and moral exhortation. He argued in books such as *The Children of Light and the Children of*

Darkness, against the philosophical pragmatism of figures such as John Dewey, that something much more spiritually profound and mysterious was involved in human selfhood than what the Darwinian biological model of adaptive or maladaptive organism/environment relationships could provide, that the human will was the site of conflicts of ultimate allegiances and rebellions against the created order, and that human selfishness and pride, and the power struggles among individuals and nations that arise as a consequence — what he called the knowledge of the "children of darkness" regarding human nature — must be understood and mobilised by the foolish liberal "children of light" if their universal conceptions of the higher law were to make any headway in the new international arena.

And so we are left with a great unanswered question. How by prophetic argument can we bring about the changes in the world that our faith in the higher law requires? To my knowledge no fully adequate answer has yet been given to this question. We know a part of the answer, but not enough of the answer. This is not in any way to belittle the moral, spiritual and political wisdom of the great teachers and leaders of human civilization, or our own hard-won insights. This wisdom is considerable and should be consulted at every turn. We need every bit of it. But neither ancient nor modern prophetic movements have been able to make the argument — as GEIG has not yet been able to make the argument — that is able to successfully bring about the fundamental changes between humans and nature, and between nations and peoples that the world so desperately needs. Surely this question must be an essential part of the future agenda of the Global Ecological Integrity Group if it is to remain true to its prophetic vocation.

CONTRIBUTORS

Prof. Vladimir Bencko, M.D., Ph.D.
Professor of hygiene and epidemiology, Institute of Hygiene and Epidemiology, First Faculty of Medicine, Charles University in Prague & General University Hospital in Prague, Czech Republic
E-mail: vladimir.bencko@lf1.cuni.cz

Trude Borch works as a senior researcher at the independent research institute Norut in Tromsø, Norway. Her educational background is in economics and social anthropology. Borch works on diverse issues related to resource management and sustainability, but specializes on the management of marine and coastal resources and conflicting interests between commercial, recreational and indigenous fishers, such as marine fishing tourism.
E-mail: trude.borch@norut.no
Web Address: www.norut.no

Klaus Bosselmann, PhD
Dr Klaus Bosselmann is Professor of Law and Director of the New Zealand Centre for Environmental Law at the University of Auckland. Chair of the IUCN Commission on Environmental Law Ethics Specialist Group.
E-mail: k.bosselmann@auckland.ac.nz

Donald A. Brown is Associate Professor of Environmental Ethics, Science, and Law at Penn State University where he teaches interdisciplinary courses on climate change and sustainable development. Before holding this position he was an environmental lawyer for the states of Pennsylvania and New Jersey and Program Manager for United Nations Organizations at the United States Environmental Protection Agency's Office of International Environmental Policy. In this position he represented the United States EPA on US delegations to the United Nations negotiating climate change, biodiversity, and sustainable development issues. His latest book is *American Heat, Ethical Problems with the US Response to Global Warming*.
E-mail: dabrown57@gmail.com

Peter G. Brown is a Professor at McGill University in the School of Environment and the Departments of Geography and Natural Resource Sciences. He is the author of *Restoring the Public Trust, The Commonwealth of Life*, and a co-author of *Right Relationship*. His most recent co-edited book is *Water Ethics*.

Helmut Burkhardt, Dr. rer. nat., Dipl.Phys.
Professor Emeritus Ryerson University, Toronto Ontario, Canada. Early research in thermonuclear fusion, now in solar energy and in the full spectrum of sustainability issues. Life member and past president of Science for Peace Canada, member of the Canadian Pugwash Group, and of the International Network of Engineers and Scientists for Global responsibility.
E-mail: helmut.burkhardt@bell.net

Sheila D. Collins, Ph.D.
Professor of Political Science & Director
Masters Program in Public Policy & International Affairs
William Paterson University of New Jersey
Web: www.wpunj.edu/cohss/departments/pol_sci/faculty/collins/index.dot

Philippe Crabbé (Environmental and Natural Resources Economics Emeritus Professor, University of Ottawa). He was the first director of the Institute of the Environment at the University of Ottawa and served on the two past Intergovernmental Panel of Climate Change assessments. He is currently working on Beneficial Environmental Practices in Agriculture and on Climate Change Adaptation (Co-Principal Investigator) in the Carribeans.

Assoc. Prof. Pavel Cudlín (pavelcu@usbe.cas.cz) graduated in Natural Sciences at the Charles University and obtained PhD at the Czechoslovak Academy of Sciences in landscape ecology. Since 1974 he has been employed in the Institute of Systems Biology and Ecology (earlier Institute of Landscape Ecology), last years as a head of the Department of Forest Ecology. He has been involved for a long time in studies on root and mycorrhiza ecology, forest ecosystem decline under global change, and biodiversity and ecosystem service valuation.

Doug Daigle
Director, CREST Program
School of the Coast & Environment
Louisiana State University

Joseph Dellapenna
Professor of Law, Villanova University, Director of the Model Water Code Project, American Society of Civil Engineers, Rapporteur of the Water Resources Law Committee of the International Law Society. Recognized by the US Supreme Court as an authority on transnational litigation.

Rose A. Dyson, Ed.D.
Consultant in Media Education
President: Canadians Concerned About Violence in Entertainment
www.C-CAVE.com
Director of Communications
Canadian Peace Research Association

J. Ronald Engel, Ph.D.
Senior Research Fellow, Center for Humans and Nature; Professor Emeritus, Meadville/Lombard Theological School
(University of Chicago affiliate)
Web Address: www.humansandnature.org

Dr Robert Goodland, Tropical ecologist
World Bank's Lead Environmental Adviser (1978-2001) wrote and persuaded the Bank to adopt most of its current mandatory Social and Environmental Safeguard Policies. Last year he was awarded IUCN's Coolidge medal for outstanding lifetime achievements in conservation
E-mail: RbtGoodland@gmail.com

Vicky Karageorgou
Phd in European Environmental Law
Lecturer in European Administrative and Environmental Law
Panteion University of Political and Social Sciences, Athens Greece
Lawyer specialized in cases in the field of environmental, planning and energy law

Dr. James R. Karr, Ecology, Environmental Science and Policy
Professor Emeritus, University of Washington, Seattle, WA.
E-mail: jrkarr@u.washington.edu

Kathryn Kintzele is Director of the Global Program for the US-based Center for Humans and Nature. She is the Deputy Chair of the Ethics Specialist Group of the IUCN Commission on Environmental Law and the Co-Chair of the Biosphere Ethics Initiative, a soft law program that seeks to incorporate ethical principles of biodiversity conservation into law and policy. Kathryn is also an Adjunct Professor at Valparaiso University School of Law.

Meidad Kissinger, Assistant professor at the Geography and Environmental Development department at the Ben Gurion University, Israel. Dr Kissinger expertise is in the field of ecological and sustainability accounting. His recent work advances ecological footprint analysis including exploring the connections between human activities in consuming regions and their dependence and pressure on 'hidden' ecosystems goods and services in other regions around the world.

Shira Kramer, MHS, PhD
President
Epidemiology International, Inc.
11019 McCormick Road, Suite 200
Hunt Valley, MD 21031 USA
E-mail: s.kramer@epiinternational.com

Mimi E. Lam is a Research Associate at the University of British Columbia Fisheries Centre in the Policy and Ecosystem Restoration in Fisheries group. She chairs the Traditional Ecological Knowledge Section of the Ecological Society of America. Lam researches the human dimensions of fisheries, articulating socio-economic and cultural values in science and policy. She co-edited The Privilege to Fish Special Feature on 'fishing rights' in Ecology and Society.
E-mail: m.lam@fisheries.ubc.ca
Web Address: http://sites.google.com/site/ferrfc/members/Dr-Mimi-Lam

Alexander Lautensach (DiplBiol. biology; MSc. molecular biology; BEd. secondary education; MScT. science education; PhD. education in environmental ethics). Assistant professor, School of Education, University of Northern British Columbia; Associate editor, Journal of Human Security; Deputy director, Human Security Institute. Research interests: environmental ethics, human security, science education, teaching and learning for sustainability, bioethics education.

Sabina W.B. Lautensach (BA policial science & anthropology; MA. international relations; PhD intercultural conflict resolution). Editor-in-chief, Journal of Human Security http://www.rmitpublishing.com.au/jhs.html . Director, Human Security Institute http://www.hsihumansecurity institute.com Research interests: social anthropology, confidence building measures, conflict resolution, human security, intercultural negotiation.

Owen McIntyre, B.A., LL.B., Ph.D.
Senior Lecturer, Faculty of Law
University College Cork
National University of Ireland

Heather McLeod-Kilmurray, PhD
Associate Professor and
Director, Centre for Environmental Law and Global Sustainability
Faculty of Law, University of Ottawa
E-mail: hmcleodk@uottawa.ca

Jack P. Manno, Ph.D.
Associate Professor of Environmental Studies
State University of New York College of
Environmental Science and Forestry
Web Address: www.esf.edu/es/faculty/manno.asp

Assoc. Prof. Jan Pokorný graduated in Natural Sciences at the Charles University. He worked on photosynthetic processes at the Agricultural University in Prague and later became head of the Třeboň Section of the Institute of Botany of the Czech Academy of Sciences. He directs ENKI, a public benefit corporation for environmental research, and has worked in several European countries as well as Africa and Australia. In 1998 he was elected to the panel of the Ramsar Convention for Central and Eastern Europe.
E-mail: pokorny@enki.cz

Robert Rattle is an independent researcher, consultant, author, and scholar based in Canada with interests in sustainable consumption, social determinants of health, Health Impact Assessment, Internet and communication technologies, and globalization.
E-mail: rrattle@alumni.uwaterloo.ca

William E. Rees, PhD, FRSC
Professor, University of British Columbia
School of Community and Regional Planning
Web Address: www.scarp.ubc.ca/profiles/faculty/William%20Rees
E-mail: wrees@interchange.ubc.ca

Michael W. Schröter, Ph.D
Legal Advisor for Environmental Affairs at Representation of Bremen
Visiting Research Fellow at New Zealand Centre for Environmental Law
E-mail: mschroeter@gmx.net

Sara Seck was appointed Assistant Professor at the Faculty of Law, University of Western Ontario in 2007. She completed a PhD in 2008 at Osgoode Hall Law School examining whether home states have obligations under international human rights and environmental laws to regulate transnational mining companies to prevent and remedy harm. Her publications have appeared in the Canadian Yearbook of International Law, the Yale Human Rights and Development Law Journal, and the Osgoode Hall Law Journal, among others.

Assoc. Prof. Josef Seják graduated in 1968 from the University of Economics, Prague. 25 years affiliated to the Institute of Economics, Czechoslovak Academy of Sciences, with specialisation on evaluations of natural and environmental resources. Since 1993 to 2003 at the Czech Environmental Institute (CEI), in 1999-2003 director. Dean of the Faculty of Environment at J.E. Purkyne Univ. in Usti n. L. in 2004-2007. Visiting lecturer at Univ. of Applied Sciences in Turku, Finland.
E-mail: sejak@fzp.ujep.cz

Colin L. Soskolne, Ph.D. (Epidemiology)
Professor
Department of Public Health Sciences
School of Public Health
University of Alberta, Edmonton, Alberta, Canada
Web Address: www.colinsoskolne.com
E-mail: colin.soskolne@ualberta.ca

Donald Spady, MD, MSc, FRCP(C)
Adjunct Associate Professor of Pediatrics & Public Health
Faculty of Medicine & Dentistry
School of Public Health, University of Alberta
E-mail: dspady@ualberta.net

Linda Te Aho, LLB, LLM
Senior Lecturer in Law
Te Piringa – Faculty of Law
University of Waikato, New Zealand
Web Address: www.waikato.ac.nz

Vanessa Timmer, Post-Doctoral Research Fellow at the University of British Columbia; Research Fellow, Earth System Governance Project; Executive Director, One Earth Initiative Society. Dr Timmer's research focuses on the intersection of sustainability, governance, systems thinking and social change.

Laura Westra, Ph.D., Ph.D. (Law)
Professor Emerita (Philosophy)
University of Windsor
Sessional Instructor, Faculty of Law
Web Address: www.ecointegrity.net
GEIG: www.globalecointegrity.net
E-mail: lwestra@interlog.com